The
PRINCETON
FIELD GUIDE *to*
PREDATORY
DINOSAURS

The
PRINCETON
FIELD GUIDE *to*
PREDATORY
DINOSAURS

GREGORY S. PAUL

Princeton University Press

Princeton and Oxford

Published by Princeton University Press
41 William Street, Princeton, New Jersey 08540
99 Banbury Road, Oxford OX2 6JX

press.princeton.edu

All Rights Reserved

ISBN 978-0-691-25316-9
ISBN (e-book) 978-0-691-25333-6

Library of Congress Control Number: 2024934500

British Library Cataloging-in-Publication Data is available

Editorial: Robert Kirk and Megan Mendonça
Production Editorial: Kathleen Cioffi
Text Design: D & N Publishing, Wiltshire, UK
Jacket Design: Benjamin Higgins
Production: Steven Sears
Publicity: Matthew Taylor and Caitlyn Robson
Copyeditor: Eva Silverfine

This book has been composed in LTC Goudy Oldstyle Pro (main text) and Acumin Pro (headings and captions)

Drawing previous page: *Avimimus portentosus*

Printed in China

10 9 8 7 6 5 4 3 2 1

CONTENTS

PREFACE

If I were, at about age twenty as a budding paleozoologist and paleoartist, handed a copy of this book by a mysterious time traveler, I would have been shocked as well as delighted. The pages would reveal a world of new flesh- and, in some cases, plant-eating dinosaurs and ideas that I barely had a hint of or had no idea existed at all. My head would spin at the revelation of the therizinosaurs, such as the wacky feathered *Beipiaosaurus*, and at the biplane flying dromaeosaurs, not to mention the little halszkaraptors with their duck-like beaks, the brow horns and atrophied arms of bulldog-faced *Carnotaurus*, or the bat-like membranous wings of scansoriopterygids. What to make of the island-dwelling *Balaur* and its weird feet? It is a particular pleasure to restore the skeleton of the once mysterious *Deinocheirus*, long known from only its colossal arms—its peculiar skeleton does not disappoint. And who would have imagined it would become possible to figure out the colors of some feathered dinosaurs? I would note the new names for some old dinosaurs, including that now there are three royal species of *Tyrannosaurus*: *T. rex, T. regina,* and *T. imperator*. And that direct evidence of head-to-head combat between the tyrant king and horned *Triceratops* would show that the latter survived. There would be the dinosaur-bearing beds with the familiar yet often exotic names: Morrison, Chinle, Wessex, Lameta, Tendaguru, Oldman, Kirtland, Djadokhta, Nemegt, Cloverly, Judith River, New Egypt, Navajo Sandstone, Horseshoe Canyon, Forest Sandstone, Santa Maria, Lowenstein, Bahariya, Hell Creek, Chaneres, Kayenta, Portland, Lufeng, and Lance. Plus there are the novel formations, at least to my eyes and ears: Yixian–Jiufotang, Tiourarén, Dinosaur Park, Longjiang, Lourinha, Barun Goyot, Two Medicine, Iren Dabasu, Anacleto, Painten, Santana, Rio Neuguen, Shishugou, Demopolis, Huincul, Aguja,

Ischigualasto, Nanxiong, Kaiparowits, Qingshan, Qiaotou, Cerro del Pueblo, Hanson, Snow Island Hill, Tiaojishan, Echkar, Elliot, Candeleros, Los Colorados, Tropic Shale, Pari Aike, Maleri, Portezuelo, Prince Creek, Majiacun, Ulansuhai, Shinekhudag, Kitadani, Calizas de la Huérguina, and Maevarano. The sheer number of new dinosaurs would demonstrate that an explosion in dinosaur discoveries and research, far beyond anything that had previously occurred and often based on new high technologies, marked the end of the twentieth century going into the twenty-first.

Confirmed would be the paradigm shift already under way in the late 1960s and especially the 1970s that observed that dinosaurs were not so much reptiles as they were near birds and often paralleled mammals in form and function. Dinosaurs were still widely seen as living in tropical swamps, but we now know that some lived through polar winters so dark and bitterly cold that low-energy reptiles could not survive. Imagine a small dinosaur shaking the snow off its feathery insulation while its body, oxygenated by a birdlike respiratory complex and powered by a high-pressure four-chambered heart, produces the heat needed to prevent frostbite.

Producing this book has been satisfying in that it has given me yet more reason to achieve more fully a long-term goal: to illustrate the skeletons of almost all predatory dinosaur species for which sufficiently complete material is available. These restorations have been used to construct the most extensive library of side-view life studies of dinosaur hunters in print to date. The result is a work that covers what is now two centuries of scientific investigation into the group of animals that ruled the continents for over 150 million years. Enjoy the travel back 175 million years.

ACKNOWLEDGMENTS

A complaint earlier this century on the online Dinosaur List by Ian Paulsen about the absence of high-quality dinosaur field guides led to the production of the first and then the second edition of this dinosaur guide. The exceptional success of those editions combined with the continuous flux of new discoveries and research led to production of a third. Many thanks to those who have provided the assistance over the years that has made this book possible, including Kenneth Carpenter, Asier Larramendi, James Kirkland, Michael Brett-Surman, Philip Currie, Robert Bakker, Scott Persons, Rinchen Barsbold, Xu Xing, Frank Boothman, Daniel Chure, David Burnham, Kristina Curry Rogers, Alex Downs, Steven and Sylvia Czerkas, David Evans, James Farlow, Jay Van Raalte, Hermann Jaeger, Frances James, Tracy Ford, Catherine Forster, Rodolfo Coria, Mark Hallett, Jerry Harris, Thomas Holtz, Peter Larson, Mark Norell, Edwin Colbert, James Madsen, Donald Baird, Wann Langston, Nicholas Hotton, Dale Russell, Halszka Osmólska, Teresa Maryanska, Lawrence Witmer, Ralph Molnar, David Varricchio, David Weishampel, Jeffrey Wilson, Nicholas Longrich, Jordan Mallon, Darren Naish, Fernando Novas, Kevin Padian, Armand de Ricqles, Timothy Rowe, Scott Sampson, John Scannella, Matthew Lamanna, Mary Schweitzer, Boris Sorkin, John Foster, Guy Leahy, Charles Martin, Octavio Mateus, Liu Jinyuan, Carl Mehling, Markus Moser, Mike Fredericks, Donald Glut, Masahiro Tanimoto, Robert Telleria, John Jackson, Michael Triebold, Ben Creisler, Dan Varner, Jens Lallensack, Saswati Bandyopadhyay, Ji Shuan, Clint Boyd, Mickey Mortimer, Tyler Greenfield, Mikko Haaramo, John Schneiderman, and many others. I would also like to thank all those who worked on this book for Princeton University Press: Robert Kirk, Kathleen Cioffi, Megan Mendonça, Eva Silverfine, and anonymous readers.

Ornithomimus? edmontonicus

The feathery predatory dinosaur *Velociraptor*

INTRODUCTION

HISTORY OF DISCOVERY AND RESEARCH

The remains of dinosaurs predatory and otherwise have been found by humans for millennia and probably helped form the basis for belief in mythical beasts, including dragons. A few dinosaur bones were illustrated in old European publications without their true nature being realized. In the West, the claim in the Genesis creation story that the planet and all life were formed just 2,000 years before the pyramids were built hindered the scientific study of fossils. At the beginning of the 1800s, the numerous three-toed trackways of avepod dinosaurs found in New England were attributed to big birds, in part because they lacked tail drag marks. By the early 1800s, the growing geological evidence that Earth's history was much more complex and extended back into deep time began to free researchers to consider the possibility that long-extinct and exotic animals once walked the globe.

Modern dinosaur paleontology began 100 years ago in the 1820s in England. A few bones and teeth of the predatory *Megalosaurus* and herbivorous *Iguanodon* were found and the findings published in 1824 and 1825. For a few decades it was thought that the bones coming out of ancient sediments were the remains of oversized versions of modern reptiles. In 1842 Richard Owen recognized that many of the fossils were not standard reptiles, and he coined the term "Dinosauria" to accommodate them. Owen had pre-evolutionary concepts of the development of life, and he envisioned dinosaurs as elephantine versions of reptiles, so they were restored as heavy-limbed quadrupeds. This restoration led to the first full-size dinosaur sculptures for the grounds of the Crystal Palace in the 1850s, including a vaguely croc-like *Megalosaurus*, which helped initiate the first wave of dinomania as the sculptures excited the public. A banquet was actually held within one of the uncompleted figures. These wonderful examples of early dinosaur art still exist.

Across the pond in New England, three-toed trackways, some quite large, were abundant on early Mesozoic slates, especially from the Connecticut Valley. In the early and mid-1900s, Edward Hitchcock, who had no idea of how very ancient the prints were, concluded that the combination of three digits and lack of tail drag marks meant these were short-tailed birds from before the biblical flood, some titanic like the newly found great moas of New Zealand.

The first complete dinosaur skeletons, uncovered in Europe shortly before the American Civil War, were those of small examples, the armored *Scelidosaurus* and the bird-like avepod *Compsognathus*. The latter showed that the theropods—named a couple of decades later by Othniel Marsh—were bipeds that looked a lot like long-tailed birds. That caused some to propose that tridactyl Mesozoic footprints were those of nonavian dinosaurs—but the trackways' consistent lack of tail drag marks did not cause paleoartists to keep the appendages off the ground for another hundred years. The modest size of these fossils limited the excitement they generated among the public.

Megalosaurid *Torvosaurus*

Found shortly afterward in the same Late Jurassic Soln-hofen island lagoonal sediments as *Compsognathus* was the "first bird," *Archaeopteryx*, complete with teeth and feathers. The remarkable mixture of avian and reptilian features preserved in this little dinobird did generate wide-spread interest, all the more so because the publication of Charles Darwin's theory of evolution at about the same time allowed researchers to put these dinosaurs in a more proper scientific context. The enthusiastic advocate of bio-logical evolution, Thomas Huxley, argued that the many similarities between *Compsognathus* and *Archaeopteryx* indi-cated a close link between the two groups.

At this time, the fossil bones action was shifting to the United States. Before the Civil War, incomplete remains had been found on the Eastern Seaboard, including the large raptorial *Dryptosaurus*. But matters really got moving when it was realized that the forest-free tracts of the West, which were then being cleared of the First Peoples, offered fossil hunting grounds that were the best yet for the re-mains of extinct titans. The westward expansion of rail-roads made this fossil hunting practical, which quickly led to the "bone wars" of the 1870s and 1880s. Edward Cope and Charles Marsh, having taken a dislike for one another that was as petty as it was intense, engaged in a bitter and productive competition for dinosaur fossils that would produce an array of complete skeletons. For the first time it became possible to appreciate the form of classic Late Jurassic Morrison dinosaurs, such as agile predatory *Allo-saurus* and *Ceratosaurus*, along with their prey of elephan-tine sauropods—*Apatosaurus*, *Brontosaurus*, *Diplodocus*, and *Camarasaurus*, the protoiguanodont *Camptosaurus*, and the bizarre plated *Stegosaurus*. Also found were more little bird-like predators, such as *Ornitholestes*. Popular interest in the marvelous beasts was further boosted, all the more so with the first museum mount of a hunting dinosaur, an *Allosau-rus* at the American Museum of Natural History in New York near the beginning of the 1900s. Interestingly, its tail is clear of the ground as it feeds on a sauropod carcass. The same was true of the *Ceratosaurus* put up at the Smithson-ian a few years later, yet this posture did not take.

By the turn of the century, discoveries began to include younger deposits, such as the Lance and Hell Creek, which produced classic dinosaurs from the end of the dinosaur era, including duck-bills, tank-like ankylosaurs, horned *Triceratops*, and the terror of their dinosaurian lives, the great *Tyrannosaurus*. When its magnificent skeleton was put up at the American Museum during the Great War, its im-pact cannot be overemphasized: it became the iconic di-nosaur fossil for decades, solidifying the idea of dinosaurs being erect-bodied tail draggers. As paleontologists moved north into Canada in the early decades of the twentieth century, they uncovered a rich collection of slightly older Late Cretaceous dinosaurs including *Albertosaurus* and *Daspletosaurus*, as well as the ratite-mimicking *Struthiomi-mus* and its fleet-footed kin. Illustrating these amazing dis-coveries were the first generation of paleoartists who, after

some early flings, adopted a generally conservative stance in which tails flopped and all feet were somehow always on the ground at the same time, even as they portrayed ancient mammals cavorting and galloping about ancient scenes.

Inspired in part by the American discoveries, paleon-tologists in other parts of the world looked for new di-nosaurs. In southeastern Africa, the colonial Germans uncovered at exotic Tendaguru supersauropods and spiny stegosaurs, although not much in the way of their killers showed up. Just one medium-sized theropod skeleton was discovered, and the lack of a skull long hid that it and other elaphrosaurs were herbivorous. In northern Africa was found a series of large trunk vertebra, *Spinosaurus*, sporting an enormous spine sail that is still perplexing di-nosaurologists. In the 1920s Henry Osborn at the Amer-ican Museum in New York dispatched Roy Andrews to Mongolia in a misguided search for early non-African hu-mans that fortuitously led to the recovery of small Late Cretaceous dinosaurs, including the "egg-stealing" *Ovirap-tor* and the similarly advanced, near-bird theropods, sickle-clawed *Velociraptor* and *Saurornithoides*. Dinosaur eggs and entire nests were found, only to be errantly assigned to a small omnivorous protoceratopsid rather than the ovi-raptorid that had actually laid and incubated them. As it happened, the Mongolian expeditions were somewhat misdirected. Had paleontologists also headed northeast of Beijing, they might have made even more fantastic, feath-ery discoveries that would have dramatically altered our view and understanding of dinosaurs, birds, and their evo-lution, but that event would have to wait another three-quarters of a century.

The mistake of the American Museum expeditions in heading northeast contributed to a set of problems that seriously damaged dinosaur paleontology as a science between the twentieth-century world wars. Dinosaurol-ogy became rather ossified, with the extinct beasts widely portrayed as sluggish, dim-witted evolutionary dead ends doomed to extinction, an example of the "racial senes-cence" theory that was widely held among researchers who preferred a progressive concept of evolution at odds with more random Darwinian natural selection. It did not help matters when artist-paleontologist Gerhard Heil-mann published a seminal work that concluded that birds were not close relatives of dinosaurs, in part because he thought dinosaurs lacked a wishbone furcula, which had just been found, but misidentified, in *Oviraptor*. The ad-vent of the Depression, followed by the trauma of World War II—which led to the loss of some important specimens on the continent as a result of Allied and Axis bombing—brought major dinosaur research to a near halt.

Even so, public interest in dinosaurs remained high. The *Star Wars–Jurassic Park* of its time, RKO Picture's *King Kong* of 1933 amazed audiences with its dinosaurs seem-ingly brought to life, with the mighty ape battling with the terrible *Tyrannosaurus* to the latter's death. Unfortunately, the very popularity of dinosaurs was giving them a circus air that convinced many scientists that they were beneath their scientific dignity and attention.

Despite the problems, discoveries continued. In an achievement remarkable for a nation ravaged by the Great Patriotic War and suffering under the oppression of Stalinism, the Soviets mounted postwar expeditions to Mongolia that uncovered the Asian version of *Tyrannosaurus*, aka *Tarbosaurus*, and the enigmatic arms of enormous-clawed *Therizinosaurus*. Equally outstanding was how the Poles, led by the renowned Halszka Osmólska, took the place of the Soviets in the 1960s, discovering in the process the famed complete skeleton of *Velociraptor* engaged in combat with *Protoceratops*. They too found another set of mysterious long arms with oversized claws, *Deinocheirus*. *Gallimimus* is a well-preserved, long-billed ornithomimid. Another leader of Mongolian paleozoology has been the Mongolian native Richen Barsbold.

In the United States, before World War II, Roland Bird studied the Cretaceous Texas trackways that suggested giant avepod dinosaurs hunted even bigger sauropods. Shortly after the global conflict, the Triassic Ghost Ranch quarry in the Southwest, packed with complete skeletons of little *Coelophysis*, provided the first solid knowledge of the beginnings of predatory dinosaurs. Also found shortly afterward in the Southwest was the closely related but much larger crested *Dilophosaurus* of the Early Jurassic. But research was slow; it would be decades before these fossils were scientifically detailed.

What really spurred the science of dinosaur research were the Yale expeditions to Montana in the early 1960s, which dug into the little-investigated Early Cretaceous Cloverly Formation. The discovery of the close *Velociraptor* relative, *Deinonychus*, finally made it clear that some predatory dinosaurs were sophisticated, energetic, agile dinobirds, a point reinforced by the realization that they and the other sickle claws, the troodontids, as well as the ostrich-like ornithomimids, had fairly large, complex brains. These developments led John Ostrom to note and detail the similarities between his *Deinonychus* and *Archaeopteryx* and to conclude that birds are the descendants of energetic, small theropod dinosaurs. Ostrom further observed that the bird- and mammal-like body design of dinosaurs meant that their newly apparent presence in Arctic regions could not be taken as evidence that those ancient polar habitats were warm even in winter but rather that dinosaurs' flesh and plant eating did not have low-level reptilian energetics.

Realizing that the consensus dating back to the original discovery that dinosaurs were an expression of the reptilian pattern was flawed, Robert Bakker in the 1960s and 1970s issued a series of papers contending that dinosaurs and their feathered descendants constituted a distinct group of archosaurs whose biology and energetics were more avian than reptilian. In the eye-popping article "Dinosaur Renaissance" in a 1975 *Scientific American*, Bakker proposed that some small dinosaurs themselves were feathered. In accord with the countless trackways, Barney Newman noted that dinosaurs, predatory included, were not dragging their tails. In the late 1980s, this author's *Predatory Dinosaurs of the World*, which covered the same basic subject as this field guide, helped set the advanced "New Look" of dinosaurs, including popularizing the white skeleton immersed in black soft tissues profile that has become the illustrative norm. It further argued that some dinosaurs, such as *Deinonychus*, *Saurornithoides*, and *Oviraptor*, were secondarily flightless and closer to birds than they were *Archaeopteryx*. At the same time, researchers from outside paleontology stepped into the field and built up the evidence that the impact of a mountain-sized asteroid was the long-sought great dinosaur killer. This extremely controversial and contentious idea turned into the modern paradigm when a state-sized meteorite crater, dating to the end of the dinosaur era, was found in southeastern Mexico. At the same time, the volcanic Deccan Traps of India were proposed as an adjunct in the demise of the nonbird dinosaurs.

All this did not happen without contention. A number of researchers objected that energy efficiency—a big topic at the time in the wake of the oil shortages—was so advantageous that only flying birds and pterosaurs, and big-brained mammals, have wastefully high metabolic rates; the dinosaurs, especially the big ones, should have stuck with the energy-saving metabolisms of reptiles. The idea that the pretty birds so many spend time watching descended from the brutish dinosaurs did not sit well with all, and some continued to argue that bird origins were among the basal archosaurs after all. Others proposed that the first protocrocodilians, which were small, lithe ground runners, were the actual close relations of birds. Very disputed were those like myself who were adorning smaller dinosaurs with feathers and other fluff when no fossils showing such had shown up. We countered that neither had scales been found on small predatory dinosaurs, so it was a matter of scientific choice until proven one way or another.

The radical and controversial concepts greatly boosted popular attention on dinosaurs, culminating in the *Jurassic Park* novels and films that sent dinomania to unprecedented heights, with the big star of the show being the *Tyrannosaurus*, which this illustrator helped design, and the overly smart "raptors." The elevated public awareness was combined with digital technology in the form of touring exhibits of robotic dinosaurs. In museums, dinosaur tails were being put up in the air, sometimes by rebuilding old mounts, including the venerable New York American Museum *Tyrannosaurus*. This time the interest of paleontologists was raised as well, inspiring the second and ongoing golden age of dinosaur discovery and research, which is surpassing that which has gone before. Assisting the work are improved scientific techniques in the area of evolution and phylogenetics, including cladistic genealogical analysis, which has improved the investigation of dinosaur relationships although it is being overused, the model having inherent limitations. New generations of artists lift the tails and get feet off the ground to represent the more dynamic gaits that are in line with the more active lifestyles the researchers now favor.

Dinosaurs are being found and named at an unprecedented rate as dinosaur science goes global, with efforts under way on all continents. In the 1970s the annual meeting of the Society of Vertebrate Paleontology might have seen a half-dozen presentations on dinosaurs; now such presentations number a couple of hundred. Especially important has been the development of local expertise made possible by the rising economies of many Second World nations, reducing the need to import Western expertise.

In South America, Argentine and North American paleontologists collaborated in the 1960s and 1970s to reveal the first Middle and Late Triassic predatory protodinosaurs, finally showing that the very beginnings of dinosaurs started among surprisingly small archosaurs. Since then, Argentina has been the source of endless remains from the Triassic to the end of the Cretaceous that include the early dinosaurs *Eoraptor* and *Herrerasaurus*; slender-skulled, long-armed dromaeosaurs; supertitanosaur sauropods that oversized avepods, such as *Giganotosaurus*, *Tyrannotitan*, and *Meraxes*, preyed on; and short-faced, stub-armed abelisaurs, including big brow-horned *Carnotaurus*. Abelisaurs also show up in India and Africa,

including Madagascar. The latter further contains some of the lighter noasaurs, among which *Masiakasaurus* has forward-curling front teeth that look good for spiking fish and other small prey. Even more intriguing are the very sophisticated Cretaceous alvarezsaurs, with delicate skulls, birdlike pelves, very long legs, and arms and hands abbreviated and strengthened into digging tools, which suggest they broke into insect colonies. First discovered in South America, they are also known from other continents. A small avepod sporting ribbon-like display feathers described by a German team under dubious legal circumstances has been returned to Brazil.

In southern Africa excellent remains of an Early Jurassic species of *Coelophysis* verified how uniform the dinosaur fauna was when all continents were gathered into Pangaea. Northern Africa has been the major center of activity as a host of sauropods and theropods have filled in major gaps in dinosaur history. Australia is geologically the most stable of continents, with relatively little in the way of tectonically driven erosion either to bury fossils or later to expose them, so dinosaur finds have been comparatively scarce despite the aridity of the continent. The most important discoveries have been of Cretaceous dinosaurs that lived close to the South Pole, showing the climatic extremes to which dinosaurs were able to adapt. Glacier-covered Antarctica is even less suitable prospecting territory, but even it has produced the Early Jurassic crested avepod *Cryolophosaurus*, and the Late Cretaceous big protobird, *Imperobatar*.

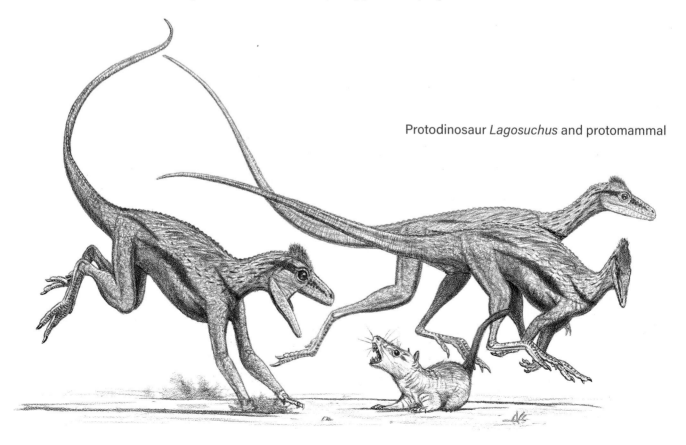

Protodinosaur *Lagosuchus* and protomammal

At the opposite end of the planet, the uncovering of a rich Late Cretaceous fauna on the Alaskan North Slope has confirmed the ability of dinosaurs, including large tyrannosaurids, to dwell in latitudes cold and dark enough in the winter that lizards and crocodilians are not found in the same deposits. Farther south, a cadre of researchers have continued to plumb the great dinosaur deposits of western North America as they build the most detailed sample of dinosaur evolution from the Triassic until their final loss. One new predatory dinosaur species after another, from podokesauroid, to ceratosaur, to allosaur, to tyrannosaur, to protobird species, is coming to light. *Tawa* was a four-toed archaic theropod, as was square-headed, buck-toothed *Daemonosaurus*. *Allosaurus* skeletons, which are now common museums displays, are now multiple species, although just how many is still under consideration—same for *Ceratosaurus*. Sorting out genera and species of gracile *Ornithomimus*, *Struthiomimus*, and *Dromiceiomimus* is proving vexing. A whole slew of fearsome tyrannosaurids from the Southwest have emerged: *Lythronax*, *Bistahieversor*, *Teratophoneus*, and *Sauronitholestes*. Farther north additional *Daspletosaurus* are being named and challenged. The same is the case for *Tyrannosaurus*, which looks like it is not just *T. rex*. Small skeletons often thought to be growing *Tyrannosaurus* are proving to be the lithe genera *Nanotyrannus* and *Stygivenator*, sporting large two-fingered hands like East Coast *Dryptosaurus*. *Nothronychus* is a big, plump-bellied therizinosaur. Sickle-clawed *Sauronitholestes* is now as completely known as *Deinonychus*. *Anzu* is a head-crested oviraptorosaur. As for eggs, complete nests assigned to the birdy troodonts have been uncovered. Not showing up yet are the nests of the big dinopredators, for reasons that are obscure.

Europe too has seen its discoveries, including the Spanish allosauroid *Concavenator*, bearing an oddly configured sail atop its hips. Romania has produced the end of Mesozoic, island-dwelling *Balaur*, which, like therizinosaurs, has gone back to four toes but otherwise is too poorly known to know what it is. In southern Germany the Solnhofen quarries continue to be worked and churn out *Archaeopteryx* remains that further detail its skeletal and feather anatomies while indicating that there are multiple species. British *Baryonyx*, along with a host of other spinosaurs from Africa and South America, *Suchomimus* and *Irritator* among them, have filled in a lot of information about what turned out to be a croc-like skulled collection. Of late a number of remains from across north Africa have been used to restore *Spinosaurus* as a peculiarly proportioned dinosaur as aquatic as a crocodile, a theory that is open to considerable question until further fossils are uncovered.

Now Mongolia and especially China have become a great frontier in dinosaur paleontology. Even during the chaos of the Cultural Revolution, Chinese paleontologists made major discoveries, including avepods small and great. As China modernized and Mongolia gained independence, Canadian and American researchers worked with their increasingly skilled and independent resident scientists, who have become a leading force in dinosaur research. It was finally realized that the oviraptors found associated with nests at the Flaming Cliffs were not eating the eggs but brooding them in a preavian manner. Since then, ring-shaped mega nests over 3 m (10 ft) across laid by very large oviraptors have come to light. Also revealing have been assorted therizinosaurs; after considerable consideration, researchers have determined that they are indeed four-toed, ponderous bird relatives with beaks for browsing and blunt teeth for chewing. Almost all of China as well as Mongolia are productive when it comes to avepods, and the genera list is formidable— dilophosaur-like head-crested *Sinosaurus*; *Limusaurus*, which shows that elaphrosaurs are toothless plant eaters; species of allosaur-like *Yangchuanosaurus*; tyrannosauroids *Guanlong* and *Dilong*; the shallow-skulled tyrannosaurid *Alioramus*; yet more velociraptorine species as well as troodonts like *Sinornithoides* and *Byronosaurus*; therizinosaurs *Jianchangosaurus*, *Alxasaurus*, *Segnosaurus*, and *Erlikosaurus*; and a host of oviraptorosaurs as per *Avimimus*, nesting *Citipati*, *Huanansaurus*, *Banji*, *Conchoraptor*, and *Oksoko*. Lately, coming from central Asia, are the cartoonish dromaeosaurs *Halszkaraptor* and *Natovenator*, which may have paddled about and swum for their prey and grabbed at it with their duck-like heads.

But that is just the start. After many decades paleontologists started paying attention to the extraordinary fossils being dug up by local farmers from Early Cretaceous Yixian–Jiufotang lake beds in the northeast of China. In the mid-1990s, complete specimens of small compsognathid theropods, labeled *Sinosauropteryx*, began to show up; their bodies were covered with dense coats of bristly protofeathers, to the delight of Ostrom and others. The feathers in illustrations of small dinosaurs dispute ended. It would even turn out that the 1 tonne *Yutyrannus* was well adorned with protofeathers. More recently it has been argued that it is often possible to determine the color of the feathers, and now even preserved scales! The Yixian–Jiufotang beds are so extensive and productive that they have become an inexhaustible source of beautifully preserved material as well as of strife, as the locals contend with the authorities for the privilege of excavating the fossils for profit—sometimes altering the remains to "improve" them—rather than for rigorous science. The feathered dinosaurs soon included the early oviraptorosaurs *Protarchaeopteryx* and *Caudipteryx*. Even more astonishing have been the Yixian–Jiufotang dromaeosaurs. The small, sickle-claw microraptorines bear fully developed wings not only on their big arms but on their similarly long legs as well. This indicates not only that dromaeosaurs first evolved as fliers but that they were adapted to fly in a manner quite different from the avian norm. A number of small land-bound dromaeosaurs are also present, along with a host of early troodonts such as *Mei* and *Sinovenator*. The therizinosaur *Beipiaosaurus* looks like a refugee from a Warner Brothers cartoon. There are new museums in China brimming with enormous numbers of undescribed dinosaur skeletons on display and in storage. Making the Yixian–Jiufotang deposits all the more interesting is that they appear to be the rare remnants of highland lakes thousands of meters above sea level, where chilling winters were similar to those at the poles at the time.

But wait—there's more from northeastern China: remains in the earlier Tiaojishan beds that predate the Solnhofen. It is therefore not surprising the fossils include the most basal known protobirds along the lines of small, winged *Anchiornis*, *Caihong*, *Serikornis*, *Eosinopteryx*, and *Pedopenna*, which may be in the same species as the prior dinobird. What has been surprising is that the Tiaojishan contains the bat-winged scansoriopterygids that look like a failed dinosaurian experiment at flight that could not compete with their feather-winged relations.

A major issue that has roiled the field of dinosaurology is the increasing private sales of fossils, sometimes legal, sometimes not. Regarding the former, Americans are free to sell any fossils found on their properties for market prices. Those were modest until the nearly complete *Tyrannosaurus* "Sue" was, after legal battles over its actual ownership involving Indigenous peoples, sold at auction for nearly US$8 million in the 1990s. It ended up at a respected institution; however, that episode led to an enormous price inflation for dinosaur skeletons and eggs, especially those of the big flesh eaters, with treasured *Tyrannosaurus* at the top of the financial pyramid. Of late the equally titanic *Tyrannosaurus* "Stan" went to an obscure Middle Eastern concern for over $30 million. After that the much smaller but more complete tyrannosauroid Bloody Mary, which some think is a juvenile *Tyrannosaurus*, sold for a few million and is now residing in a museum. Almost all these affairs involved court cases and legal fees. With prices so high, specimens are at risk of going into inaccessible collections that limit their scientific utility, while professional researchers with limited science funding may be denied access to private lands. The elevated value of dinopredator fossils has led to extensive poaching and smuggling from countries that ban private possession and export of fossils, such as China and Mongolia, to the degree that dinosaur hunters in central Asia have to deal with poachers who follow their expeditions in the field. A *Tarbosaurus* specimen put up for sale in the United States was sent back to Mongolia after legal action. These problems have led to calls for the United States to adopt laws that require all significant specimens to go into qualified institutions. That proposal is politically not practical. And if not prospected by private collectors, many specimens would remain in the ground and be subject to loss to erosion sooner or later. Resolution of these issues escapes ready solution. And any federal or state controls are moot on Indigenous peoples' lands.

On a global scale, the number of dinosaur trackways that have been discovered is in the many millions. This is logical in that a given dinosaur could potentially contribute only one skeleton to the fossil record but could make innumerable footprints. In a number of locations, trackways are so abundant that they form what have been called "dinosaur freeways." Some of the trackways were formed in a manner that suggests the predators worked in packs. A few may record the attacks of predatory theropods on herbivorous dinosaurs. Tail drag marks remain barely existent.

The history of dinosaur research is not just one of new ideas and new locations; it is also one of new techniques and technologies. The turn of the twenty-first century has seen paleontology go high tech with the use of computers for processing data and high-resolution CT scanners to peer inside fossils without damaging them. Skeletal and life restorations are being generated in 3-D digital format—whether this has improved the process of restoring prehistoric organisms is another matter. Dinosaurology has also gone microscopic and molecular in order to assess the lives of dinosaurs at a more intimate level, telling us how fast they grew, how long they lived, and at what age they started to reproduce. Bone isotopes are being used to help determine dinosaur diets and to state that some dinosaurs were semiaquatic. And it turns out that feather pigments can be preserved well enough to restore original colors. Meanwhile the *Jurassic World* franchise helps sustain popular interest in the group even as it presents an obsolete, prefeather image of the birds' closest relations.

The evolution of human understanding of dinosaurs has undergone a series of dramatic transformations since they were scientifically discovered almost 200 years ago. This is true because dinosaurs are a group of "exotic" animals whose biology was not obvious from the start, unlike fossil mammals or lizards. It has taken time to build up the knowledge base needed to resolve their true form and nature. The latest revolution is still young. When I was a youth, I learned that dinosaurs were, in general, sluggish, cold-blooded, tail-dragging, slow-growing, dim-witted reptiles that did not care for their young. The idea that some were feathered and that birds are living descendants was beyond imagining. Dinosaurology has matured in that it is unlikely that a reorganization of similar scale will occur in the future, and we now know enough about the inhabitants of the Mesozoic to have the basics well established. Predatory dinosaurs will not return to be hydrophobes, ever unwilling to pursue fleeing prey into the water, and dinosaurs' tails will not be chronically plowing through Mesozoic muds. Dinosaurs are no longer so mysterious. Even so, the research is nowhere near its end. To date, over 200 valid predatory dinosaur genera encompassing over 300 species have been discovered and named. This probably represents at most a quarter, and perhaps a much smaller fraction, of the species that have been preserved in sediments that can be accessed. And, as astonishingly strange as many of the dinosaurs uncovered so far have been, there are equally odd species waiting to be unearthed. Reams of work based on as-yet-undeveloped technologies and techniques will be required to provide further details about both dinosaur biology and the world in which they lived. And although a radical new view is improbable, there will be many surprises.

WHAT IS A PREDATORY DINOSAUR?

To understand what a predatory dinosaur—dinopredator for short—is, we must first start higher in the scheme of animal classification. The Tetrapoda are the vertebrates adapted for life on land—amphibians, reptiles, mammals, birds, and the like. Amniota comprises those tetrapod groups that reproduce by laying shelled eggs, with the proviso that some have switched to live birth. Among amniotes are two great groups. One is the Synapsida, which includes the archaic pelycosaurs, the more advanced therapsids, and the mammals, which are the only surviving synapsids. The other is the Sauropsida, much the same as the Reptilia, all of which but some early forms belong to the great Diapsida, which included most of the ancient marine reptiles, which are detailed in *The Princeton Field Guide to Mesozoic Sea Reptiles*. Surviving diapsids include turtles, the lizard-like tuataras, true lizards and snakes, crocodilians, and birds. The Archosauria is the largest and most successful group of diapsids and includes crocodilians and dinosaurs. Birds are literally flying dinosaurs.

Archosaurs also include the basal forms informally known as thecodonts because of their socketed teeth, themselves a diverse group of terrestrial and aquatic forms that include the ancestors of crocodilians and the winged pterosaurs (covered in *The Princeton Field Guide to Pterosaurs*), although those fliers were not intimate relatives of dinosaurs and birds.

The great majority of researchers now agree that the dinosaurs are monophyletic in that they shared a common ancestor that made them distinct from all other archosaurs, much as all mammals share a single common ancestor that renders them distinct from all other synapsids. This consensus is fairly recent—before the 1970s it was widely thought that dinosaurs came in two distinct types that had evolved separately from thecodont stock, the Saurischia and Ornithischia. It was also thought that birds had evolved as yet another group independently from thecodonts. Dinosauria is formally defined as the phylogenetic clade that includes the common ancestor of *Triceratops* and birds and all their descendants; for coverage of dinosaurs that are not predatory see *The Princeton Field Guide to Dinosaurs*. Because different attempts to determine the exact relationships of the earliest dinosaurs produce somewhat different results, there is some disagreement about whether the most archaic, four-toed theropods were dinosaurs or lay just outside the group. This book includes them, as do most researchers. Just before the dinosaurs were the little protodinosaurs, which although a small group that did not last very long included predators, omnivores, and herbivores; along with dinosaurs, the protodinosaurs form the dinosauriformes. These, in turn, form with pterosaurs and their probable lagerpetid relations either the dinosauromorphs or ornithodires, the terminology not being entirely stable at this time.

In anatomical terms, one of the features that most distinguishes dinosaurs centers on the hip socket. The head of the femur is a cylinder turned in at a right angle to the shaft of the femur and fits into a cylindrical, internally open hip socket. This allows the legs to operate in the nearly vertical plane characteristic of the group, with the feet directly beneath the body. You can see this system the next time you have chicken thighs. In protodinosaurs the femur head is not as strongly cylindrical, and the inside of the hip socket is not entirely open. The ankle is a simple fore-and-aft hinge joint that also favors a vertical leg posture; this is a feature of the dinosauromorphs also seen in pterosaurs. Dinosaurs were "hind-limb dominant" in that they were either bipedal or, even when they were quadrupedal, most of the animal's weight was borne on the legs, which were always built more strongly than the arms. Predatory protodinosaurs were predominantly bipedal but could

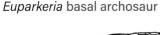

Euparkeria basal archosaur

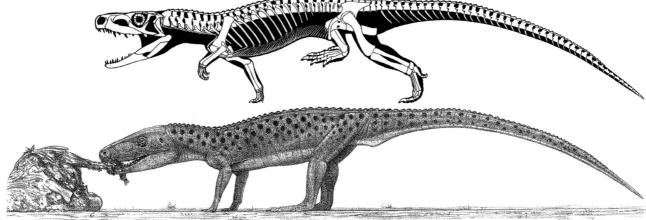

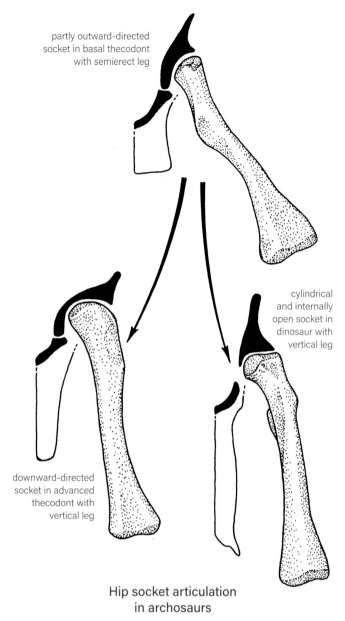

partly outward-directed
socket in basal thecodont
with semierect leg

cylindrical
and internally
open socket in
dinosaur with
vertical leg

downward-directed
socket in advanced
thecodont with
vertical leg

**Hip socket articulation
in archosaurs**

feeding in the water like moose or fishing cats, at most a few became strongly amphibious in the manner of hippos, much less marine like seals and whales. The only strongly aquatic dinosaurs are some birds. The occasional statement that there were marine dinosaurs is therefore very incorrect—these creatures of Mesozoic seas were various forms of reptiles that had evolved over the eons.

The situation within dinosaurs is interesting because it has become chaotic. For over a century, dinosaurs have generally been segregated into the two classic groups: the Saurischia, which include the largely herbivorous sauropodomorphs and the usually predaceous theropods, birds included; and the vegetarian Ornithischia, which despite the name have nothing to do with birds. But while ornithischians are clearly a real, monophyletic clade, the saurischian group has always been weak. Part of the problem is that the earliest species of a given group are all pretty similar to one another, and it is difficult to parse out the few, still subtly different characters that show they are at the base of one of the greater, more sophisticated groups but not the other. Add to that the incompleteness of the fossil record, and we are missing most of the information needed to sort out the situation. As a result, some researchers have removed the theropods from the Saurischia and more closely allied them with ornithischians in the Ornithoscelida, in which case those dinosaurs not in the latter can be placed in Paxdinosauria rather than Saurischia, which has long had its much different meaning. But there are problems with that scheme too, and it is possible that the herbivorous sauropodomorphs and ornithischians form a joint clade, the Ornithischiformes or Phytodinosauria, they being another example of unstable terminology. Others think the old Saurischia and Ornithischia are real clades. All possibilities remain viable at this time; the situation may not be resolvable with the data on hand, if ever, and this volume does not take a firm stand on the issue.

Getting to the predatory dinosaurs in particular, these are usually called theropods. But some of the early flesh-consuming dinosaurs, the herrerasaurs among them, may well not be theropods, which is the clade that includes birds. They are informally called basal dinosaurs because their phylogenetic position is uncertain. As for the classic name Theropoda, the tag has long been awkward in that the name means beast footed, which is effectively meaningless. The name also fails to do the job of elucidating the key feature of the group, the functionally and sometimes literally three-toed pes, that characterizes the vast majority of the clade, including of course birds: the three long central toes are usually the primary load-bearing foot support in predatory dinosaurs beyond a few of the earliest theropods, the outer toe being absent and the inner usually at most small, sometimes reversed, and other times entirely lost. Among theropods, those with four complete toes are limited to those very few basal examples and to the much later therizinosaurs and the poorly known *Balaur*, which

probably progress on all fours as well. Predatory dinosaurs were obligatory bipeds, with hands unsuitable for ground locomotion. The feet of predaceous dinosauriformes were digitigrade, with the ankle held clear of the ground. Dinosauriformes shared a trait also widespread among archosaurs in general, the presence of large and often remarkably complex sinuses and nasal passages.

Aside from the above basic features, dinosaurs, even when we exclude birds, were an extremely diverse group of animals, rivaling mammals in this regard. Dinosaurs ranged in form from nearly birdlike types such as the sickle-clawed dromaeosaurs to rhino-like horned ceratopsians to armor-plated stegosaurs to elephant- and giraffe-like sauropods and dome-headed pachycephalosaurs. They even took to the skies in the form of birds. However, dinosaurs were limited in that they were persistently terrestrial. Although some dinosaurs may have spent some time seem to have reevolved the strong yet not retroverted inner toe. A name that better characterizes the tridactyl main body of the group while recognizing its inclusion of birds is Avepoda, which applies to all of the nearly all Theropoda that had a strongly reduced inner toe or descended from such. Not that all avepods walked on as many as three toes. In the deinonychosaurs, the second-most inner toe is modified to be extendable in order to bear enlarged sickle claws that trackways confirm were held clear of the ground, but that toe is still large. A living avepod dinosaur with just two toes is the fast ostrich, in which only the central toe is big, the next outer toe being small and providing some lateral support.

The avepods closest to birds are usually tagged maniraptors and paravians. The contents of these groups have become disputed and confused over the years, and they are connected with the hypothesis that protobirds somehow never lost flight. Airfoilans are the dinosaurs, including birds, that had arm-borne airfoils of some sort or descended from such dinosaurs; aveairfoilans are airfoilans with basal avian-grade shoulder girdles, and typically sported feathered forelimb airfoils, and their descendants.

Because birds are dinosaurs in the same way that bats are mammals, the dinosaurs aside from birds are sometimes referred to as "nonavian dinosaurs." This usage can become awkward, and in general in this book dinosaurs that are not birds are, with some exceptions, referred to simply as dinosaurs.

Dinosaurs seem strange, but that is just because we are mammals biased toward assuming the modern fauna is familiar and normal and past forms are exotic and alien. Nor were dinosaurs part of an intuitively expected evolutionary progression that was necessary to set the stage for mammals culminating in humans. That said, dinopredators tend to be not super strange among the Dinosauria, most of them sporting a somewhat to very birdy look with a large tail usually tagged on—the biggest exception are the therizinosaurs with their super saber claws and bloated bellies. What dinosaurs do show is a parallel world, one in which mammals were permanently subsidiary. The dinosaurs show what largely diurnal land animals that evolved straight from similarly day-loving ancestors should actually look like. Modern mammals are much more peculiar, having evolved from nocturnal beasts that came into their own only after the entire elimination of nonavian dinosaurs. While dinosaurs dominated the land, small nocturnal mammals were abundant and diverse as they are in our modern world. If not for the events that brought the Age of Dinosaurs to a quick end, dinosaurs would probably still be the global norm.

Major predatory dinosaur group names

DATING DINOSAURS

How can we know that dinosaurs lived in the Mesozoic, first appearing in the Late Triassic, over 230 million years ago, and then disappearing at the end of the Cretaceous, 66 million years ago?

As gravels, sands, and silts are deposited by water and sometimes wind, they build up in sequence atop the previous layer, so the higher in a column of deposits a dinosaur is, the younger it is relative to dinosaurs lower in the sediments. Over time, sediments form distinct stratigraphic beds that are called formations. For example, *Apatosaurus, Brontosaurus, Diplodocus, Barosaurus, Stegosaurus, Camptosaurus, Allosaurus,* and *Ornitholestes* are found in the Morrison Formation of western North America, which was laid down in the Late Jurassic, from 150 to 156 million years ago. Deposited largely by rivers over an area covering many states in the continental interior, the allosaur- and ceratosaurs-dominated Morrison Formation is easily distinguished from the marine Sundance Formation, lying immediately below, as well as from the similarly terrestrial Cedar Mountain Formation above, which contains very different sets of dinosaurs. Formed over millions of years, the Morrison is subdivided into lower (older), middle, and upper (younger) levels. So a fossil found in the Sundance is older than one found in the Morrison, a dinosaur found in the lower Morrison is older than one found in the middle, and a dinosaur from the Cedar Mountain is younger still.

Geological time is divided into a hierarchical set of names. The Mesozoic is an era—preceded by the Paleozoic and followed by the Cenozoic—that contained the three progressively younger periods called the Triassic, Jurassic, and Cretaceous. These are then divided into Early, Middle, and Late, except that the Cretaceous is split only into Early and Late despite being considerably longer than the other two periods (this was not known when the division was made in the 1800s). The periods are further subdivided into stages. The Morrison Formation, for example, began to be deposited during the last part of the Oxfordian and continued through the entire Kimmeridgian. Lasting 40 million years, the Cedar Mountain Formation spans all or part of seven stages of nearly all the Early Cretaceous into the beginning of the Late Cretaceous.

The absolute age of recent fossils can be determined directly by radiocarbon dating. Dependent on the ratios of carbon isotopes, this method works only on bones and other organic specimens going back up to 50,000 years, far short of the dinosaur era. Because it is not possible to date Mesozoic dinosaur remains directly, we must instead date the formations in which the specific species are found. This is viable because a given dinosaur species lasted only a few hundred thousand to a few million years.

The primary means of absolutely determining the age of dinosaur-bearing formations is radiometric dating. Developed by nuclear scientists, this method exploits the fact that radioactive elements decay in a very precise manner over time. The main nuclear transformations used are uranium to lead, potassium to argon, and one argon isotope to another argon isotope. This system requires the presence of volcanic deposits, which initially set the nuclear clock. These deposits are usually in the form of ashfalls, similar to the one deposited by Mount Saint Helens over neighboring states, that leave a distinct layer in the sediments. Assume that one ashfall was deposited 144 million years ago and another one higher in the sediments 141 million years ago. If a dinosaur is found in the deposits in between, then we know that the dinosaur lived between 144 and 141 million years ago. As the technology advances and the geological record is increasingly better known, radiometric dating is becoming increasingly precise. The further back in time one goes, the greater the margin of error and the less exactly the sediments can be dated. In many cases, the dating is well set and not likely to change—the end of the Mesozoic happened just a dash over 66 million years ago. In other cases, less favorable geological circumstances leave matters less precisely settled; when the second edition of this book was prepared, the shift from the Oxfordian to the Kimmeridgian stage was calculated to have occurred less than 155 million years ago; however, it was just updated to over 157 million, a value that may change based on further research.

Because it was not possible to date geological deposits accurately when the time divisions were mapped and named in the 1700s and 1800s, they later proved to be very irregular in the amount of time covered by each division; for instance, the Norian stage is about 10 times longer than the Hettangian stage.

Volcanic deposits are often not available, and other methods of dating must be used. Doing so requires biostratigraphic correlation, which can in turn depend in part on the presence of "index fossils." Index fossils are organisms, usually marine invertebrates, that are known to have existed for geologically only brief periods of time, just a few million years at most. Assume a dinosaur species is from a formation that lacks datable volcanic deposits. Also assume that the formation grades into marine deposits laid down at the same time near its edge. The marine sediments contain organisms that lasted for only a few million years or less. Somewhere else in the world, the same species of marine life was deposited in a marine formation that includes volcanic ashfalls that have been radiometrically dated to between 83 and 81 million years. We can then conclude that the dinosaur in the first formation is also 83 to 81 million years old.

A number of predatory dinosaur-bearing formations lack both volcanic deposits and marine index fossils. It is often not possible to date the dinosaurs in these deposits accurately. It is only possible to broadly correlate the level of development of the dinosaurs and other organisms in the formation with faunas and floras in better-dated formations, and this produces only approximate results. This situation is especially common in central Asia. The reliability of dating therefore varies. It can be very close to the actual value in formations that have been well studied and contain volcanic deposits; these can be placed in specific parts of a stage. At the other extreme are those formations that, because they lack the needed age determinants and/or have not been sufficiently well examined, can only be said to date from the early, middle, or late portion of one of the periods, an error that can span well over 10 million years. North America currently has the most robust linkage of the geological time scale with its fossil dinosaurs on Earth.

THE EVOLUTION OF PREDATORY DINOSAURS AND THEIR WORLD

Dinosaur predators and their prey appeared in a world that was both ancient and surprisingly recent—it is a matter of perspective. The human view that the time of Mesozoic dinosaurs was remote in time is an illusion that results from our short life span. A galactic year, the time it takes our solar system to orbit the center of the galaxy, is 200 million years. Only one galactic year ago the dinosaurs had just appeared on planet Earth. When dinosaurs first emerged, our solar system was already well over four billion years old, and 95% of the history of our planet had already passed. A time traveler arriving on Earth when dinosaurs first existed would have found it both comfortingly familiar and marvelously different from our time.

As the moon slowly spirals out from the Earth because of tidal drag, the length of each day grows. When dinosaurs initially evolved, a day was about 22 hours and 45 minutes long, and the year had around 385 days; when they went largely extinct, a day was up to 23 hours and over 30 minutes, and the year was down to 371 days. The moon would have looked a little larger and would have more strongly masked the sun during eclipses—there would have been none of the rare annular eclipses in which the moon is far enough away in its elliptical orbit that the sun rings the moon at maximum. The "man on the moon" leered down upon the dinosaur planet, but the prominent Tycho crater was not blasted into existence until toward the end of the Early Cretaceous. There were no other big galaxies closer to ours than is Andromeda today, but on occasion star-packed globular clusters may have passed fairly close to our solar system. As the sun converts an increasing portion of its core from hydrogen into denser helium, it becomes hotter by nearly 10% every billion years, so the sun was about 2% cooler when dinosaurs first showed up and around a half percent cooler than it is now when most went extinct.

At the beginning of the great Paleozoic era over half a billion years ago, the Cambrian Revolution saw the advent of complex, often hard-shelled organisms. Also appearing were the first simple vertebrates. As the Paleozoic progressed, first plants and then animals, including tetrapod vertebrates, began to invade the land, which saw a brief Age of Amphibians in the late Mississippian followed by the classic Age of Reptiles in the Pennsylvanian and much of the Permian. By the last period in the Paleozoic, the Permian, the continents had joined together into the C-shaped supercontinent Pangaea, which straddled the equator and stretched nearly to the poles north and south. Lying like a marine wedge between the terrestrial arms of northern Laurasia and southern Gondwana was the great Tethys tropical ocean; all that is left of that is the modest Mediterranean. Seventy percent of the world was dominated by the enormous Panthalassic superocean, whose vast expanse was marred only by occasional islands; the Pacific is its somewhat lesser descendant. With the majority of land far from the oceans, most terrestrial habitats were harshly semiarid, ranging from extra hot in the tropics to sometimes glacial at high latitudes. The major vertebrate groups had evolved by that time. Among synapsids, the mammal-like therapsids, some up to the size of rhinos, were the dominant large land animals in the Age of Therapsids of the Late Permian. These were apparently more energetic than reptiles, and those living in cold climates may have used fur to conserve heat. Toward the end of the period, the first archosaurs appeared. These lowslung, vaguely lizard- or crocodilian-looking creatures were a minor part of the global fauna. The conclusion of the Permian was marked by a great extinction that is attributed to the massive volcanism then forming the enormous Siberian Traps—there is no evidence of a major meteorite impact at the time—which so heavily contaminated the atmosphere in multiple ways that the global environment was severely disrupted chemically and weatherwise. In some regards the Permian–Triassic extinction exceeded that which killed off the terrestrial dinosaurs 185 million years later.

At the beginning of the first period of the Mesozoic, the Triassic, the global fauna was severely denuded. As it recovered, the few remaining therapsids enjoyed a second evolutionary radiation and again became an important part of the wildlife. And again, they never became truly enormous or tall, although a few late examples were as big as small elephants. This time they had serious competition, as the archosaurs also underwent an evolutionary explosion, first expressed as a wide variety of thecodonts, some of which reached a tonne in mass. While one group evolved into aquatic, armored crocodile mimics, some quite large, and others became armored land herbivores, most were terrestrial predators that moved on erect legs achieved in a manner different from dinosaurs. The head of the femur did not turn inward; instead, the hip socket expanded over the femoral head until the shaft could be directed downward. Some of these erect-legged archosaurs were nearly bipedal. Others became toothless plant eaters. One branch of the thecodonts is still with us, the crocodilians that began to appear in the Triassic as small, lightly built land runners that spun off aquatic forms. It is now realized that in many respects the Triassic thecodonts filled the lifestyle roles that would later be occupied by dinosaurs. Even so, these basal archosaurs never became gigantic or very tall. Many thecodonts had a very undinosaurian feature: their ankles were in the form of crocodilians, being complex, door-hinge-like joints in which a tuber projecting from one of the ankle bones helped increase the leverage of the muscles on the foot, rather as in mammals. In the Norian, a modest extinction event cut back somewhat on the diversity of therapsids and thecodonts. In the Late Triassic, the membrane-winged, long-tailed pterosaurs show up in the fossil record fully formed, since their evolution had started earlier. Their go-to mode of escape from predators dinosaurian and otherwise was to take flight. That pterosaurs had the same kind of simple-hinge ankle seen in dinosaurs is a reason that the two groups appear to be related, forming the Ornithodira. The energetic pterosaurs were insulated; we do not yet know whether other nondinosaurian archosaurs, including protodinosaurs and early dinosaurs, were also covered with thermal fibers, but the possibility is substantial.

By the middle Triassic quite small predatory archosaurs were evolving that exhibited a number of features of dinosaurs. Although the hip socket was still not internally open, the femoral head was turned inward, allowing the legs to operate in a predominantly vertical plane. The ankle was the simple fore-aft hinge that it remains in birds. The skull was lightly constructed. At first recorded as skeletal remains from the early Late Triassic of South America—trackways from elsewhere suggest they appeared earlier in the period—protodinosaurs have since been found on other continents. Some, such as *Lagosuchus* and *Lewisuchus*, were predaceous, others herbivorous. These bipedal/quadrupedal early near dinosaurs would survive only until the Norian, at least the plant-eating examples, as they were displaced by their descendants. Protodinosaurs show that predatory dinosaurs started out as little creatures; big theropods did not descend from the big basal archosaurs.

From small things big things can evolve, and very quickly. In the Carnian stage of the Late Triassic the fairly

The Early Late Triassic *Lewisuchus admixtus* and *Lagosuchus talampayensis* prey

large-bodied, small-hipped, four-toed first wave of dino-predators were on the global stage, including herrerasaurs, *Tawa*, and odd-headed *Daemonosaurus*. These fully bipedal early dinosaurs with cylindrical hip joints dwelled in a world still dominated by complex-ankled archosaurs and would not last beyond the Norian or maybe the Rhaetian stage, perhaps because these early dinosaurs did not have an aerobic capacity high enough to vie with their yet newer dinosaurian competitors. The Norian saw the appearance of the great group that is still with us, the bird-footed ave-pod theropods, whose enlarging hips and beginnings of the avian-type respiratory system imply a further improvement in aerobic performance and thermoregulation. In the Trias-sic, all the avepods represented by skeletal remains are po-dokesauroids, as per *Coelophysis*, which did not exceed 200 kg (400 lb) as far as is known; a few trackways record dino-predators about twice that heft. Some splay-toed tridactyl trackways hint at avepods more derived than podokesau-roids. That the latter's thigh muscle anchoring pelvic ilial plates were not yet as enlarged as those of later predatory dinosaurs means their locomotion was not as advanced. Not yet appearing as far as we know on Triassic dino-predators were bone-based display devices, although soft tissue organs cannot be ruled out. At about the same time, the first members of one of the grand groups of herbiv-orous dinosaurs are recorded in the fossil record, the small-hipped, semibipedal prosauropods with lightly built heads and slender teeth best suited for tender vegetation such as ferns. Prosauropods were well armed with big hooked thumb claws and could kick and lash out with their big tails. All these new dinosaurs gave thecodonts and therap-sids increasing competition as they rapidly expanded in di-versity and, in the case of the plant eaters, size also. Just

15 or 20 million years after the evolution of the first little protodinosaurs, prosauropods weighing 2 tonnes had de-veloped. In only another 10 million years, prosauropods as big as elephants were extant, becoming the first truly gigantic land animals. These long-necked dinosaurs were also the first herbivores able to browse at high levels, many meters above the ground. The first of the beaked herbiv-orous ornithischians arrived in the Carnian. These little semibipeds were not common, and they, as well as small theropods and prosauropods, may have dug burrows as refuges from the predator-filled world. By the last stage of the Triassic, dinosaurs predaceous and herbivorous were becoming the ascendant land animals, although they still lived among a number of thecodonts and some therapsids. From the latter, at this time, evolved the first mammals. Mammals and dinosaurs have therefore shared the planet for over 200 million years—but for 140 million of those years, mammals remained small and correspondingly sub-ject to going down the gullets of avepods.

Because animals could wander over the entire super-continent with little hindrance from big bodies of water, faunas tended to exhibit little difference from one region to another. And with the continents still collected together, the climatic conditions over most of the supercontinent remained harsh. It was the greenhouse world that would prevail through the Mesozoic. The carbon dioxide level was two to ten times higher than it is currently, boosting temperatures to such highs—despite the slightly cooler sun of those times—that even the polar regions were relatively warm in winter. The low level of tectonic activity meant there were few tall mountain ranges to capture rain or interior seaways to provide moisture. Hence, there were great deserts, and most of the vegetated lands

The Late Triassic *Tawa hallae*

were seasonally semiarid, but forests were located in the few regions of heavy rainfall and groundwater created by climatic zones and rising uplands. It appears that the tropical latitudes were so hot and dry that the larger dinosaurs, with their high energy budgets, could not dwell near the equator and, except for tenuous coastal strips, were restricted to the cooler, wetter, higher latitudes. The flora was in many respects fairly modern and included many plants with which we would be familiar. Wet areas along watercourses were the domain of rushes and horsetails. Some ferns also favored wet areas and shaded forest floors. Other ferns grew in open areas that were dry most of the year, flourishing during the brief rainy season. Large parts of the world may have been covered by fern prairies, comparable to the grasslands and shrublands of today. Tree ferns were common in wetter areas. Even more abundant were the fern-like or palm-like cycadeoids,

similar to the cycads that still inhabit the tropics. Taller trees included water-loving ginkgoaceans, of which the maidenhair tree is the sole—and, until widely planted in urban areas, the nearly extinct—survivor. Dominant among plants were conifers, most of which at that time had broad leaves rather than needles. Some of the conifers were giants rivaling the colossal trees of today, such as those that formed the famed Petrified Forest of Arizona. Flowering plants were completely absent.

The end of the Triassic about 200 million years ago saw another extinction event. A giant impact had occurred in the hard, ancient rocks of southeastern Canada, but it was a few million years before the specific time of the extinction. Coincident with the extinction were intense volcanics tied to the formation of the budding Atlantic Ocean. The heavy dusting of the atmosphere with reflective aerosol debris cooled off the planet to the degree that minor glaciations

The Early Jurassic
Coelophysis rhodesiensis and
Massospondylus carinatus

occurred at high latitudes. The thecodonts and therapsids suffered the most: the former were wiped out with the exception of crocodilians, and only scarce remnants of the latter survived along with furry mammal relatives. The more energetic, sometimes insulated pterosaurs and dinosaurs were able to deal with the big chills without much difficulty and sailed through the crisis into the Early Jurassic with little disruption, so much so that avepods such as *Coelophysis* remained common and little changed, as did prosauropods. For the rest of the Mesozoic, dinosaurs, including those carnivorous, would enjoy almost total dominance on land except for some substantial semiterrestrial crocodilians; otherwise there simply were no competitors above a few kilograms in weight. Such extreme superiority was unique in Earth's history. The Jurassic and Cretaceous combined were the Age of Dinosaurs.

For reasons that are obscure, delicate paired fore-and-aft head crests over the snout were a repeated display feature of a variety of avepods in much of the Jurassic. Atypical was *Cryolophosaurus* with a transversely broad crest above the orbits—that it lived near the South Pole shows that some dinosaurs were using their fast-burning metabolisms to cope with chilly weather. Appearing in the Early Jurassic were the averostran avepods whose big hips heralded a more powerful set of leg muscles for better locomotion powered by improved respiration. Podokesauroids as per *Dilophosaurus* and averostrans upscaled to hundreds of kilograms and maybe a tonne or so, posing a greater threat to the herbivorous dinosaurs. Regarding those, a literally big novelty of the Early Jurassic was the advent of sauropods that had descended from prosauropods. Sauropods independently evolved birdlike respiration and large pelves in parallel to averostrans. Elephantine in both size and body and leg form, with very long necks and tails added, the sauropod's columnar legs meant these were the first dinosaurs not able to achieve a full run in order to try to outpace the fleet-footed avepods. So they had to stand and fight with their powerful tails and hands bearing stout thumb claws, the latter when rearing up like angry bears. As the Jurassic progressed, the small-hipped Triassic-grade podokesauroids and prosauropods proved unable to compete with their more modern, bigger-hipped relations and were gone by the end of the Early Jurassic. It is likely that the loss of podokesauroids in favor of averostrans and the prosauropods in favor of sauropods was finalized by another bout of mass volcanism, this time in southern Gondwana in the late Early Jurassic, which is associated with tough conifers becoming the dominant trees during the rest of the Jurassic into the Cretaceous. Although some avepods were getting moderately large, the much more gigantic adult sauropods enjoyed a period of relative immunity from attack. Ornithischians remained uncommon, and one group was the first set of dinosaurs to develop armor protection against the flesh-eating avepods. Another group of ornithischians was the small, chisel-toothed, semibipedal heterodontosaurs, which established that fiber coverings

had evolved in some small dinosaurs by this time, if not earlier. These could either run away from attackers or fight with beaks and fang-like front teeth. On the continents, crocodilians remained small and fully or partly terrestrial.

Farther west, the supercontinent was beginning to break up, creating African-style rift valleys along today's Eastern Seaboard of North America that presaged the opening of the Atlantic. Even so, Europe remained an Indonesia-like archipelago of islands immediately northeast of North America, as it had been for a long while. More importantly for dinosaur faunas, the increased tectonic activity in the continent-bearing conveyor belt formed by the mantle caused the ocean floors to lift up, spilling the oceans onto the continents in the form of shallow seaways that began to isolate different regions from one another, encouraging the evolution of a more diverse global wildlife. The expansion of so much water onto the continents also raised rainfall levels, although most ecosystems remained seasonally semiarid. The increased rain helped cool the continents some, allowing big dinosaurs to roam about at all latitudes. The moving land masses also produced more mountains able to squeeze rain out of the atmosphere.

Beginning 175 million years ago, the Middle Jurassic began the long Age of Sauropods; they first matched medium-sized whales in bulk and trees in height, and by the end of the stage the height maximum had topped off at six to seven stories on animals of three dozen tonnes. A few sauropods had tail clubs to whack at yangchuanosaur avepods and each other with. Also appearing were the first small, armored stegosaur ornithischians that also introduced tail spikes for dealing with the dinopredators. Even smaller were the little ornithopods, the beginnings of a group of ornithischians whose dental batteries gave them great evolutionary potential. Although the increasingly sophisticated tetanuran, avetheropod, and coelurosaur avepods evolved and featured highly developed avian-type respiratory systems, for reasons that are not obvious they continued to fail to produce true giants, those yet known reaching barely a tonne at most, no larger than the largest mammalian land carnivores. The often stocky megalosaurs were a frequent feature of the later Jurassic of Europe and North America. There is tenuous evidence that flowering plants were present by the middle of the Jurassic, but if so they were not yet common.

The Late Jurassic, which began 160 million years ago, was the apogee of two dinoherbivore groups, the sauropods and the stegosaurs. Sauropods would never again be so diverse. Some rapidly enlarged to 50 to 75 tonnes, and a few may have greatly exceeded 100 tonnes, rivaling the biggest baleen whales. But it was a time of growing danger for the sauropods: avepods had finally evolved hippo-sized megalosaurs, Asian yangchuanosaurs, and big brow-horned Euro–American allosaurs larger than any mammalian land predators and able to tackle the colossal herbivores. With the appearance of the megaavepods began the Age of Averostran Avepods, which would last through

23

the rest of the Mesozoic. However, yangchuanosaurs did not make it through the Late Jurassic. North American Late Jurassic nose-horned ceratosaurs were not as hefty, but their oversized teeth gave them exceptional slash and bite power. The now rhino- and sometimes elephant-sized stegosaurs were also at their most diverse. Some developed spectacular plate arrays; these dinosaurs could not run and had to fight to survive. But the other group of big armored dinosaurs, the short-legged ankylosaurs sans tail clubs, was beginning to develop. Also entering the fauna were the first fairly large ornithopods, sporting thumb spikes with which to poke at avepods. Asia saw the development of small semibipedal ceratopsians with powerful beaks to bite avepods that were trying to mug them.

The still long-armed ancestors of tyrannosaurs seem to have been developing in the latter Jurassic, and assorted gracile, small-game-catching coelurosaurs were numerous. These coelurosaurs included the haplocheirids, which may have been the ancestors or close to it of the strange alvarezsaurs, or alternatively the swift ornithomimosaurs, of the coming Cretaceous. Appearing on the paleo scene were the first known avepods shifting away from dining on flesh toward plants, elaphrosaurs whose arms and hands were dramatically reduced and which used long legs to flee their carnivorous cousins. But it is the advent of the highly bird-like and partly arboreal airfoilans at the end of the mid-Jurassic going into the late part of the period that was a major event that survives to today. Initial dinosaur flight appears to have come in two versions. One experiment was the bat-winged gliding scansoriopterygids of Asia; these apparently soon disappeared, perhaps because of competition from the bird-winged aveairfoilans. More importantly for the long-term future of dinosaurs of the air were Chinese deinonychosaur anchiornids, the earliest dinosaurs known to have had large feathers on their arms as well as legs. This was the time of the first avepod size squeeze, in which dinosaurs usually on the large side spun off diminutive forms of just a fifth of a kilogram (half a pound) as part of dinosaurs taking wing. A few million years later, when Europe was still a nearshore extension of northeastern North America, the first "bird," the deinonychosaur *Archaeopteryx*, was extant. Preserved in lagoonal deposits on the northwestern edge of the then great Tethys Ocean, it had a combination of very large arms and long, asymmetrical wing feathers, indicating that it was developing the early stages of powered flight. Also found in the deposits is fragmentary *Alcmonavis*, the first avian to have an expanded central finger base to better support the outer flight feathers. The advent of the little aveairfoilans also heralded the first major increase in dinosaurian mental powers, as brain size and complexity rose to the lower avian level. Pterosaurs, which retained smaller brains, remained small bodied, and most still had long tails. Although some crocodilians were still small runners competing with the avepods for prey, the highly amphibious crocodilians of the sort we are familiar with were appearing—but not

particularly large; that there were no big freshwater armored archosaurs in the Jurassic is another paleomystery. Although small, mammals were undergoing extensive evolution in the Jurassic. Many were insectivorous or herbivorous climbers, but some were burrowers in part to escape avepods and others had become freshwater-loving swimmers. Dinosaurs predatory and otherwise did not yet show signs of being specifically adapted for life in the water, for reasons that remain opaque. Something that did not happen in the Triassic/Jurassic was the evolution of predaceous dinosaurs with really big hips bearing really long legs for maximal speed performance.

During the Middle and Late Jurassic, carbon dioxide levels were incredibly high, with the gas making up between 5% and 10% of the atmosphere, 10 to 20 times twenty-first-century levels (and about twice preindustrial norms). As the Jurassic and the classic Age of Sauropods ended, the incipient North Atlantic was about as large as today's Mediterranean. Vegetation had not yet changed dramatically from the Triassic. What happened to the fauna at the end of the Jurassic is not well understood because of a lack of deposits. Some researchers think there was a major extinction, in part because new data are showing that the very different Early Cretaceous dinofaunas were up and running closer to the Jurassic–Cretaceous (J/K) boundary than was thought, maybe even a little before, indicating a sudden and fast turnover of the type that is usually evidence of a major global disruption. There is not yet obvious geological evidence of a big bad event at the time. Whatever happened, ceratosaurs, megalosaurs, allosaurs, and elaphrosaurs were victims of the times.

The epic Cretaceous began some 145 million years ago. This 80-million-year-long period would see an explosion of dinosaur meat eating and other evolution that surpassed all that had gone before as the continents continued to split, the south Atlantic began to open, India detached from Antarctica and then Madagascar as it began to head toward Asia, and seaways crisscrossed the continents. Greenhouse conditions became less extreme as carbon dioxide levels gradually edged downward, with a quick sharp drop near the end of the Valanginian that led to glaciers appearing even at mid-latitudes in the Hauterivian cold snap. However, carbon dioxide levels did not drop to the modern preindustrial level. Early in the Cretaceous, the warm Arctic oceans generally kept conditions up there balmy even in winter. At the other pole, continental conditions rendered winters frigid enough to form permafrost. General global conditions were a little wetter than they were earlier in the Mesozoic, but seasonal aridity remained the rule in most places, and true rain forests continued to be scarce at best.

While sauropods remained abundant and often enormous to the end of the Mesozoic, they were markedly less diverse than before. To a fair extent the Cretaceous was the Age of Ornithischians. Ornithopods small and especially large flourished. Thumb-spiked-wielding iguanodonts soon became common herbivores in the Northern Hemisphere.

Their well-developed plant-pulping dental batteries may have been a key to their success. A few evolved tall sails formed by their vertebral spines. Among ceratopsians, the small Asian parrot-beaked-armed psittacosaurs first proliferated, and their relatives, the big-headed protoceratopsids, which probably battled predators with the fierceness of suids, appeared in the same region. So did the first of the dome-headed pachycephalosaurs that could butt the flanks of tormentors with their solid-topped heads. Stegosaurs, however, soon departed from the planet. Even so, the dinosaurs tended to add new groups without losing much of the old ones, building up their diversity over the Mesozoic. In the place of stegosaurs, the low-slung and extremely fat-bellied armored ankylosaurs became a major portion of the global fauna, their plates and spikes providing protection from the Laurasian, sometimes titanic, carcharodontosaur allosauroids and the snub-nosed, short-armed, heavy-limbed abelisaurs down in Gondwana. The carcharodontosaurs sometimes evolved back sails, while the abelisaurs were prone to feature head adornments, but the paired crests of Jurassic avepods were long gone for reasons not

The Late Jurassic *Allosaurus fragilis*

understood. Another group of giant theropods was the croc-snouted spinosaurs, apparently adapted to catch fish as part of their diet. Bone isotopes indicate that spinosaurs were semiaquatic like hippos, making them the first dinosaur group to at long last become fairly water loving.

It was among the smaller theropods that dinosaur evolution really went wild in the Early Cretaceous. Relatives of the earlier elaphrosaurs and the big abelisaurs, the similarly southerly noasaurs were lightly built toothed predators with a propensity toward fishing. Coming along evolution wise were the not yet gigantic Northern Hemisphere tyrannosauroids, heavily feathered, up to a tonne, and still large armed and three fingered. And the first of the herbivorous ostrich-mimicking ornithomimosaurs were present. But the focus of events was among the nearly avian little aveairfoilans. As revealed by the spectacular volcanic highland lake deposits featuring snowy winters of northeastern China, deinonychosaurs developed into an array of flying and flightless forms, with the latter probably secondarily flightless descendants of the fliers. The famous sickle-clawed dromaeosaurs appear to have begun as small aerialists with two sets of wings, the normal ones on the arms and an equally large set on the hind legs. From these appear to have evolved bigger bodied—some a few hundred kilograms—but smaller armed terrestrial dromaeosaurs that hunted larger game. The other major sickle-clawed Cretaceous deinonychosaur group, the more lightly built and swifter-running, big-eyed troodonts, also thrived. The flightless troodonts may have evolved from the more aerially capable anchiornids of the prior geoperiod.

At the same time, not only had birds themselves descended from deinonychosaur dinosaurs, but the Chinese deposits show they had already undergone a spectacular evolutionary radiation by 125 million years ago, with some species flying about in enormous flocks. Some retained teeth, others were toothless. Some had long tails, most did not. None were especially large. By this time the big bird split had occurred. While many of these early birds were basal forms, two great groups had evolved. On one hand were the enantiornithines that became the most abundant avians of the Mesozoic. On the other were then less numerous euornithines, which are still around as our modern dinosaurs that grace our skies. Among the basal birds were the toothed, long-skulled, and long-tailed herbivorous jeholornithiforms.

It is possible that they were the ancestral type for the enigmatic, potbellied, land-bound plant chewing therizinosaur dinosaurs that sported tellingly similar skulls. And the parrot-beaked omnivoropterygid basal birds in turn bear a striking resemblance to the similarly deep-headed and short-tailed protarchaeopterygid and caudipterid omnivorous oviraptorosaurs from the same formations. Therefore, there is reason to propose that the short-tailed oviraptorosaurs with their many other bird-style features were yet another group of secondarily flightless dinobirds, ones more advanced than the long-tailed archaeopterygian-dromaeosaur-troodont deinonychosaurs and the therizinosaurs. The conventional view held by most researchers is that flightless therizinosaurs and oviraptorosaurs happened to be merely convergent with the flying jeholornithiforms and omnivoropterygids, respectively. Among oviraptorosaurs, the Asian caudipterids were, along with troodonts, the first dinosaurs running about on the very long and gracile legs that suggest the highest-grade land animal speeds.

Pterosaurs, now short tailed and consequently more dynamic fliers, were becoming large as they met increasing competition from the air-capable dinosaurs in the form of birds. Also fast increasing in size were the freshwater crocodilians, some gigantic, making them an increasing threat even for big predatory dinosaurs coming to water to drink or for other purposes, the avepods being vulnerable to being pulled underwater and drowned. Some large crocodilians were semiterrestrial and able to compete with large avepods on land as well as in the water. Still scampering about were a few small running crocodilians. Some carnivorous mammals were big enough, about a dozen kilograms, to catch and consume the smallest dinosaurs and their babies. Even gliding mammals were cutting through the air by this time.

During the late Early Cretaceous flowering plants began to become an important portion of the global flora. The first examples were small shrubs growing along shifting watercourses, where their ability to rapidly colonize new territory was an advantage. Others were more fully aquatic, including water lilies.

In the Late Cretaceous, which began 100 million years ago with the continental breakup well under way, the

The Early Cretaceous *Deinonychus antirrhopus*

Panthalassic colossus became the almost-as-great Pacific, and interior seaways often covered vast tracts of land. As carbon dioxide levels continued to drop, the dark Arctic winters became cold enough to match the conditions seen in today's high northern forests, and glaciers crept down high-latitude mountains. Mammals were increasingly modern but still small. Pterosaurs, marine and terrestrial, became gigantic to a degree that stretches credulity. Toward the end of the Cretaceous, the freshwater-loving azhdarchids sported wings of 11 m (over 35 ft) and easily outweighed ostriches. These monsters of the air were competitors and threats to avepods, and vice versa, though they could not stand up to the bigger hunting dinosaurs. Small running crocodilians continued to offer some competition to small and juvenile avepods, while a few even became herbivorous. As for the conventional freshwater crocodilians, in some locales they become colossi, up to 12 m (close to 40 ft) long and approaching 10 tonnes, as large as the biggest flesh-eating avepods. Although these monsters fed mainly on fish and smaller tetrapods, they posed a real threat to all but the largest dinosaurs when they came to large bodies of water to drink or swam across bodies of

basal pygostilian
Confuciusornis

ornithurian
Ichthyornis

enantiornithine
Sinornis

enantiornithine
Yixianosaurus

ornithurian
Hesperornis

enantiornithine
Longipteryx

Cretaceous birds

water. The hazard should not be exaggerated, however, because these supercrocs do not appear to have been very numerous in many locations, were absent at higher latitudes, and did not last all that long. Even so, their existence may have discouraged the evolution of profoundly aquatic dinosaurs, none of which went fully aquatic, much less marine.

Although sauropods soon became limited to the titanosaurs, that group proliferated across most of the globe, being especially diverse in the Southern Hemisphere and wrapping up the 150 million years that made them the most successful herbivore group in Earth's history. Sauropods disappeared from North America for part of the Late Cretaceous, only to reappear in the drier southwestern regions toward the end. Some sauropods were armored; this may have in part been a means to protect the juveniles against the increasing threat posed by a growing assortment of predators. A few titanosaurs were rather small ground grazers that looked like meat on the table for megaavepods. Others were land titans perhaps as big as the biggest whales. Titanosaurs, supersized and otherwise, were subject to attack from abelisaur and carcharodontosaur allosauroid theropods, some of the latter matching bull elephants in bulk. However, all the allosauroids went out of existence in the early Late Cretaceous—why is not apparent. Perhaps even larger were the African sail-backed spinosaurs of the early Late Cretaceous; this group too did not make it to the end of the Mesozoic for reasons not yet explained. Claims that the last but still pneumatic-bodied spinosaurs were as highly aquatic as modern crocodilians are excessive.

The ultrawide-bodied ankylosaurs continued their success, especially in the Northern Hemisphere. One group of the armored herbivores developed tail clubs with which to deter and, if necessary, damage their meat-hungry enemies. Clubless nodosaurids, some of which could use big shoulder spines to charge at large avepods, also persisted. The stand and fight with thumb spikes iguanodonts faded from the scene to be replaced by the duck-billed hadrosaurs whose only defense was kicking and running. The most common herbivores in much of the Northern Hemisphere, hadrosaurs may have been adapted in part to browse on the flowering ground cover that was beginning to replace the fern prairies. Small ornithopods, not all that different from the bipedal ornithischians that had appeared back near the origins of the dinosaurs, continued to dwell over much of the globe. In the Northern Hemisphere the protoceratopsids, small in body and big in parrot-beak weapon heads, were common in many locales. It was from this stock that some of the most spectacular dinoherbivores evolved—the rhino- and elephant-sized ceratopsids whose oversized heads sported horns, neck frills, great parrot-like beaks, and slicing dental batteries with which to try to fend off the tyrannosaurs. These dinosaurs were limited largely to the modest-sized stretch of North America that lay west of the interior seaway.

Birds, some still toothed, most belonging to the abundant egg-burying enantiornithines, the rest euornithines, continued to thrive. The enantiornithines were a rather diverse lot, including wading, swimming, fishing, insect-hunting, fruit- and seed-feeding, and raptorial hunting forms, with a large portion being arboreal. Few, if any, became particularly large. The ornithurian euornithines were much less common and not as broadly diverse, but some of the toothed versions expanded out onto the oceanic airs, a first for dinosaurs. A group of those, the hesperornitheans, lost their wings and became loon-like foot-propelled divers up to 2.5 m long, the first fully marine dinosaurs not to be matched until the later penguins. By the Late Cretaceous the classic short-armed coelurosaurs were no longer extant, unable to deal, it seems, with all the advanced avepods—and the juveniles of the big species—about. The small predatory theropods consisted of the intelligent and sickle-clawed swift troodonts and leaping dromaeosaurs, some of which were still able to fly. One dromaeosaur group, the newly found halszkaraptors, lost their rod-stiffened tails seemingly in favor of a more watery life, pursuing prey with their duck-like snouts. Also successful were the short-tailed nonpredatory aveairfoilans, those being the deep-headed omnivorous avimimid, caenagnathid, and oviraptorid oviraptorosaurs, many of which exhibited dramatic cassowary-style head crests, as well as the small-headed, big-clawed herbivorous therizinosaurs. In both groups some species became quite large, *Therizinosaurus* being elephant size. So did some ornithomimosaurs—the once enigmatic *Deinocheirus* proving to be a duck-headed five tonner. The Late Cretaceous saw the appearance of the trio of big vegetarian avepods among the therizinosaurs, oviraptorosaurs, and ornithomimosaurs. Among the latter group and fairly common in the Northern Hemisphere, the very long- and slender-legged ornithomimids with their oversized hips became perhaps the fastest of all dinosaurs—the fleetest modern ratites perhaps excepted—in their effort to dodge attack from also swift tyrannosaurs. They were closely matched speedwise by the little colonial bug-eating alvarezsaurs, which when not dashing hither and yon used their short and amazingly stout arms and reinforced hands to break into insect nests. Also swift among the oviraptorosaurs were Asian avimimids.

Culminating more than 150 million years of Mesozoic megaavepod history were the great tyrannosaurids, the most sophisticated and powerful of the gigantic dinopredators. Reduced arms with just two fingers, enormous long legs for speed, skulls sculpted for hunting the big game, and somewhat reduced tails gave them their exceptionally handsome looks even when they grew as big as elephants. But their global impact is not to be overstated: the classic tyrannosaurids came into existence only some 15 million years before the end of the Mesozoic and were limited to Asia and North America. Apparently they occasionally wandered, along with other theropods, hadrosaurs, and ankylosaurs, back and forth across the polar Bering land bridge, where some became specialized for the sharp winter climate. Farther south in central Asia, more temperate climes included winter snows. In North America a size race occurred as tyrannosaurids, ceratopsids, ankylosaurs, and pachycephalosaurids reached unprecedented sizes for their groups. It had been thought that the great mass expansion did not occur until the last three million years of the late Maastrichtian, but it is being realized that some American tyrannosaurids reached elephantine bulk in the late Campanian and early Maastrichtian. That such occurred in the American southwest is not surprising because giant titanosaur sauropods dwelled there. Why more northerly tyrannosaurids become so big is less clear in the absence of similarly bulky prey. The size boost became more general approaching the close of the Cretaceous, resulting in the classic *Tyrannosaurus*, *Triceratops*, *Ankylosaurus*, and *Pachycephalosaurus* fauna; the edmontosaur hadrosaurs got bigger too. This size increase looks like the result of a predator–prey arms race, an expansion of the resource base as the retreating interior seaway linked the eastern and western halves of the continent into a larger land area, or a combination of both. The common view that *Tyrannosaurus* was a super predator represented by only one species, juvenile and adult, in its habitat is being overturned by growing evidence for a number of tyrannosauroid taxa of varying sizes, the smaller *Nanotyrannus* and especially *Stygivenator* examples sporting arms as long or actually longer than those of adult *Tyrannosaurus*. The striking diversity of the regions' tyrannosaurs was due to the reunification of the entire continent, allowing the faunas from east and west to intermix for the first time in tens of millions of years as the once great interior seaway drained away. Not known is whether tyrannosaurs were so diverse across much of North America or such diversity was limited to regions where the ranges of a number of the predators happened to overlap. Not getting larger were ornithomimids, some of which were actually small. Not doing especially well as the Late Cretaceous progressed were dromaeosaurs and troodonts, perhaps owing to competition from the small tyrannosauroids and the juveniles of their giant parents. *Tyrannosaurus* juveniles were on the rare side, so it was the more common *Stygivenator*s and *Nanotyrannus* that were the main competitors for the other small avepods in that ecosystem. On a European island, the peculiar, so far skull-less *Balaur* of unclear relationships reevolved four functional toes, the inner two bearing sickle claws that it may or may not have used for hunting, while apparently losing the flight of its predecessors. In much of the rest of the world the old-style nonaverostran abelisaurs still ruled the megaavepod roost. They paralleled tyrannosaurids in their reduced arms and long legs configured for high speeds but did not become titanic.

The first known big bird was the flightless late Late Cretaceous *Gargantuavis*. As big as today's similarly continental ratites, just what kind of avian it was is not yet known; there is a possibility that it was a ground-bound enantiornithine. Also appearing at this time were the first neornithes; these initially water-loving toothless forms were the start of modern birds. Aside from their toothless bills on highly flexible heads, they brooded their eggs in the modern avian manner.

With India on its own as it tectonically sailed north, closing off the Tethys Ocean, by the end of the Cretaceous the continents had moved far enough apart that the world was assuming its modern configuration. At the terminus of the period a burst of uplift and mountain building had helped drain many of the seaways, although Europe remained an island archipelago. Flowering plants were fast becoming an ever more important part of the flora. Grasses were about, although not extensive, and none of the nonpredaceous avepods were well adapted for grazing them. Tall therizinosaurs could browse on an assortment of hardwood trees. Conifers remained dominant, however, and classic closed canopy rain forests still did not exist.

Then things went catastrophically wrong.

EXTINCTION

The mass extinction at the end of the Mesozoic is generally seen as the second most extensive in Earth's history, after the one that ended the Paleozoic. However, the earlier extinction did not entirely exterminate the major groups of large land animals. At the end of the Cretaceous all dinosaurs including most but not all birds, the only major land and aerial animals, were lost, leaving only a few flying birds as survivors of the group. Among the birds, all the toothed forms, plus the major Mesozoic bird branch, the enantiornithines, as well as the flightless birds of the time, were also destroyed. So were the last of the superpterosaurs and the most gigantic of the crocodilians.

It is difficult to exaggerate how remarkable the loss of the dinosaurs was. If they had repeatedly suffered the elimination of major groups and experienced occasional diversity squeezes in which the Avepoda was reduced to a much smaller collection that then underwent another evolutionary radiation until the next squeeze, then their final loss would not be so surprising. But the opposite is the case. A group that had thrived for over 175 million years over the entire globe, rarely suffering the sudden destruction of a major subgroup and usually building up diversity in form and species over time as it evolved increasing sophistication, was in short order completely expunged. The small dinosaurs went with the large ones, predators along with herbivores and omnivores, and intelligent ones along with those with reptilian brains. It is especially notable that even the gigantic forms did not suffer repeated extinction events. Giant avepods were always a diverse and vital group for most of the reign of dinosaurs. In contrast,

Latest Cretaceous
Tyrannosaurus imperator,
Triceratops horridus, and
Quetzalcoatlus

many of the groups of titanic Cenozoic mammals appeared, flourished relatively briefly, and then went extinct. Predatory dinosaurs appear to have been highly resistant to large-scale extinction. A reason for that may be the way they reproduced. Big avepods were fast breeders that laid large numbers of eggs, making them weed species that enjoyed high population recovery and expansion potential as long as at least some of the many eggs and/or juveniles survived a severe crisis and could continue the species. In contrast, big mammals reproduce slowly, dropping a calf only once a year or less, and that offspring then needs the careful attention of a nursing parent to grow up. That time investment leaves slow-breeding species highly vulnerable to elimination if the ecosystem temporarily becomes toxic. Those flesh-eating dinosaurs that largely fed on vegetarian

dinosaurs were of course doomed if the latter, which were also fast breeders, were exterminated. The smaller carnivorous dinosaurs, able to hunt assorted lesser prey, should have been more trophically flexible and thus able to survive in a suddenly disturbed world. Rendering dinosaur near elimination still more remarkable is that one group of avepod dinosaurs, the birds, did survive, as well as aquatic crocodilians, lizards, snakes—the latter had evolved by the Late Cretaceous—amphibians, and mammals that proved able to weather the same crisis.

It has been argued that dinosaurs were showing signs of being in trouble in the last few million years before the final extinction. Whether they were in decline has been difficult to verify or refute even in those few locations where the last stage of the dinosaur era was recorded in the geological record, such as western North America where the fossil data are the best. For example, western North American hadrosaurs sported an array of crested forms during the Campanian and early Maastrichtian, but by the end of the latter they were down to one flat-skulled genus. Over the same period ceratopsids went from a diversity of centrosaurs and chasmosaurs to just one species. On the other hand, *Tyrannosaurus* seems to have gone from one species to two. That the big herbivore decline was not just a western North American occurrence is demonstrated by the apparent decline in the variety of dinosaurian eggs in China in the last two million years of the Mesozoic. Even if these declines did occur, they were modest, and other regions of the globe may not have seen a decrease. At the Cretaceous/ Paleocene (K/Pg) boundary (formerly the Cretaceous/Tertiary [K/T] boundary), the total population of juvenile and adult dinosaurs should have roughly matched that of similar-sized land mammals before the advent of humans, numbering in the billions and spread among many dozens or a few hundred species on all continents and many islands.

A changing climate has often been offered as the cause of the dinosaurs' demise. But the climatic shifts at the end of the Cretaceous were neither strong nor greater than those already seen in the Mesozoic. And dinosaurs inhabited climates ranging from tropical deserts to icy winters, so yet another change in the weather should not have posed such a lethal problem. If anything, reptiles should have been more affected. Mammals consuming dinosaur eggs are another proposed agent. But dinosaurs had been losing eggs to mammals, and for that matter lizards and the like, for nearly 175 million years, and so had reptiles and birds without long-term ill effects. The spread of diseases as retreating seaways allowed once-isolated dinosaur faunas to intermix is not a sufficient explanation because of the prior failure of disease to crash the dinosaur population, which was too diverse to be destroyed by one or a few diseases and which would have developed resistance and recovered its numbers. Also unexplained is why other animals survived.

The solar system is a shooting gallery full of large rogue asteroids and comets that can create immense destruction. There is widespread agreement that the K/Pg extinction was caused largely or entirely by the impact of at least one meteorite, a mountain-sized object that fell from the northeast at a steep angle and formed a crater 180 km (over 100 mi) across, located on the Yucatán Peninsula of Mexico. The evidence strongly supports the object being an asteroid rather than a comet, so speculations that a perturbation of the comet-filled Oort cloud as the solar system traveled through the galaxy and its dark matter are at best problematic. The explosion of 100 teratons surpassed the power of the largest H-bomb detonation by a factor of 20 million and dwarfed the total firepower of the combined nuclear arsenals at the height of the Cold War. The blast and heat generated by the explosion wiped out the fauna in the surrounding vicinity, and enormous tsunamis, either arriving directly from the impact area or locally induced by the massive worldwide earthquakes emanating from the massive hit, cleared off coastlines of what was left of the interior seaway and across the Atlantic and even the vast Pacific. The cloud of high-velocity debris ejected at near-orbital velocities into near space glowed blazing hot as it reentered the atmosphere in the hours after the impact, creating a world-wrapping pyrosphere that may have been searing enough to bake animals to death as it ignited planetary wildfires. A fossil site exactly at the K/Pg divide in North Dakota appears to record some of these events, including remains of dinosaurs possibly drowned by the local flooding and fish with impact debris in their gills. The latter has been used to determine the impact was a springtime disaster. The initial impact would have been followed by a solid dust pall that, by reflecting sunlight, plunged the entire world into a dark, cold winter lasting for years, as well as causing severe air pollution and acid rain. As the aerial particulates settled, the climate then flipped as enormous amounts of carbon dioxide—released because the impact happened to hit a tropical marine carbonate platform and dug deeply into it at a steep angle (a more glancing impact or one elsewhere might not have had such serious effects)—created what some reconstruct as an extreme greenhouse effect that baked the planet for many thousands of years. Such a combination of agents appears to solve the mystery of the annihilation of the dinosaurs. The modern-style birds that got through, albeit barely, may have done so because they brooded their eggs solely with body heat, protecting them from the dreadful climate, and had flexible kinetic bills that aided feeding on whatever was left to eat. Even so, this scenario has its issues.

As big as the asteroid was, it was the size of a mere mountain and was dwarfed by the planet it ran into. It is not certain whether the resulting pyrosphere was as planetarily lethal as some estimate. Even if it was, heavy storms covering a small percentage of the land surface should have shielded a few million square kilometers, in total equaling the size of India, creating numerous scattered refugia. In exposed locations, dinosaurs that happened to be in burrows, caves, or deep gorges or in fresh waters not subject to big wave action, should have survived the pyrosphere, like the lucky crocodilians and amphibians. So should many of the eggs buried in covered nests. Birds and amphibians, which are highly sensitive to environmental toxins, survived the acid rain and pollution. Because dinosaurs were rapidly reproducing "weed species" whose self-feeding young could survive without the care of the parents, at least some dinosaur populations should have made it through the crisis, as did some other animals, rapidly recolonizing the planet as it recovered. Recent work challenges the extremity of the asteroid winter and the following hot house.

A major complicating factor is that massive volcanism occurred at the end of the Cretaceous as enormous lava flows covered 1.5 million square kilometers (~580,000 square miles), a third of the Indian subcontinent. Had this not been going on, the extinction could be blamed on the impact, whereas if the asteroid impact had not occurred, then the volcanism could be charged. Nor is it likely that bouts of intense volcanics were a regular occurrence. As it is, the extinction waters are muddied—similar to how the atomic bombings of Japan and the massive Soviet attack at the same time mean that the surrender cannot simply be assigned to the former. It has been proposed that the massive air pollution, including severe greenhouses resulting from carbon dioxide fluxes after chills from the sunlight-reflecting aerosol hazes, vented from the repeated supereruptions of the Deccan Traps damaged the global ecosystem so severely in so many ways that dinosaur populations collapsed in a series of stages, perhaps spanning tens or hundreds of thousands of years. This hypothesis is intriguing because extreme volcanic activity was responsible for the great Permian–Triassic extinctions, and other infrequent mass eruptions appear to have caused serious losses including in the Mesozoic. Although the K/Pg Deccan Traps were being extruded before the Yucatán impact, and in the process were possibly degrading the global fauna and flora, evidence indicates that the impact—which generated earthquakes of magnitude 9 over most of the globe (11 on the exponential scale at the impact site)—may have greatly accelerated the frequency and scale of the eruptions. If this is correct, then the impact was responsible for the extinction not just via its immediate, short-term effects but by sparking a level of extra-intense supervolcanism that prevented the recovery of the dinosaurs. It is also possible that the Yucatán impactor was part of an asteroid set that hit the planet repeatedly, further damaging the biosphere. Even so, the combined impact–volcanic hypothesis does not fully explain why dinosaurs failed to survive problems that other continental animals did.

Although extraterrestrial impact(s), perhaps indirectly linked with volcanism, is the leading explanation, the environmental mechanisms that destroyed all the nonflying dinosaurs while leaving many birds and other animals behind remain incompletely understood.

AFTER THE AGE OF AVEPOD DINOSAURS

Perhaps because trees were freed from chronic assault by the towering sauropods and therizinosaurs, dense forests, including rain forests, finally appeared. In the immediate wake of the extinction there were no large land animals, and only large freshwater crocodilians could make a living feeding on fish. The loss of so many dinosaurs led to a second, brief Age of Reptiles as superboa snakes as long as the biggest theropods and weighing over a tonne quickly evolved in the tropics, which also sported big freshwater turtles being attacked by enormous crocodilians. By 40 million years ago, about 25 million years after the termination of large dinosaurs, some land and marine mammals were evolving into giants rivaling the latter. What no mammals did was produce carnivores that came at all close to matching the elephant-sized carcharodontosaurids and tyrannosaurids. The biggest mammalian omnivores and carnivores were creodonts and bears of about a tonne, the greatest cats topped off at only 400 kg (900 lb).

Modern birds

swan

roadrunner

A great evolutionary irony is that the sole survivors of the Dinosauria, the ornithurine birds, greatly benefited from the elimination of competition from other avian relations and dinosaurs at the beginning of the Cenozoic. With the enantiornithines and toothed euornithines gone, the beaked ornithurines were set free to proliferate into a growing array of feathered creatures of a tremendous variety of adaptative forms and species. Ornithurines are split into the paleognaths and the vastly more abundant and diverse neognaths. Members of both groups lost flight and became small and large land runners and, in the case of neognaths, marine swimmers. In South America, the often big phorusrhacids evolved into deep-hooked beaked hunters in competition with marsupial predators and even briefly moved into a North America dominated by placentals. Fast-running paleognathous ratites have proven able to make a fair go of it in the face of a modern mammalian fauna on continents. On some big islands, ratites weighing over a third of a tonne were still roaming about until human arrivals did them in just a thousand years ago. Ancient penguins got up to a sixth of a tonne. But the main story of Cenozoic dinosaurs was their governance of the daylight skies, while the night was dominated by mammalian fliers, the bats. The largest fliers have sported wings spanning 6–7 m (20–23 ft) and weighed 50–70 kg (100–150 lb). At the other end of the size spectrum hummingbirds get down to 2 g (0.07 oz) and span just 6 cm (2.4 in), the smallest dinosaurs. The late Cenozoic bird population, maybe in the area of 200 billion, spread over 12.5 thousand species. The greatest success story of modern flying dinosaurs? The marvelous diversity and numbers of the little but sophisticated passerine songbirds that fill field guides.

BIOLOGY

GENERAL ANATOMY AND LOCOMOTION

Predatory dinosaur heads have ranged from remarkably delicately constructed to fairly massively built. In all examples the nasal passages or the sinuses or both were very well developed, a feature common to archosaurs in general. All predatory dinosaurs have had large openings in the skull in addition to the nostrils and orbits—one immediately in front of the orbits that contains the sinuses, another directly behind the orbits for jaw muscles, and another above the latter on the skull roof for the same purpose. Unlike mammals with their extensive facial musculature, dinosaurs, like reptiles and birds, lacked facial muscles, so the skin was directly appressed to the skull. This feature makes dinosaur heads easier to restore than those of mammals. The external nares are always located well forward in the nasal depression. The skin covering the large openings probably bulged gently outward to accommodate sinuses or jaw muscles, depending on what filled them. At the back of some large averostran avepod skulls, the top of the braincase seems to form a prominent transverse crest, but this actually anchored the powerful dorsal neck muscles from behind and the upper ends of jaw muscles on their front surfaces.

Among amphibians, tuataras, lizards, and snakes, the teeth tend to be set close to one another along fairly sharp-rimmed jaws, with the upper teeth always outside those on the lower jaw. The mouth is sealed, and the teeth are covered by nonmuscular lips that cannot be pulled back to bare the teeth when the mouth opens. So even when the mouth is agape, the teeth often remain entirely covered by lips, as well as gums; this is so in the biggest flesh-eating monitor lizards as well as herbivorous iguanas. This arrangement appears to be true of most theropods and sauropods as well—one fossil of the latter appears to have a patch of lip tissue preserved astride some teeth. Thus, the artistic tendency to show teeth entirely or partly exposed in dinopredators as well as sauropodomorphs may be an error as glaring as it is common. An exception among avepods would be the spinosaurs, which have a more crocodilian arrangement in which at least the front teeth are widely

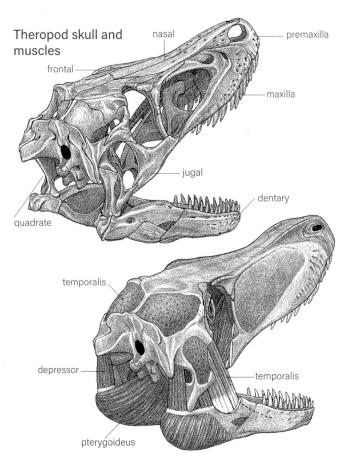

Theropod skull and muscles

nasal — premaxilla

frontal

maxilla

jugal

dentary

quadrate

temporalis

depressor

temporalis

pterygoideus

spaced in separate sockets on rounded jaw rims, with some lower teeth outside the uppers, so spinosaurs' snaggy front teeth may have been lipless and exposed when the jaws were closed. Daemonosaurs have large, protruding front teeth that may have been bared. The same may have been true of the forward-tipped teeth at the front of the lower jaws of masiakasaurs. Some aveairfoilan avepods evolved beaks. In some of those, including therizinosaurs, the beak was limited to the front of the mouth; in others, the beak displaces all the teeth.

Teeth were often worn to varying degrees at the tips and serrations, sometimes heavily so, chipped, or badly broken and became increasingly nonfunctional. Set in sockets, all dinosaur teeth were constantly replaced through life in the manner of reptiles. Those of predatory dinosaurs lasted from just a couple of months to a couple of years. Size was not an important factor in the differing rates. Tyrannosaurs had the slowest shedding timing, dromaeosaur and troodont teeth lasted about a year, ceratosaur and allosaur teeth were on a schedule of about three months, and some abelisaurs had the fastest

turnover. After the dinosaur died, teeth often slipped partly out of their sockets, exposing the roots and making them look longer than they were in life. Teeth ranged from blunt, leaf-shaped dentition suitable for crunching plants to serrated blades adapted to piercing and dicing flesh. Like the teeth of today's carnivores, those of predatory theropods were never razor sharp, as is often illogically claimed, because they would have injured the mouths they were in; one can run a finger firmly along the serrations without harm. With teeth only good for puncturing and slicing flesh, food could not be chewed and was bolted down in chunks, or small prey was swallowed whole. *Tyrannosaurus* teeth were stouter than others and may have allowed some bone crushing. Toothed avepods that were omnivorous or herbivorous sported blunter, more leaf-shaped teeth suited for pulping plant material. Because dinosaurs were not lizards or snakes, they lacked flickering tongues. Dinosaurs had well-developed hyoids, suggesting that the tongues they supported were similarly developed. In predatory theropods the tongue was probably simple and inflexible.

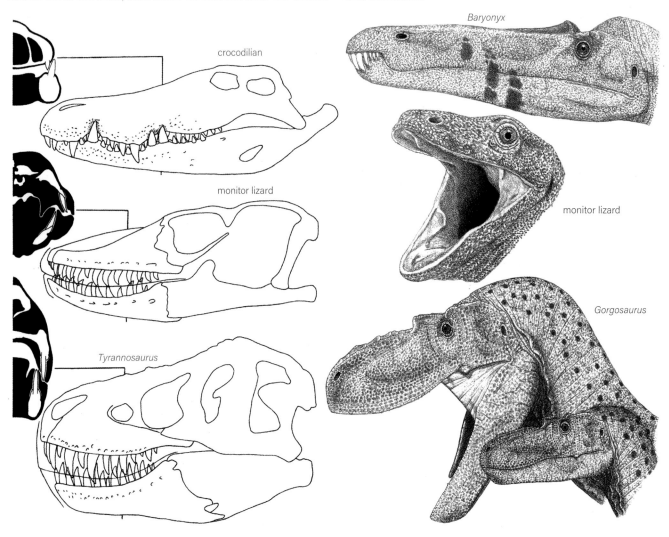

crocodilian

monitor lizard

Tyrannosaurus

Baryonyx

monitor lizard

Gorgosaurus

Archosaur lip anatomy

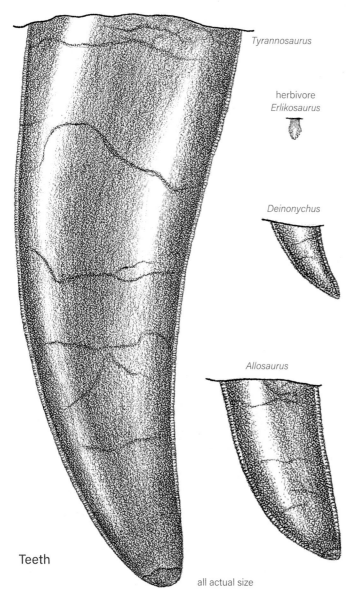

Tyrannosaurus

herbivore
Erlikosaurus

Deinonychus

Allosaurus

Teeth

all actual size

and a nictitating membrane, so the same was presumably true in the dinosaurs.

The dinosaur outer ear is a reptilian-style, deep, rather small depression between the quadrate and jaw-closing muscles at the back of the head. The eardrum was set in the depression and was connected to the inner ear by a simple stapes rod. The orientation of the semicircular canals of the inner ears is being used by researchers to determine the posture of dinosaur heads. The situation may, however, be more complicated, reducing the reliability of this method. In living animals, the relationship between the orientation of the semicircular canals and the normal carriage of the head is not all that uniform, even between individuals of a species. That animals position their heads in different ways depending on what they are doing does not help. It seems that the posture of the semicircular canals is determined at least in part by the orientation of the braincase with the rest of the skull and does not reflect the orientation of the head as well as has been thought. The heads of predatory dinosaurs would have been held close to horizontal, or tilted down a little.

Although the absolute size and exact shape of dinosaur muscles cannot be exactly determined, their form and relative size among the different groups can be approximated. The exceptionally complex limb muscles of living mammals are the heritage of the unusual history of the early members of the group. Dinosaurs retained the simpler muscle patterns of reptiles, which are still seen in birds.

The necks of many predatory dinosaurs tend to articulate in a distinctive birdlike S curve. The beveling of the vertebrae is especially strong in some avepods, including allosaurids, tyrannosaurids, and some dromaeosaurs, so much so they may not have been able to straighten out their necks entirely. Others had straighter necks. If anything, animals tend to hold their necks more erect than the articulations indicate. There has been a tendency to make dinosaur necks too short by placing the shoulder girdle too far forward. It was standard for predatory dinosaurs that were not birds to have 10 cervical vertebrae, other dinosaurs had fewer, others more. The flexibility of predatory dinosaur necks ranged from modest in the stouter-necked, big examples, to fairly high in longer-necked examples, but no dinosaur had the special adaptations—including up to over two dozen cervicals in long-necked birds—that make avian necks exceptionally mobile. Contributing to the greater flexibility of bird necks is the absence of overlapping ribs. The ribs did overlap one another via a slender tapering posterior rod in the majority of predatory dinosaurs, a widespread archosaur condition seen in crocodilians. Presumably the overlap strengthened the neck without preventing flexion: the overlapping ribs could slide along each other. The loss of the overlap in ornithomimids and most airfoilans indicates further increases in neck flexibility.

The bulging neck muscles of stout-necked predaceous dinosaurs were tremendously powerful, being part of

In some large dinopredators the eyes were in the upper part of the orbit. Bony sclerotic eye rings that helped support the large orbs often show the actual size of the eye, both in total and indirectly in that the diameter of the inner ring tends to closely match the area of the visible eye when the eyelids are open. Predatory dinosaurs had large eyes, yet relative eye size decreases as animals get bigger. Although the eyes of giant theropods were very large, they looked small compared with the size of their heads. Even the eyes of ostriches, the biggest among living terrestrial animals, do not appear that large on the living bird. In the predatory daylight raptors, a bony bar running above the eyeball provides the fierce "eagle look." Interestingly, the flesh-eating Mesozoic dinosaurs lacked this bar. Whether the pupils of dinosaur eyes were circular or slit shaped is not known. The latter are most common in nocturnal animals, and either may have been present in different species. The eyes of birds and reptiles are protected by both lids

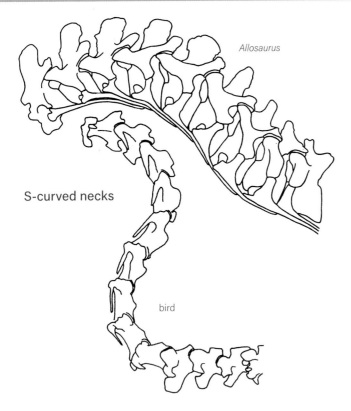

Allosaurus

S-curved necks

bird

the flesh-penetrating and ripping, killing, and feeding mechanism. These were best developed in the great tyrannosaurids, *Tyrannosaurus* most of all. The longer, less massively constructed necks of other, usually smaller, meat-eating species were less massively muscled. Long, solid vertebrae, browsing giraffe necks are not heavily muscled despite needing to carry a large head, to the degree that the bulges of each vertebra are externally visible. The elongated necks of the similarly herbivorous ornithomimosaurs and therizinosaurs held up much smaller heads and were highly pneumatic, so they should not have been similarly lightly muscled. Exactly how is not clear. There has been little investigation of the soft tissues of the long necks of ratites and swans. X-rays hint that their trachea and esophagus may not be restricted to the underside of the neck vertebrae between the cervical ribs, but if that is true in living birds, how this may apply to their nonavian relations is obscure. The upper muscles that hold the head and neck up are stronger than those that lower them, and the former form a subtle bulge contour over the latter, which runs the length of the neck.

The normal vertebral count of predatory dinosaur trunks was 13. These articulated in a dorsally gentle convex arc with the sacral vertebrae of the hips, which numbered from two in protodinosaurs, was usually five or six, and a few more in some derived examples as the pelvic ilium they supported got larger. The nature of the vertebral articulations, in larger examples often reinforced by ossification of interspinal ligaments, means that dinosaurs had stiffer backs than lizards, crocs, most mammals, and

for that matter protodinosaurs, although theropod trunk vertebrae were not normally fused the way they often are in birds. Because the trunk was not highly flexible, the superficial muscles were not well developed, which is why there is not much meat to be found on the ribcages of chickens, unlike the tasty ribs of bovids and suids. As in most vertebrates, including lizards, crocodilians, and birds, the front ribs are strongly swept back in articulated predatory dinosaur skeletons; they are not vertical as they are in many mammals. The belly ribs tended to be more vertical, but this condition is variable. In a few large avepods, the neural spines along some or much of the animal's midline became exceptionally tall, producing dorsal fins. Such are seen in some archaic "mammal-like reptiles," mammals like bison, and occasional sauropods and ornithopods. These probably carried sail-like thin tissues to help produce a dramatic top profile. The abelisaur avepods had an unusual opposite feature. The transverse processes that project sideways from the vertebrae—normally set well below the tops of the tall neural spines—were angled upward enough to be about as high as the neural spines, resulting in flat-topped trunks and necks. A similar condition is common on the tail bases of many basal theropods, being taken to an extreme in most abelisaurs. Seemingly too subtle to be a display feature, presumably this arrangement had some function regarding the action of the muscles they supported, although this is not well understood.

As typical of meat eaters, the bellies and hips of the flesh-eating theropods were narrow, reflecting the small size of their digestive tracts when empty as well as their athletic form. Because intestinal tracts of herbivores are much more capacious, and normally filled with plant material undergoing gradual digestion, the abdomens and hips of plant-consuming therizinosaurs were extremely broad.

The protodinosaur pelvis was rather small and shallow as it is in lizards and crocodilians. A striking features of the predatory dinosaur pelvis from early on was how deep it usually was, primarily because the narrow vertical pubes was so long and, to a certain extent, so were the ischial rods to the aft. Aside from sauropodomorphs in which pelvic depth was not so extreme, there was nothing else like it. In derived avepods the pubis was often tipped with a large boot; what this was all about is not known. The animals could sit on the pubic and ischial tips with their main trunks tilted up a bit clear of the ground, but other creatures don't need this peculiar head-scratching scheme, and ground impressions of resting avepods do not necessarily show an imprint of the pubis; that of the ischium is seen in in some. Locomotion may be somehow involved but how has not been elucidated. An exception to the deep pubis was seen in the little, bat-winged scansoriopterygids, presumably to flatten the hips to aid climbing. The orientation of the pubis ranged from being procumbent, which was common in basal forms, to becoming usually vertical, or retroverted in the early herrerasaurs and especially the aveairfoilans. The later retroversion was another way to

Resting theropod impressions

make the trunk shallower for climbing while helping reduce the frontal air drag of the body by making it shallower; the retroversion also compensated for the reduction of the tail by shifting the center of mass of the belly aft. In derived aveairfoilans, including birds, the pubis closely parallels the ischium; any trace of the boot is lost. In the plant-digesting therizinosaurs, increasing the volume available for the big fermenting gut would have been another reason for swinging the pubis back.

When hungry, the gut in front of the deep pelvis would have been somewhat hollowed out in dinopredators like in canids and cats on the hunt for flesh to gorge on and fill their highly expandable stomachs. Unlike most dinosaurs, birds included, the predators retained the gastralia common to other archosaurs, these being a series of flexible bony rods in the skin of the chest and belly. Each segment of the gastralia was made of multiple pieces, making the complex array flexible and allowing the abdomen to dramatically expand and contract, depending on how full the digestive tract was with the body parts of the last meal. This flexibility is especially important in big predators, which tend to binge on the newest kill and then fast until the hunger induced by an empty stomach compels the animal to go to the intense effort and considerable danger of another hunt in a gorge and fast cycle. In most well-articulated specimens, the gastralia are bowed convex outward, which makes the dinosaur's bellies look plump: however, this appearance is the result of the post-mortem bloating of carcasses. A recently found, exceptionally well-preserved skeleton of a tyrannosaur that did not balloon before preservation shows the expected hollow set of gastralia. That said, abdominal air sacs may have filled out some the space of the gut even when the dinosaurs were hungry more than in nonpneumatic mammals. When packed full after enjoying a full meal the belly would have been rotund. These variations between extremes would have been less so in lesser carnivorous dinosaurs that tend to pick up small game items here and there, keeping their digestive tracts partly full most of the time. Not being predaceous, ornithomimid gastralia would have normally been more filled out because their somewhat larger digestive systems were somewhat larger and usually contained digesting foodstuffs; this

seems borne out by articulated skeletons. The gastralia baskets of therizinosaurs were more rigid, probably because these sophisticated herbivores always kept their bulging abdomens full of fermenting fodder. These structures were present in pterosaurs and prosauropods but not sauropods and most ornithischians.

In most predatory dinosaurs the hip vertebrae and tail were in much the same line as the trunk vertebrae. Because tail drag marks are extremely rare among the immense number of trackways, the old-style convention of persistently tail-dragging dinosaurs cannot be correct, which is why pretty much all the skeletons that were mounted that way have been reposed, and the *Tyrannosaurus* tail does not drag in *Jurassic Park*. Only while rearing up while walking could the tail drag in typical predatory dinosaurs, which the great abundance of their trackways show was very infrequent. The therizinosaur hip and tail were flexed upward relative to the trunk vertebrae. This position allowed the trunk to be held strongly pitched up while the hips and tail remained horizontal, increasing the vertical browsing reach of the head while the dinosaur retained the ability to move on its hind legs. Even small- and medium-sized dinosaurs without caudal specializations may have been able to briefly bounce on their tail occasionally while lashing out with the hind limbs. The dinosaurs' tails were highly flexible along most of their length in most examples. In deinonychosaurs, flight needs tended to stiffen the tails, to the degree that in most sickle-clawed dromaeosaurs much of the tail was stiffened to at least some degree by ossified rods. In compensation, the tail base was hyperflexible, especially vertically.

The great majority of predaceous dinosaurs had long reptilian tails, made proximally deep by long chevrons beneath the centra and broad by transverse processes on the sides of the centra. The resulting flanks of this front portion of the tail anchored the great caudofemoralis muscle. Connected to the back surface of the femur, this was a powerful leg retractor that in addition to forming a major contour to the tail base made a big contribution to the power stroke. In slender-tailed deinonychosaurs the chevrons were shallow like in long-tailed pterosaurs, while some therizinosaurs and the oviraptorosaurs opted for short, birdlike appendages, sometimes ending in fused pygostyle tips. In these small-tailed forms the caudofemoralis is no longer a critical leg muscle, its function being replaced by enlarged muscles anchored on the underside of the ilium behind the hip joint. In birds the aft ilium is often elongated to accommodate these leg retractors.

In most dinosaurs, including most predatory examples, the shoulder joint is placed a little forward of the frontmost ribs. This differs from mammals, in which the shoulder joint is astride the side of the chest. In the bipedal predatory dinosaurs that never used their arms for ground locomotion, including birds, the shoulder girdle is fixed in place, partly by a furcula that braces both scapula blades. Furculas are usually considered to be cofused clavicles, but

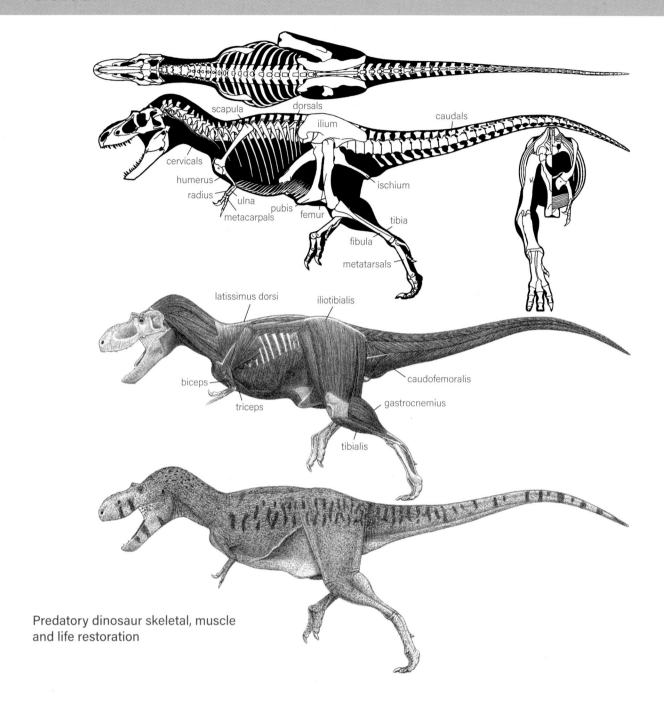

cervicals
scapula
dorsals
ilium
caudals
humerus
radius
ulna
pubis
metacarpals
femur
ischium
tibia
fibula
metatarsals

latissimus dorsi
iliotibialis
biceps
triceps
caudofemoralis
gastrocnemius
tibialis

Predatory dinosaur skeletal, muscle and life restoration

they may instead be modified interclavicles that displaced the clavicles. Scapula blades were broad until the avetheropods, when they become strap-like in the avian manner. The scapula blade and the attached short coracoid was typically subvertical, as is the norm for tetrapods. The exception was the birdlike airfoilans, aveairfoilans most of all, whose scapula blades are horizontal, while the coracoid is vertical or somewhat retroverted, as well as somewhat elongated. This arrangement places the shoulder joint high on the shoulders and next to the first ribs. The same scheme is found in the other flying archosaurs, the pterosaurs—bats have a completely different arrangement. The sternum is neither large nor ossified in most avepods until the aveairfoilans. In basal aveairfoilans we know it became larger because the space between the coracoid and the beginning of the gastralia became longer, but the sternals were still not bony. Later they were ossified as squarish plates, although deep keels did not develop until well into avian evolution.

In flying birds the shoulder joint faces sideways and upward so the arms can be held out to the side and raised vertically for flapping. In many dinopredators the arms could also be swung laterally to help grapple with prey. But even in winged protobirds like *Archaeopteryx*, the arms could not be directed straight up. The short, extremely stout and powerful arms of alvarezsaurs could also swing well out to

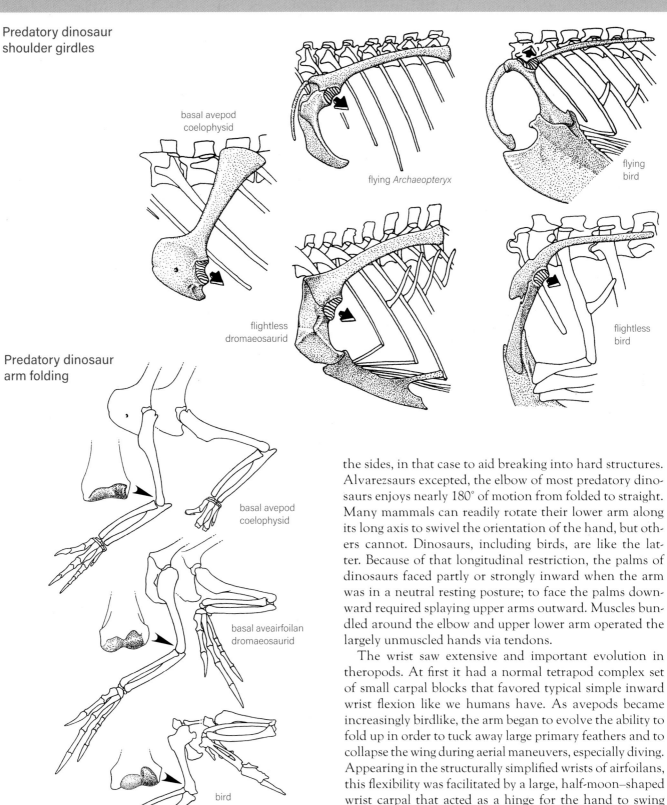

Predatory dinosaur shoulder girdles

basal avepod coelophysid

flying *Archaeopteryx*

flying bird

flightless dromaeosaurid

flightless bird

Predatory dinosaur arm folding

basal avepod coelophysid

basal aveairfoilan dromaeosaurid

bird

the sides, in that case to aid breaking into hard structures. Alvarezsaurs excepted, the elbow of most predatory dinosaurs enjoys nearly 180° of motion from folded to straight. Many mammals can readily rotate their lower arm along its long axis to swivel the orientation of the hand, but others cannot. Dinosaurs, including birds, are like the latter. Because of that longitudinal restriction, the palms of dinosaurs faced partly or strongly inward when the arm was in a neutral resting posture; to face the palms downward required splaying upper arms outward. Muscles bundled around the elbow and upper lower arm operated the largely unmuscled hands via tendons.

The wrist saw extensive and important evolution in theropods. At first it had a normal tetrapod complex set of small carpal blocks that favored typical simple inward wrist flexion like we humans have. As avepods became increasingly birdlike, the arm began to evolve the ability to fold up in order to tuck away large primary feathers and to collapse the wing during aerial maneuvers, especially diving. Appearing in the structurally simplified wrists of airfoilans, this flexibility was facilitated by a large, half-moon–shaped wrist carpal that acted as a hinge for the hand to swing back toward the lower arm. Complete wing folding did not become fully developed until advanced birds.

A very distinctive feature of the theropod hand that it shared with some other dinosaurs was the big-clawed, inwardly directed thumb weapon. Also distinctive was the emphasis placed on the inner three fingers, with the second the longest. The outer two fingers were at best reduced,

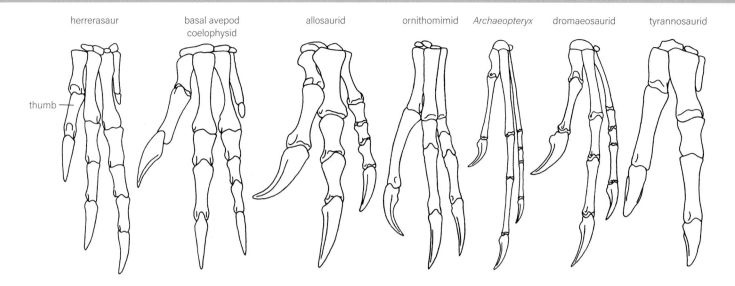

herrerasaur · basal avepod coelophysid · allosaurid · ornithomimid · *Archaeopteryx* · dromaeosaurid · tyrannosaurid

thumb

Predatory dinosaur hands

and the hand was very nearly or completely tridactyl by the avetheropods. Although considerable movement was possible, the dinosaur's fingers were not as flexible and supple as those of humans. As some avepods evolved toward the avian condition, finger flexion was further reduced to better support enlarging hand feathers. The fingers were padded by slender pads that bulged some at the joints. Avepod arms underwent a wide array of radical evolutionary changes in differing clades. Four times among predaceous nonaveairfoilans, in ceratosaurs–abelisaurs, carcharodontosaurids, *Gualicho*, and tyrannosaurids, the arms became reduced down to limited if any functionality. In each case, the nature of the alterations was very distinctive in accord with the evolutionary derivation of the group, with the fingers being reduced to two in *Gualicho* and tyrannosaurs. Severe arm reduction is also typical of neoflightless birds. Also sporting short arms were alvarezsaurs, but in this case they were constructed like those of body builders, probably for digging, with the thumb a massive digit tipped with a big broad claw. The other fingers were very atrophied to the point of being of limited or no use, in some only two fingers were left, and in others essentially only the thumb was left, making them the only known one-fingered dinosaurs. The arms of the nonpredaceous ornithomimosaurs and therizinosaurs became long, large clawed, and adapted for feeding or defense. The real upscaling of dinosaur arms occurred in aveairfoilans, as they were enlarged for climbing and increasingly adapted for flight, until the hand was the fused, flattened, clawless primary feathers support of birds. Pterosaur and bat wrists and hands could hardly be more different, including super elongated fingers to support their wing membranes, which are also attached to the legs.

In the hindlimb the ilium plate of the upper pelvis started out small in basal dinosauriforms, so their thigh muscles were rather narrow. The plate got somewhat larger and the thigh correspondingly broader in the early theropods and avepods and continued to expand in size until it was large in most averostrans, increasing the breadth, bulk, and power of the thigh muscles in the manner inherited by birds. Among the biggest plates seen were on the extra-fast ornithomimids and tyrannosaurs, and also are seen in birds that spend a lot of time on the ground. The upper edge of the ilium was and is a major visible contour in living dinosaurs. Also expanding the size of the leg muscles was a large, forward-projecting cnemial crest at the knee end of the tibia. This crest helps anchor the big, bulging drumstick of muscles ensheathing the upper shank, which many enjoy when consuming fowl. This big bundle of muscles operates the nearly muscle-free feet via tendons.

It is difficult to restore the precise posture of dinosaur limbs because in life the joints were formed by thick cartilage pads similar to those found on store-bought chickens, which are immature. That dinosaurs normally retained thick cartilage pads in their limb joints throughout their entire lives, no matter how fast or big they became, is a poorly understood difference between them and adult birds and mammals, which have well-ossified limb joints. The manner in which dinosaurs grew up and matured may explain the divergence. In terms of locomotory performance, cartilage joints do not seem to have done dinosaurs any more harm than they do big running birds that still have cartilage joints when fully grown but not yet mature, and they may have had advantages in distributing weight and stress loads. The poor ossification of dinosaur limb joints hinders restoring their posture; even so, some basics can be determined.

The dinosaurian cylindrical hip socket did not allow the legs to sprawl out to the sides in most dinosaurs. As a result, the limbs were close to vertical in the fore-and-aft plane. The correspondingly narrow-gauge fossil trackways record the hindprints falling on the midline or very close to it, like in birds and erect-limbed mammals, humans included. Pterosaurs with ball-and-socket hip joints were more variable in limb posture as their trackways affirm; bats are sprawlers. This does not, however, mean that the erect legs of dinosaur-birds work in a perfectly vertical plane; that happens in few if any animals. The femur is somewhat everted outward, especially when swinging forward to clear the gut, all the more so when the abdomen was filled with a recent big kill. The knee was correspondingly bowed out somewhat, the opposite of humans in which it is bowed in a dash—we do not have bellies in front of our knees to clear. The shank sloped inward some to the slightly inward-bent ankle, and the upper foot was vertical or a little sloped outward in fore or aft view.

There was a notable exception to strictly erect avepod legs. The small, early microraptorine dromaeosaurs sported extra wings on their hindlimbs that, in order to work, required the legs to splay horizontally out to the sides. To accommodate that, the femoral head was rounded and the hip socket partly closed off and more upwardly oriented than in the rest of the dinosaurs. This reversion to the protodinosaur condition allowed the legs to operate in both the vertical and horizontal planes and in between. With no other dinosaurs having leg wings, this was not to be repeated.

In the fore-and-aft plane all predatory dinosaurs, like most terrestrial tetrapods, have had strongly flexed hip, knee, and ankle joints that provided the springlike limb action needed to achieve a full run, in which all feet were off the ground at some point in each complete step cycle. In addition, the ankle remained highly flexible, allowing the long foot to push the dinosaur into a ballistic stride. This was true of even the most gigantic 10 tonne avepods; they did not have the columnar limbs of sauropods or elephants that prevent them from running, rather, their limbs were similar to the fairly flexed limbs of giant running rhinos. The knee joints of dinosaurs with flexed limbs were not fully articulated if they were straightened. Humans have vertical legs with straight knees because our vertical bodies place the center of gravity in line with the hip socket. In bipedal dinosaurs, because the head and body were held horizontally and were well forward of the hips, the center of gravity was ahead of the hip socket even with the long tail acting as a counterbalance, so the femur had to slope strongly forward to place the feet beneath the center of gravity. As a result, the theropod femur did not retract much past vertical at the end of the limb stroke even when running at top speed, unlike mammals, including humans, in which the femur swing further aft. This arrangement is taken to an extreme in short-tailed birds, whose femur is nearly horizontal when they are walking in order to place the knees and feet far enough forward; when running, the femur of birds swings more strongly backward, but not to full vertical, to fully utilize the power of the big thigh muscles. The theropod knee could be nearly folded up. Like the elbow, the door-hinge-like roller ankle had nearly 180° of motion from folded to straight. Mesozoic trackways show avepods sometimes walking pigeon toed, sometimes not. The walking gait of the narrow-bodied beasts would have been smoothly striding. The exception would seem to be the fat-bellied therizinosaurs, but their trackways show the same narrow-gauge gait as the other theropods rather than the extremely pigeon-toed waddling of plump-bodied geese. There is no anatomical reason to think any dinosaur hopped, and not a single of the enormous sample of trackways shows otherwise.

As per the dinosaur-avian norm, theropod feet are digitigrade like cats and dogs, with the ankle held clear of the ground while the entire weight-bearing toes are pressed to the ground. In plantigrades, as per pterosaurs, bats, bears, and people, the entire foot up to the ankle rests flat; in

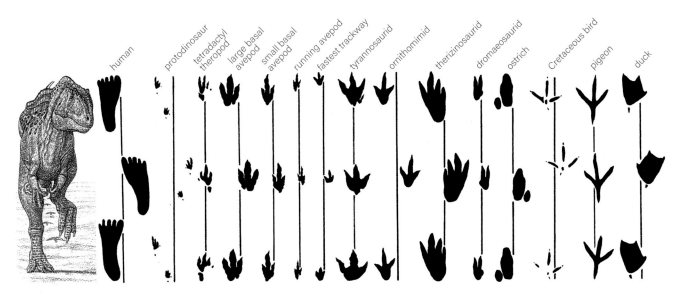

Predatory dinosaur and other trackways

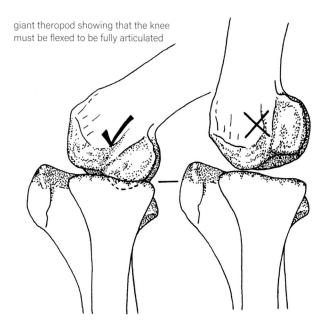

giant theropod showing that the knee
must be flexed to be fully articulated

Knee flexion

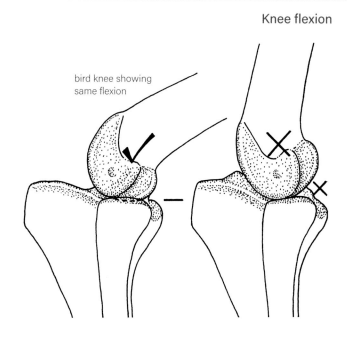

bird knee showing
same flexion

unguligrades only the tips of the toes make ground contact, as in horses and deer. The ostrich is an odd bird in that only the front sections of the toes are on the ground. The load-bearing toes, four in protodinosaurs and first theropods and therizinosaurs, three in most dinosaurs, two in sickle-clawed deinonychosaurs, are underlain by modest padding, which became relatively broader as did the toe bones with greater size and with larger padding under the main base of the foot. The details of this padding are regularly recorded, with some distortion, in fossil footprints. The three central toes of avepods are usually, but not always, not widely divergent until birds, when they splay out much more. In the standard tridactyl and less common bidactyl avepods, the inner toe is clear of the ground or close to it. Usually it subparallels the other toes, and it can be wondered why the hallux was retained, it not appearing to have much if any of a function. It obviously did have functions in peculiar *Balaur*, being enlarged and hyperextendable, bearing a big claw as a weapon, and perhaps being used for climbing. In more arboreal and aerial airfoilans, the hallux became more reversible and distally placed for climbing and perching as well as handling prey items. In some birds the hallux is fully reversed and all the way down the foot, giving it full contact with the ground as well as excellent perching purchase. Even more extreme are those birds that have also reversed the fourth toes, such as roadrunners and woodpeckers: these are sort of tetradactyl. At the opposite extreme are those avepods, often fleet runners, that have entirely lost the inner toe; these include the ornithomimids and their mimics the ratites. These are the most truly tridactyl avepods, matched in this particular point by the big ornithopod dinosaurs and three-toed perrisodactyl ungulates like rhinos. With just two toes, ostriches are closet to two- and one-toed ungulates. The second toes became weaponized in deinonychosaurs, which

may or may not include *Balaur*, and a few birds with an enlarged claw, the saber claw of the cassowary being the premiere living example of that. Paid very little attention is the outermost toe, five, which is reduced to a mere short splint immediately below the ankle, which was retained in Mesozoic avepods other than some birds to the end of the period despite it having no apparent use. Perhaps it provided some leverage to the foot-extending muscles in the drumstick. The little bone is lost in most birds, perhaps to save as much extraneous weight as possible among the fliers—in contrast, the outer toe of early pterosaurs is a long multisection splint that helps support the trailing edge of the membrane between the legs; in the rest of the pterosaurs it too is reduced to a splint of obscure function if any. Pterosaurs had four complete toes, bats keep all five.

The speed at which a fossil trackway was laid down can be approximated—with emphasis on the approximated—by correlating the stride length of the trackway with the length of the articulated leg from the hip joint to the base of the foot. The latter can be estimated from the length of the foot, which is four to one in a surprising array of animals—humans and theropods of all sizes and most types included. Mammals and birds of all sizes tend to walk at speeds around 3–7 km/h (average 3 mph). Note that squirrels will bound rapidly, halt for a moment, then bound some more, stop, and so on. Humans and their dogs typically move at a similar overall pace, as do elephants. This consistency is because the cost of locomotion per given distance scales closely to available aerobic power as size increases, so being big does not provide a major advantage. A typical walking speed recorded by trackways of tridactyl dinosaurs is around 5–7 km/h (4 mph).

Perhaps the fastest of dinosaurs are ostriches, which are credited with speeds of over 60 km/h (over 40mph); among Mesozoic avepods this may have been approached

Predatory dinosaur feet

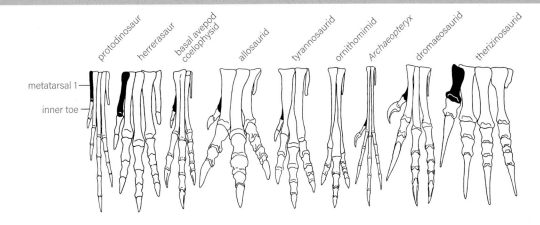

or matched by the nonpredatory ornithomimids. Also looking swift were small troodonts, perhaps as much to escape being prey as to catch it, and insectivorous alvarezsaurs for defense. On the offense were smaller and juvenile tyrannosaurids, built for high speed. About a third or more of the mass of running animals like ostriches consists of leg muscles. Adult animals do not spend much time using these to burn lots of energy per unit time and exhausting themselves running about without a compelling reason, so fossil speed trackways are not surprisingly scarce. The highest Cretaceous dinosaur land speed calculated so far is in the area of 40–50 km/h (~30 mph), made by a 100 kg (200 lb) probable avepod. It is not a surprise that small- and medium-sized dinosaurs with long, slender, flexed legs were able to run at speeds comparable to those of similar-sized ground birds and galloping mammals. Difficulties arise when trying to estimate the top speeds of flexed-limbed dinosaurs weighing many tonnes. Computer analysis has calculated that *Tyrannosaurus* could reach a top speed ranging from no better than that of a similar-sized elephant, 25 km/h (15 mph), up to the 40 km/h of a sprinting athlete. Because big-hipped, birdlike *Tyrannosaurus* was much better adapted for running than are elephants, and even by large avepod standards is exceptionally speed adapted with its enormous pelvis and running birdlike limbs, it is unlikely that it was similarly slow as proboscideans. Its anatomy is more in line with it being twice as fast or so as elephants, able to approach or match galloping rhinos and nonthoroughbred horses. Lesser tyrannosaurids may have been a dash swifter than *Tyrannosaurus*; other less speed adapted comparably titanic-sized avepods were likely to have been somewhat slower. Computer analyses are of at best limited use because they are hard pressed to fully simulate important aspects of animal locomotion, including the energy storage of prestretched elastic leg tendons and the resonant springlike effect of the torso and tail. If the latter proves true, then other giant running dinosaurs may have used special adaptations to move faster than our computer models are indicating. If marine mammals were known only from fossils, it would be logically concluded that being high metabolic rate endotherms none would have been able to hold their

breaths long enough to dive very deep into high-pressure water that would certainly kill them, and computer simulations would be cited as confirmation of that probable fact. Yet some whales and seals can dive thousands of meters because of soft tissue physiological adaptations not preserved in fossils. The core and unavoidable problem with simulations is that they are simulating a simulation, not observing the actual performance of a living animal, so the results are no better than what was put into them and may not reproduce the actual parameters of a living creature. The outcomes can never be scientifically tested and verified.

Avepods pursuing or fleeing not only needed to run, they needed to be able to maneuver, whether when moving or fighting in place. A way to improve turning performance is to be more compact fore and aft, which lowers the distal inertial mass, like a spinning skater pulling in their arms to increase the revolutions per minute. Most predatory dinosaurs were fairly elongated, limiting turn ability. Tyrannosaurids' shorter trunks and tails combined with their reduced arms would have enhanced turning during their assaults on their prey. Ornithomimids had a similar trunk and tail form to enhance their knack for dodging their tyrannosaurid attackers. The very short tails of oviraptorosaurs and most therizinosaurs would have been good for defensive agility. Another way to better turning is to pull the tail up when turning, somewhat like pulling in the arms. All long-tailed theropods could do that, the deinonychosaurs best of all because of their extra-mobile tail bases. Deinonychosaurs could additionally use the combination of a stiffened distal tail with a flexible base to quickly use the appendage as a dynamic stabilizer and turn inducer when maneuvering on the ground or in the air, like pterosaurs with similar tails. Also boosting agility would have been arm and tail feather arrays that could be used as turn-inducing airfoils when at speed.

Ratites are competent swimmers, more so than most mammals, which lacking pneumatic bodies are nearly awash when swimming and therefore have to work to keep their nostrils above the water. Having similar birdy limbs, and most being pneumatically buoyant like their

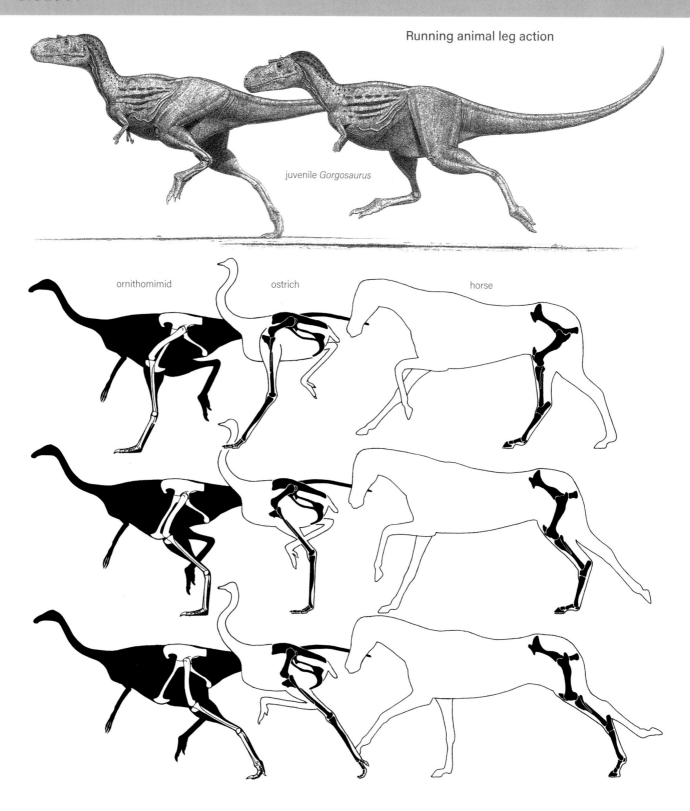

Running animal leg action

juvenile *Gorgosaurus*

ornithomimid ostrich horse

avian relations, all predatory dinosaurs were comparably adept in water. Perhaps even more adept were those with long, proximally deep tails able to add a sculling propulsion—*Ceratosaurus* appears exceptional in this regard—which excludes the slender, stiffer-tailed deinonychosaurs and the short-tailed therizinosaurs and oviraptorosaurs,

which like surface-swimming birds relied on the legs alone. Water-tolerant dinosaurs is supported by a number of examples of tridactyl-footed trackways that appear to record the animals polling along the bottom of waters shallow enough for them to reach bottom. Being such good swimmers, predatory dinosaurs could readily cross broad

Swimming predatory dinosaurs

Dilophosaurus

bottom "polling" trackways

watercourses, all the way to marine islands that they could see in the distance. Spinosaurs have been presented as specialist swimmers in the crocodilian pattern, in part because they were less buoyant than the avepod norm. But they were not nonpneumatic, so they could not have been as deep diving as crocs.

One anatomical item most predaceous dinosaurs did not have was substantial fat deposits, pursuit hunters needing to remain lean to maximize speed and agility. A possible exception would have been theropods living in climates featuring cold winters, to provide some insulation and help thermally tide over the season—although finding herbivores to kill or scavenge in winters that often weaken or dispatch plant eaters is not necessarily difficult. The slow herbivorous therizinosaurs were candidates for carrying more fat.

SKIN, FEATHERS, AND COLOR

Most dinosaurs are known from their bones alone, but we know a surprising amount about dinosaur body coverings from a rapidly growing collection of fossils that

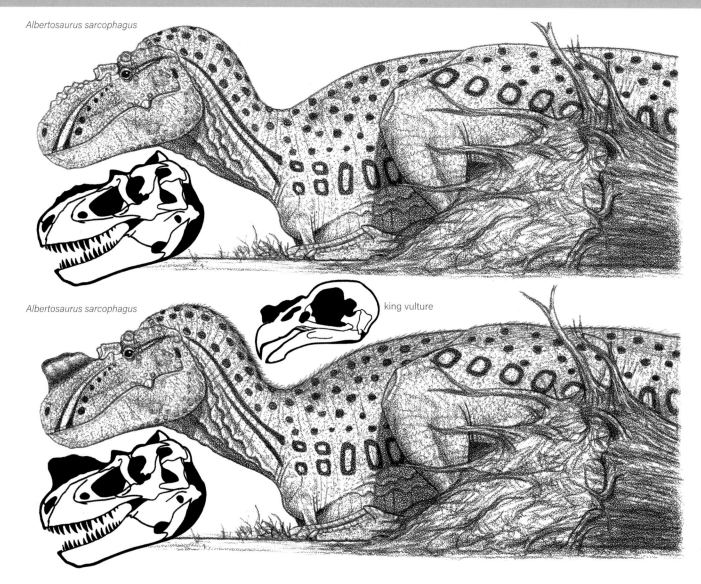

Albertosaurus sarcophagus

Albertosaurus sarcophagus

king vulture

Appearance alternatives

record their integument. It has long been known that large, and some small, dinosaurs were covered with mosaic-patterned scales. These can be preserved as impressions in the sediments before the skin rotted away, but in numerous cases traces of keratin are still preserved. Footprints sometimes preserved the shape of the bottom scales as well as the foot pads. Lizard-like overlapping scales were not common among dinosaurs, although scales like those on the tops of some bird feet may have been present in birdlike dinosaurs. Dinosaur mosaic scales were commonly semihexagonal in shape, with larger scales surrounded by a ring of smaller scales, forming rosettes that were themselves set in a sea of small scales. These scales were often flat, but some were more topographic, ranging from small beads on up. Because dinosaur scales were usually not large, they tend to disappear from visual resolution when viewed from a dozen feet or more away. However, in some cases the center scale in a rosette was a large, projecting, subconical scale; these were often arranged in irregular rows. On a given dinosaur the size and pattern of the scales varied depending on their location: those low on the body were often smaller than those higher up.

Dinosaurian soft crests, combs, dewlaps, wattles, and other soft display organs may have been more widespread than we realize. In particular, on some snouts, such as those of tyrannosaurs, the nasal bones bore a long, narrow midline rugosity that is usually restored as being somewhat further enlarged by a shallow, hard keratin sheath. The king vulture also has a modest nasal rugosity, which anchors a very prominent fleshy caruncle. It cannot be ruled out that the dinopredators with similar snout tops had comparable features. A throat pouch has been found under the jaws of an ornithomimosaur theropod, and these may have been more widespread. Bony armor was very scarce on dinopredators, they doing the attacking that drove other dinosaurs to be armored. Plates were covered with hard keratin. The exception was *Ceratosaurus*, which sported a midline

The feathered theropod *Sinosauropteryx*

dorsal row of small bony adornments; these would have been ensheathed by keratin.

Feathers have long been known on the fossils of birds preserved in fine-grained lake or lagoon bottom sediments, including winged *Archaeopteryx*. After a fossil drought that lasted until the end of the last century, over the last three decades a growing array of small avepod dinosaurs flightless and winged have been found covered with bristle protofeathers or fully developed pennaceous feathers in the Chinese Yixian–Jiufotang beds. Some researchers have claimed that the simpler bristles are really degraded internal collagen fibers. This idea is untenable for a number of reasons, including the discovery of pigmentation—either visible to the naked eye or in microscopic capsules—in the fibers that allows their actual color to be approximated. Some small, nonflying theropods also had scales at least on the tail and perhaps legs. This suggests that the body covering of small dinosaurs was variable—ostriches lack feathers on the legs, and a number of mammals from a small bat through a number of suids as well as humans, rhinos, and elephants are essentially naked.

Ironically, some paleoartists are going too far with feathering dinosaurs, giving many the fully developed aeroshells in which contour feathers streamline the head, neck, and body of most flying birds. But most Mesozoic dinosaurs did not fly, and like those birds whose ancestors lost flight long ago, flightless dinosaurs would have had shaggier, irregular coats for purposes of insulation and display. Also, modern birds have the hyperflexible necks that allow many but not all fliers to strongly U-curve the neck to the point that the head and heavily feathered neck aerodynamically merge with the body. Dinosaurs and even early flying dinobirds like *Archaeopteryx* and microraptors could not do this, so their less-flexible necks stuck out ahead of the shoulders like those of a number of modern long-necked flying birds.

Because fibers covered basal ornithischians as well as avepods, it is a good scientific bet that dinosaur insulation evolved once, in which case the filaments were all protofeathers. The absence to date of protofeathers in Triassic and Early Jurassic protodinosaurs and basal dinosaurs is the kind of negative evidence that is no more meaningful than their lack of fossil scales; this absence of evidence long led to the denial of insulation in any dinosaurs and is likely but not certainly to be settled by the eventual discovery of insulation in basal examples. However, it cannot be ruled out that insulation evolved more than once in dinosaurs. A question is why dinofur and feathers appeared in the first place. The first few bristles must have been too sparse to provide insulation, so their initial appearance should have been for nonthermoregulatory reasons. One highly plausible selective factor was display. As the bristles increased in number and density to improve their display effect, they became thick enough to help retain the heat generated by the increasingly energetic archosaurs.

A number of researchers argue that the pigment organelles of feathers preserve well and their shape varies according to color, so they are being used to restore the actual colors of feathered dinosaurs. Although some researchers have challenged the reliability of this method, it appears to be largely sound, so this book uses the colors determined by this technique—doing so maximizes the probability of achieving correct coloration, whereas not doing so essentially ensures incorrect results. It appears that the feathers of some dinosaurs were, as might be expected, iridescent, using refraction rather than pigmentation to achieve certain color effects. The hypothesis offered by some researchers, that the differing scale patterns on a particular dinosaur species correspond to differences in coloration, is plausible, but some reptiles are uniformly colored regardless of variations in scales. Dinosaur scales were better suited to carry bold and colorful patterns like those of reptiles, birds, tigers, and giraffes than is the dull gray, nonscaly skin of big mammals, and the color vision of dinosaurs may have encouraged the evolution of colors for display and camouflage. Dinosaurs adapted to living in forested areas may have been prone to using greens as stealth coloring. On the other hand, big reptiles and birds tend to be earth tinged despite their color vision. Small dinosaurs are the best candidates for bright and/or bold

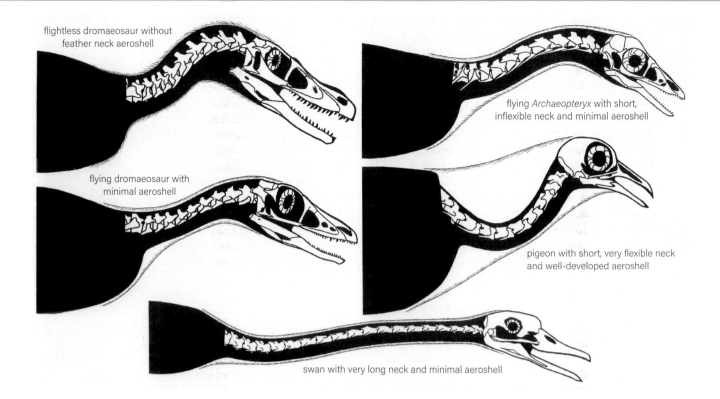

flightless dromaeosaur without feather neck aeroshell

flying dromaeosaur with minimal aeroshell

flying *Archaeopteryx* with short, inflexible neck and minimal aeroshell

pigeon with short, very flexible neck and well-developed aeroshell

swan with very long neck and minimal aeroshell

color patterns like those of many but not all small lizards and birds. On yet another hand, because humans lack vision in the ultraviolet range, we miss seeing a lot of the coloration of many animals, so a number of reptiles and especially birds that look drab to us—including genders that look bland and much the same—feature dramatic ultraviolet color patterns, often for sexual purposes. Predaceous dinosaurs of all sizes may have used specific color displays for intraspecific communication or for startling predators. Crests and taller neural spines would be natural bases for vivid, even iridescent, display colors, especially in the breeding season. Because dinosaur eyes were bird- or reptile-like, not mammal-like, they lacked white surrounding the iris. Flesh-eating dinosaurs' eyes may have been solid black or brightly colored, like those of many reptiles and birds.

A number of birds are ensconced in poisonous feathers containing neurotoxins—in a few body tissues may be contaminated. These toxins can afflict those that touch the birds, or worse, consume them; the noxious birds are themselves immune to the chemicals. The origin of these poisons appears to be the bird's diets, including insects. The toxicity probably deters predation while suppressing parasites. Some small-feathered Mesozoic archosaurs that hunted insects may well have been similarly toxic.

RESPIRATION AND CIRCULATION

The hearts of turtles, lizards, and snakes are three-chambered organs incapable of generating high blood pressures. The lungs, although large, are internally simple structures with limited ability to absorb oxygen and exhaust carbon dioxide and are operated by rib action. Even so, at least some lizards apparently have unidirectional airflow in much of their lungs, which aids oxygen extraction. Crocodilian hearts are incipiently four chambered but are still low pressure. Their lungs are internally dead end, but they too seem to have unidirectional airflow, and the method by which they are ventilated is sophisticated. Muscles attached to the pelvis pull on the liver, which spans the full height and breadth of the rib cage, to expand the lungs. This action is facilitated by an unusually smooth ceiling of the rib cage that allows the liver to easily glide back and forth, the presence of a rib-free lumbar region immediately ahead of the pelvis, and, at least in advanced crocodilians, a very unusual mobile pubis in the pelvis that enhances the action of the muscles attached to it.

Birds and mammals have fully developed four-chambered, double-pump hearts able to propel blood in large volumes at high pressures. Mammals retain fairly large dead-end lungs, but they are internally very intricate, greatly expanding the gas-exchange surface area, and so are efficient despite the absence of one-way airflow. The lungs are operated by a combination of rib action and the vertical, muscular diaphragm. The presence of the diaphragm is indicated by the existence of a well-developed, rib-free lumbar region, preceded by a steeply plunging border to the rib cage on which the vertical diaphragm is stretched.

It is widely agreed that all dinosaurs probably had fully four-chambered, high-capacity, high-pressure hearts. Their respiratory complexes appear to have been much more diverse.

It is difficult to reconstruct the respiratory systems of ornithischians because they left no living descendants and because their rib cages differ not only from those of all living tetrapods but among different ornithischian groups—the absence of skeletal pneumaticity shows they did not have birdlike breathing, and their lungs were probably dead end, perhaps similar to mammals.

Restoring the respiratory complexes of saurischians, especially theropods, is much more straightforward because birds are living members of the group and retain the basic theropod system. Birds have the most complex and efficient respiratory system of any vertebrate. Because the lungs are rather small, the chest ribs that encase them are fairly short, but the lungs are internally intricate so they have a very large gas-exchange area. The lungs are also rather stiff and set deeply up into the strongly corrugated ceiling of the rib cage. The lungs do not dead end; instead, they are connected to a large complex of air sacs whose flexibility and, especially, volume greatly exceed those of the lungs. Some of the air sacs invade the pneumatic vertebrae and other bones, but the largest sacs line the sides of the trunk; in most birds the latter air sacs extend all the way back to the pelvis, but in some, especially flightless examples, they are limited to the rib cage. The chest and abdominal sacs are operated in part by the ribs; the belly ribs tend to be extra long in birds that have well-developed abdominal air sacs. All the ribs are highly mobile because they attach to the trunk vertebrae via well-developed hinge articulations. The hinging is oriented so that the ribs swing outward as they swing backward, inflating the air sacs within the rib cage and then deflating the sacs as they swing forward and inward. In most birds the movement of the ribs is enhanced by ossified uncinate processes that form a series along the side of the rib cage. Each uncinate process acts as a lever for the muscles that operate the rib to which the process is attached. In most birds the big sternal plate also helps ventilate the air sacs. The sternum is attached to the ribs via ossified sternal ribs that allow the plate to act as a bellows on the ventral air sacs.

In those birds with short sternums, the flightless ratites, and in active juveniles, the sternum is a less important part of the ventilation system.

The system is set up in such a manner that most of the fresh inhaled air does not pass through the gas-exchange portion of the lungs but instead goes first to the air sacs, from where it is injected through the entire lungs in one direction on its way out. Because this unidirectional airflow eliminates the stale air that remains in dead-end lungs at the end of each breath and allows the blood and airflow to work in opposite, countercurrent directions that maximize gas exchange, the system is very efficient. Some birds can sustain cruising flight at levels higher than Mount Everest and equaling those of jet airliners.

Neither the first theropods nor prosauropods show clear evidence that they possessed air sacs, and aside from their lungs therefore being dead-end organs or close to it, little is known about their respiration. In the first avepod

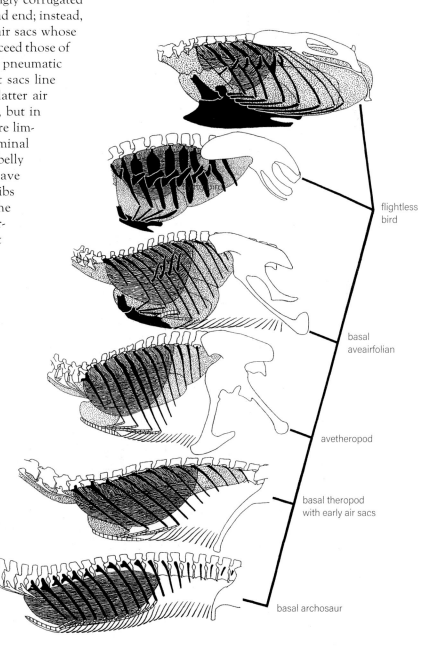

flightless bird

basal aveairfolian

avetheropod

basal theropod with early air sacs

basal archosaur

Respiratory complexes of archosaurs

theropods some of the vertebrae were pneumatic, indicating the presence of air sacs. Also, the hinge jointing of the ribs increased, indicating that they were probably helping to ventilate the lungs by inflating and deflating trunk air sacs. As theropods evolved, the hinge jointing of the ribs further increased, as did the invasion of the vertebrae by air sacs until reaching the hips. Also, the chest ribs began to shorten, probably because the lungs, set up into a corrugated ceiling of ribs, were becoming smaller and stiffer as the air sacs did more of the work. By this stage the air-sac complex was probably approaching the avian condition, and airflow in the lungs should have been largely unidirectional. The sternum was still small, but the gastralia may have been used to help ventilate the ventral belly air sacs. Alternatively, the air sacs were limited to the rib cage as they are in at least some flightless birds—the extra-long belly ribs of birds with big abdominal air sacs are absent in theropods. In many aveairfoilan theropods the ossified sternum was as large as it is in ratites and juvenile birds and was attached to the ribs via ossified sternal ribs, so the sternal plate was combining with the gastralia to inflate and deflate the air sacs. Also, ossified uncinate processes are often present, indicating that the bellows-like action of the rib cage was also improved. At this stage the respiratory complex was probably about as well developed as it is in some modern birds.

The very few researchers who think birds are not dinosaurs deny that theropods breathed like birds. Some propose that theropod dinosaurs had a crocodilian liver-pump system. Aside from theropods not being close relatives of crocodilians, they lacked the anatomical specializations that make the liver-pump system possible—a smooth rib cage ceiling, a lumbar region, and a mobile pubis. Instead, some of the theropods' adaptations for the avian air-sac system—the corrugated rib cage ceiling created by the hinged rib articulations, the elongated belly ribs—would have prevented the presence of a mobile liver. Proponents of the avepod liver pump point to the alleged presence of a deep liver within the skeletons of some small theropods. The fossil evidence for these large livers is questionable, and in any case, predators tend to have big livers, as do some birds. The existence of a crocodilian liver-pump lung ventilation system in dinosaurs can be ruled out.

The highly pneumatic sauropods show strong evidence that they too independently evolved a complex air-sac system that probably involved unidirectional airflow and approached, but did not fully match, the sophistication and efficiency of those of birds. For instance, because sauropods lacked gastralia, the air sacs should have been limited to the rib cage.

Mammal red blood cells lack a nucleus, which increases their gas-carrying capability. The red blood cells of reptiles, crocodilians, and birds retain a nucleus, so those of dinosaurs should have as well.

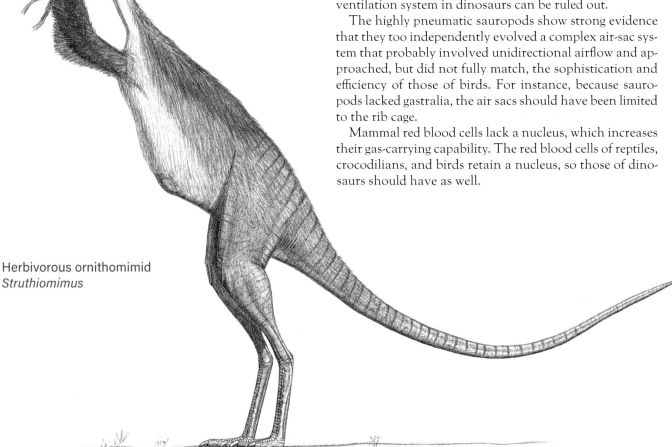

Herbivorous ornithomimid
Struthiomimus

DIGESTIVE TRACTS

In a number of avepod specimens, gastroliths, or gizzard stones, are preserved within the rib cage, often as stone bundles. This does not appear to be a general dinopredator attribute, stone bundles having not been turning up in the articulated ribcages of flesh eaters prior to aveairfoilans. They do appear in the herbivorous elaphrosaurs, ornithomimids, and oviraptorosaurs and the protobird anchiornids. The stones could be used directly to help mash and grind up plant materials as they were squeezed and manipulated by powerful gizzard muscles and/or to help mix the foodstuffs up like agitator balls in spray paint cans.

The digestive tracts of meat, fish, and insect consuming dinosaurs were relatively short, simple systems that quickly processed the easily digested chunks of flesh bolted down whole or in substantial pieces after the simple puncturing and scissoring action of the serrated teeth. Coprolites attributable to large theropods often contain large amounts of undigested bone, confirming the rapid passage of food through the tract. A few preserved remains indicate that nonvolant predatory dinosaurs retained a somewhat larger digestive tract than do birds with their overall lighter complexes, starting with a large crop. Most of the herbivorous avepods did not have tooth batteries or saw-edged beaks to chew with; their mouths merely cut off bits of vegetation to be quickly swallowed. Not being highly sophisticated herbivores, ornithomimid and oviraptorosaur abdomens were not highly capacious. Ratites carry a combination of gastroliths, gut flora, ingested fodder, and feces amounting to about a tenth of body mass, and such was probably true of the similar-grade omni/herbivorous ornithomimids and oviraptorosaurs. It was the therizinosaurs among avepods that went for high-capacity digestive complexes that broke down tough plant material, presumably via fermentation accomplished by gut bacteria. In most herbivores this occurs in the hindgut. Among such full-blown herbivores, the gut contents can be up to a fifth of total mass. In the highly specialized ruminant ungulates, cattle included, the microbial breakdown begins in the foregut in a special chamber. Hoatzins sport a comparable system in which the crop is used to initiate microbial processing—thus their tag the "stink bird" is due to the methane they put out. Therizinosaurs may have carried a ruminant-like, possibly smelly digestive complex. The cheeks that appear to have been present on therizinosaurs should have allowed them to mash food a little with their modestly developed dental batteries before swallowing. Hoatzins are unusual among avepods in doing a little chewing with their gently serrated beaks.

SENSES

The large eyes and well-developed optical lobes characteristic of most dinosaurs indicate that vision was usually their primary sensory system, as it is in nearly all birds. Reptiles and birds have full color vision extending into the ultraviolet range, so dinosaurs very probably did too. The comparatively poorly developed color vision of most mammals is a heritage of the nocturnal habits of early mammals, which reduced vision in the group to such a degree that eyesight is often not the most important of the senses. Reptile eyesight is about as good as that of well-sighted mammals, and birds tend to have very high-resolution vision, both because their eyes tend to be larger than those of reptiles and mammals of similar body size and because they have higher densities of light-detecting cones and rods than do mammals. The cones and rods are also spread at a high density over a larger area of the retina than in mammals, in which high-density light cells are more concentrated at the fovea (so our sharp field of vision covers just a few degrees). Some birds have a secondary fovea. Day-loving raptors can see about three times better than people, and their sharp field of vision is much more extensive, so birds do not have to point their eye at an object as precisely as mammals to focus on it. Birds can also focus over larger ranges, 20 diopters compared with 13 diopters in young adult humans. The vision of the bigger-eyed predatory dinosaurs may have rivaled this level of performance. The dinosaurs' big eyes have been cited as evidence for both daylight and nighttime habits. Large eyes are compatible with either lifestyle—it is the (in this case unknowable) structure of the retina and pupil that determines the type of light sensitivity.

Birds' eyes are so large relative to the head that they are nearly or entirely fixed in the skull, so looking at specific items requires turning the entire head. The same was likely to have been true of smaller-headed avepods. Theropods with larger heads should have had more mobile eyeballs that could scan for objects without rotating the entire cranium. The eyes of most dinosaurs faced to the sides, maximizing the area of visual coverage at the expense of the binocular view directly ahead. Some birds and mammals—primates most of all—have forward-facing eyes with overlapping fields of vision, and in at least some cases, vision includes a binocular, stereo effect that provides depth perception. Tyrannosaurid, ornithomimid, and many aveairfoilan theropods had partly forward-facing eyes with overlapping vision fields. Whether vision was truly stereo in any or all of these dinosaurs is not certain; it is possible

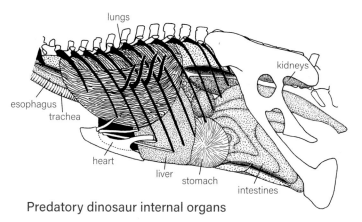

Predatory dinosaur internal organs

(labels: lungs, kidneys, esophagus, trachea, heart, liver, stomach, intestines)

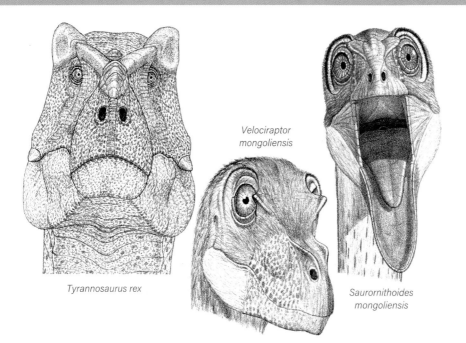

Predatory dinosaurs with forward facing eyes

Velociraptor mongoliensis

Tyrannosaurus rex

Saurornithoides mongoliensis

that the forward-facing eyes were an incidental but perhaps beneficial side effect of the expansion of the back of the skull to accommodate larger jaw muscles in tyrannosaurids, *Tyrannosaurus* most of all.

Most birds have a poorly developed sense of smell, the result of the lack of utility of this sense for flying animals as well the lack of space in heads whose snouts have been reduced to save weight. Exceptions are some vultures, which use smell to detect rotting carcasses hidden by deep vegetation, and grub-hunting kiwis. As nonfliers with large snouts, many reptiles and mammals have very well-developed olfaction, sometimes to the degree that it is a primary sensory system, canids being a well-known example. Dinosaurs often had extremely well-developed, voluminous nasal passages with abundant room at the back for large areas of olfactory tissues. In many dinosaur brains the olfactory lobes were large, verifying their effective sense of smell. Herbivorous dinosaurs probably had to be approached from downwind to avoid their sensing and fleeing from an attack. Among predatory dinosaurs, tyrannosaurs and dromaeosaurs had especially excellent olfaction, useful for finding both live prey and carcasses.

Mammals have exceptional hearing, in part because of the presence of large, often movable outer ear pinnae that help catch and direct sounds into the ear opening and especially because of the intricate middle ear made up of three elements that evolved from what were once jaw bones. In some mammals hearing is the most important sense, bats and cetaceans being the premier examples. Reptiles and birds lack fleshy outer ears, and there is only one inner ear bone. The combination of outer and complex inner ears means that mammals can pick up sounds at low volume. Birds partly compensate by having more auditory sensory cells per unit length of the cochlea, so sharpness of hearing and discrimination of frequencies are broadly similar in birds and mammals. Where mammalian hearing is markedly superior is in high-frequency sound detection. In many reptiles and birds the auditory range is just 1–5 kHz; owls are exceptional in being able to pick up from 250 Hz to 12 kHz, and geckos go as high as 10 kHz. In comparison, humans can hear 20 kHz, dogs up to 60 kHz, and bats 100 kHz. At the other end of the sound spectrum, some birds can detect very low frequencies: 25 Hz in cassowaries, which use this ability to communicate over long distances, and just 2 Hz in pigeons, which may detect approaching storms. It has been suggested that cassowaries use their big, pneumatic head crests to detect low-frequency sounds, but pigeons register even deeper bass sounds without a large organ.

In the absence of fleshy outer and complex inner ears, dinosaur hearing was in the reptilian-avian class, and they could not detect very high frequencies. Nor were the auditory lobes of dinosaur brains especially enlarged, although they were not poorly developed either. Nocturnal, flying, rodent-hunting owls are the only birds that can hear fairly high-frequency sounds, so certainly most and possibly all dinosaurs could not hear them either. Oviraptorosaurs had hollow head crests similar to those of cassowaries, hinting at similar low-frequency sound detection abilities. The big ears of large dinosaurs had the potential to capture very low frequencies, allowing them to communicate over long distances. It is unlikely that hearing was the most important sense in any dinosaur, but it was probably important for detection of prey and predators, and for communication, in all species.

VOCALIZATION

No living reptile has truly sophisticated vocal abilities, which are best developed in crocodilians. Some mammals do, humans most of all. Birds use the syrinx in the chest to generate sounds; when this scheme evolved among dinosaurs is not yet clear. A number of birds have limited vocal performance, but many have evolved a varied and often very complex vocal repertoire not seen among other vertebrates outside of people. Songbirds sing, and a number of birds are excellent mimics, to the point that some can imitate artificial sounds such as bells and sirens, and parrots can produce understandable humanlike speech. Some birds, swans particularly, possess elongated tracheal loops in the chest that they use to produce high-volume vocalizations. Cassowaries call one another over long ranges with very low-frequency sounds, and so do elephants. Birds possess the intricate voice boxes needed to help form complex vocalizations. Among dinosaur fossils only an ankylosaur skull includes a complete voice box. The complicated structure of the armored dinosaurs' larynx suggests vocal performance at an avian level, perhaps high-end performance, and such may have been true of other dinosaurs, including those predatory. The long trachea of long-necked dinosaurs such as therizinosaurs should have been able to generate powerful low-frequency sounds that could be broadcast over long ranges. Vocalization is conducted through the open mouth rather than through the nasal passages, so complex nasal passages acted as supplementary resonating chambers. It is doubtful that any nonavian dinosaur had vocal abilities to match the more sophisticated examples seen in the most vocally advanced birds and mammals. Although we will never know what dinosaurs sounded like, and the grand roars of dinosaur movies are not likely, there is little doubt that the Mesozoic forests, prairies, and deserts were filled with their voices.

GENETICS

As more fossils are found in different levels of geological formations, the evidence is growing that dinosaurs enjoyed high rates of speciation that boosted their diversity at any given time. And over time, via a rapid turnover of species, most did not last for more than a few hundred thousand years before being replaced by new species one way or another. The same is true of birds, which have more chromosomes than slower-evolving mammals. Dinosaurs presumably had the same genetic diversity as their direct avian descendants, which may have been a driving force behind their multiplicity.

DISEASE AND PATHOLOGIES

Planet Earth has long been infested with a toxic soup of diseases and other dangers that put dinopredators at high

Predatory dinosaurs in conflict,
Velociraptor and *Saurornithoides*

risk. The disease problem was accentuated by the global greenhouse effect, which maximized the tropical conditions that favored disease organisms, especially bacteria and parasites. Biting insects able to spread assorted diseases were abundant during the Mesozoic; specimens have been found in amber and fine-grained sediments. Reptile and bird immune systems operate somewhat differently from those of mammals; in birds the lymphatic system is particularly important. Presumably the same was true of their dinopredator ancestors. Wild living animals are prone to be loaded with arrays of resident multicellular parasites in the form of arthropods and worms on their skins, amidst their integument, in internal cavities, including, especially, the digestive tract, and within internal tissues. In some cases, these parasites can become debilitating.

The skeletons of predaceous dinosaurs often preserve pathologies, sometimes numerous in a given individual. Some appear to record internal diseases and disorders. Fused vertebrae are fairly common. Also found are growths that represent benign conditions or cancers. Most pathologies are injuries caused by stress or wounds; the latter often became infected, creating long-term, pus-producing lesions that affected the structure of the bone. Injuries tell us a lot about the activities of dinosaurs. Some dinosaur skeletons are so afflicted with serious defects that one of them very probably killed the beast, especially if it was immature.

The predaceous dinosaurs are, not surprisingly, especially prone to show signs of combat-related injury that derived from hunting, disputes over carcasses, and intraspecific conflicts, as well as everyday accidents. One *Allosaurus* individual shows evidence of damage to its ribs, tail, shoulder, feet, and toes as well as chronic infections of a foot, finger, and rib. The tail injury, probably caused by a kick or fall, occurred early in life. Some of the injuries, including those to the feet and ribs, look severe enough that they may have limited its activities and contributed to its death. A wound in another *Allosaurus* tail appears to have been inflicted by the spike of a stegosaur. The famous *Tyrannosaurus* "Sue" had problems with its face and tail as well as a neck rib, finger, and fibula. The head and neck wounds appear to have been caused by other *Tyrannosaurus* and in one case had undergone considerable healing. Other researchers have suggested that infections caused some of the injuries. The sickle-claw-bearing toes of dromaeosaurs and troodonts frequently show signs of stress damage. Some fossils show signs of injury from falls.

BEHAVIOR

BRAINS, NERVES, AND INTELLIGENCE

Assessing brain power is complicated because many factors are involved. One that has long been used is the mass of the brain relative to total body mass at a given size. Within the context that brains of a given performance level tend to become smaller relative to the body as size increases—elephant brains are many times absolutely larger than those of people while being many times smaller relative to body weight, and we are overall more intelligent—relatively bigger brains are likely to produce higher cognition. Also important is brain structure, with birds and mammals having more complex schemes, including large forebrains. Adding to the complications is the neural density factor. Reptiles have much lower neural density relative to brain mass than do mammals and birds, and the latter are markedly higher in this regard than mammals. The last point helps explain why birds with absolutely small brains, such as crows and parrots, achieve levels of thinking comparable to those of some far larger-headed primates. Avian brains are also markedly more energy efficient, their neurons requiring less glucose to process information. Big brains packed with lots of neurons can correlate with metabolism in that low-energy animals cannot produce enough metabolic power to operate high-cognition brains, which require a high metabolism. Less clear is whether energetic animals automatically have similarly energetic brains. In particular, it is not known whether reptilian brains can have high neural densities even if the animals run at high metabolic rates.

The brains of the great majority of dinosaurs were reptilian both in size relative to the body and in structure. There was some variation in size compared with body mass: the giant tyrannosaurids had unusually large brains for dinosaurs of their size and so did the duck-billed hadrosaurs they hunted. However, even the diminutive brains of sauropods and stegosaurs were within the reptilian norm for animals of their great mass.

Taken at face value, the small, fairly simple brains common to most dinosaurs seem to indicate that their behavioral repertoire was limited compared with those of birds and mammals, being more genetically programmed and stereotypical. But if dinosaurs are presumed to have been stable-temperature endotherms via high metabolic rates, then it is possible, albeit by no means certain, that their neural densities were in the mammalian or, since dinosaurs include birds, even in the avian range. This has led to estimates that bigger-brained dinosaurs such as tyrannosaurs were as smart as the cleverest birds, as well as primates other than humans, and may have used simple, crafted nonstone/nonmetal tools. "Crafting" implies modifying an object in some manner to make it usable, rather than just picking up a rock and using it to smash open a hard-shelled item. Crafting can be as simple as stripping leaves and side branches off a twig to make it into a probe or lever. But this is by no means certain when it comes to the majority of dinosaurs, whose thinking organs were reptilian in form. That the energetic dinosaurs had the low neural densities of reptiles unable to sustain high levels of activity appears unlikely, but their simple brains may have precluded the neural concentrations of birds. It is therefore not possible to reliably assess the intelligence of dinosaurs with reptilian-form brains at this time, and it may never be doable. Even if big theropods were not supersmart, it is pertinent that even small-brained animals can achieve remarkable levels of mental ability. Fish and lizards can retain new information and learn new tasks. Many fish live in organized groups. Crocodilians care for their nests and young. Social insects with tiny neural systems live in organized colonies that rear the young, enslave other insects, and even build large, complex architectural structures. It is not unthinkable that dinosaurs up to the biggest sauropods could use sticks and leafy branches to scratch themselves if they could reach close enough to their bodies with their mouths, use heavy sticks to knock down otherwise unreachable choice food items, or build leafy branch piles over water holes to protect them when not in use, as elephants do.

The major exception to dinosaurian reptile brains appeared in the birdlike aveairfoilans. Their brains were proportionally larger, falling into the lower avian zone, as did their complexity. It is possible if not probable that neural densities were approaching if not at the avian level. It may be that the expanded and upgraded brains of aveairfoilans evolved at least in part in the context of the initial stages of dinosaurian flight. Presumably the bigger-brained dinosaurs were capable of more sophisticated behavior than other dinosaurs. Use of very simple tools is plausible, all the more so because many small nonavian avepods that had supple-fingered hands may have been able to manipulate devices, in addition to using their mouths in ways similar to tool-using birds. On the other hand, use of crafted tools in wild birds is not extensive, and it may well be that no Mesozoic dinosaur did this. If any did, tool utilization may have occurred in the context of prying open hard-shelled food items or probing insect holes in search of prey. The insectivorous alvarezsaurs might have been

Possible predatory dinosaur pack trackways

tyrannosaurids

dromaeosaurids

Tenontosaurus

Deinonychus

especially prone to the latter, but their stout and powerful arms and hands were much better suited for bursting open insect colonies than holding tools.

The enlarged spinal cavity in the pelvic region of many small-brained dinosaurs was an adaptation to better coordinate the function of the hind limbs and is paralleled in big ground birds.

SOCIAL ACTIVITIES

Land reptiles do not form organized groups. Birds and mammals often do, but many do not. Most big cats, for instance, are solitary, but lions are highly social. Some, but not all, deer form herds.

The presence of a number of individuals of a single species of theropod in association with the skeleton of a potential prey animal has been cited as evidence that dino-predators sometimes killed and fed in packs. It is, however, often difficult to explain why so many theropods happened to die at the same time while feeding on a harmless carcass. It is more probable that the theropod skeletons represent individuals killed by other theropods in disputes over feeding privileges, an event that often occurs when large carnivorous mammals and reptiles compete over a kill.

Trackways are the closest thing we have to motion pictures of the behavior of fossil animals. A significant portion of the trackways of a diverse assortment of predatory dinosaurs are solitary, indicating that the maker was not part of a larger group. There are examples of multiple trackways of dinopredators that lie close together on parallel paths. In some cases, this may be because the track makers were forced to follow the same path along a shoreline even if they were moving independently of one another. But such parallel trackways are common enough to suggest some dinosaurian predators moved in small packs. That said, the big majority of avepod footprints track animals moving on their own as they patrolled or traveled along fresh- and saltwater shorelines.

As for the herbivorous avepods, they may have been prone to move in pods, flocks, and herds. Flocking birds almost always fly in single-species groups. On open ground where a lot of species dwell, herbivores such as wildebeest, zebras, ostriches, elephants, and gazelles often form collective herds, each taxon bringing its own best predator-detection system into the mix. One can imagine therizinosaurs, ornithomimids, and other predator-wary dinosaurs of assorted sizes doing the same, but multispecies bone beds indicating that dinosaurs did this have yet to be uncovered, so perhaps they did not.

REPRODUCTION

It has been suggested that some predatory dinosaur species exhibit robust and gracile morphs that represent the two sexes. It is difficult either to confirm or deny many of these claims because it is possible that the two forms represent different species. Males are often more robust than females, but there are exceptions. Female raptors are usually larger than the males, for instance, and the same is true of some whales. Attempts to use the depth of the chevron bones beneath the base of the tail to distinguish males from females have failed because the two factors are not consistent in modern reptiles. Head-crested oviraptorosaurs may be males if they are not mature individuals of both sexes. On the other hand, among cassowaries it is the females that have somewhat larger crests. This is

Predatory dinosaur head displays

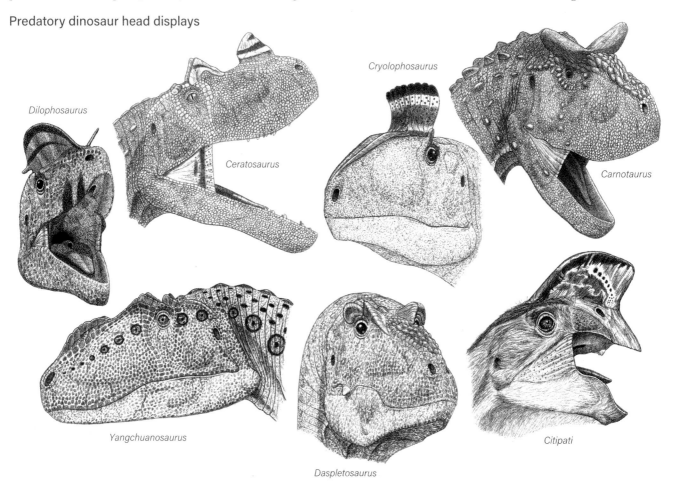

Dilophosaurus

Ceratosaurus

Cryolophosaurus

Carnotaurus

Yangchuanosaurus

Daspletosaurus

Citipati

atypical for crested birds, in which males have markedly larger display structures; that male cassowaries do a lot of the parenting may have something to do with their crest dimensions. The robust form of *Tyrannosaurus* has been tentatively identified as the female based on the inner bone tissues associated with egg production in birds, but the distribution of the stout and gracile morphs of this genus over stratigraphic time is more compatible with different species than with sexes. What is more suggestive of boys and girls is variation in size of the orbital bosses of *Tyrannosaurus* at a given geological level, although ontogeny may also be a factor. While preparing these guides, I realized that a pair of fairly complete *Allosaurus* skulls from the same quarry share features indicating they are a distinct species, yet they come in two forms, a deep and a shallow overall shape, providing perhaps the best evidence of dinosaurian gender identity yet observed.

Reptiles and some birds and mammals, including humans, achieve sexual maturity before reaching adult size, but most mammals and extant birds do not. Females that are producing eggs deposit special calcium-rich tissues on the inner surface of their hollow bones. The presence of this tissue has been used to show that a number of dinosaurs began to reproduce while still immature in terms of growth. The presence of still-growing aveairfoilans brooding their nests confirms this pattern. Most or all predatory dinosaurs probably became reproductive before reaching full adult size.

The marvelous array of head and body crests, hornlets, bosses, and feathers evolved by assorted predatory dinosaurs shows that many were under strong sexual selective pressure to develop distinctive display organs and weapons to identify themselves to other members of their species and to succeed in sexual competition. The organs we find preserved record only a portion of these visual devices—those consisting of soft tissues and color patterns are largely lost. How these organs were used varied widely. Females used display organs to signal males of the species that they were suitable and fertile mates. Males used them both to intimidate male rivals and to attract and inseminate females.

Healthy animals in their reproductive prime are generally able to dedicate more resources to growing superior-quality displays better able to attract similar quality mates. Many dinosaurs probably engaged in intricate ritual display movements and vocalizations during competition and courtship, using display organs when they had them; these behavioral displays have been lost to time. The head and body display surfaces of many dinosaurs were oriented to the sides, so they had to turn themselves to best flaunt their display. This orientation was true of the avepods with the paired crests atop their snouts, which were popular, for reasons not known, in the Jurassic in podokesauroids, basal averostrans, and early tyrannosauroids. These crests did not make it into the Cretaceous as far as is known. The transversely flattened ceratosaur nasal horn was another

side view display organ, as were the hornlets and bosses common among a variety of big avepods. Large subtriangular hornlets above the orbits were a frequent feature, continuing into most tyrannosaurids. Then, just before the final extinction, the very similar *Tarbosaurus* and *Tyrannosaurus* entirely dropped the hornlets in favor of the bosses just aft of them. Why is unknown. In *Tyrannosaurus*, the earliest of the known species, *T. imperator*, bore distinctive long and rather low rugose spindle bosses, followed by the newly recognized and prominent upright discs of *T. rex*, which lived alongside *T. regina*, sporting its own boss form in a classic species identification pattern not yet seen in the rest of Theropoda. The transverse head crest of *Cryolophosaurus*, and the stout sideway projecting horns and domes of some abelisaurs, provided unusual frontal displays.

Many birds, including flightless examples that still have large arm-born feathers, use their arm and tail arrays for intraspecific display, and such would have been probable in basal aveairfoilans. Those with the capability to fly, as per archaeopterygians and microraptorines, could have done so while on the wing, twirling in the air as they showed off to one another. Simply fluffing up and bristling head and body feathers would have been a common means of close-up display.

While intraspecific competition is often fairly pacific to avoid casualties, it can be forceful and even violent in animals that bear weapons. Male hippos and lions suffer high injury and mortality from members of their own species, and the same may have been true of predatory theropods as males battled with sharp teeth and claws, as evidenced by the large number of wounds inflicted by such on theropod skulls.

In reptiles and birds, the penis and the testes are internal, and this was the condition in dinosaurs. Most birds lack a penis, but whether any more basal dinosaurs shared this characteristic is unknown. Presumably copulation was a quick process that occurred with the female lowering her shoulders and swinging her tail aside to provide clearance for the male, which reared behind her on two legs or even one leg while placing his hands on her back to steady them.

At least some dinosaurs from theropods to sauropods to ornithopods produced hard-shelled eggs like those of birds rather than the softer-shelled eggs of reptiles, including crocodilians, and mammals. The evolution of calcified shells may have precluded live birth, which is fairly common among reptiles and is absent in birds. On the other hand, eggs of prosauropods and protoceratopsids appear to have been soft shelled, indicating that there was considerable variation in the feature in dinosaurs, perhaps even within subgroups. If so, that could help explain why remains of dinosaur eggs are surprisingly scarce through much of the Mesozoic. Even so, a growing collection of eggs and nests is now known for a variety of Late Jurassic and Cretaceous dinosaurs great and small. Firmly identifying the producer of a given type of egg requires the presence of intact eggs within the articulated trunk skeleton

Intraspecies combat

Ceratosaurus

Daspletosaurus

and in their nests shows that they were formed and deposited in pairs as in reptiles, rather than singly as in birds. Even small reptiles lay small eggs relative to the size of the parent's body, whereas birds lay proportionally larger eggs. The eggs of small dinosaurs are intermediate in size between those of reptiles and birds. It is interesting that no known Mesozoic dinosaur eggs matches the size of the gigantic 12 kg (25 lb) eggs laid by the flightless elephant birds, which, as big as they got at nearly 400 kg (800 lb), was dwarfed by many dinosaurs. The largest Mesozoic dinosaur eggs discovered so far weighed 5 kg (11 lb) and probably belonged to 1-tonne-plus oviraptors.

Two basic reproductive stratagems are known as r-strategy and K-strategy: K-strategists are slow breeders that produce few young; r-strategists produce large numbers of offspring that offset high losses of juveniles. Rapid reproduction has an advantage. Producing large numbers of young allows a species to quickly expand its population when conditions are suitable, so r-strategists are "weed species" able to rapidly colonize new territories or promptly recover their population after it has crashed for one reason or another. As far as we know, predatory dinosaurs of all sizes were r-strategists that typically laid large numbers of eggs in the breeding season, although dinosaurs isolated on predator-free islands might have been slow breeders. This reproductive strategy may explain why dinosaurs laid smaller eggs than birds, most of which produce a modest number of eggs and provide the chicks with considerable parental at-

or identifiable embryo skeletons within the eggs, as well as adults found atop their nests in brooding posture. Because each dinosaur group produced distinctive types of eggshells and shapes, the differences can be used to further identify their origin, although the producers of many types remain obscure. While the eggs of some herbivorous dinosaurs were near-perfect spheres, as far as is known those of the meat eaters were elongated, very much so in oviraptorosaurs, and strongly tapered in troodonts. The surface texture of the egg was crenulated in some and bumpy in others. The arrangement of eggs within dinosaur bodies tention. One r-strategist bird group is the big modern ratites, which produce numerous eggs. This is in contrast to the big island ratites that laid only one to a few oversized eggs a year because the young were not at risk of being snarfed up by predators, until humans liquidated the populations just a thousand years ago in part by eating the giant eggs. Elephant-sized avepods were very different in this respect from same-sized mammals, which are K-strategists that produce few calves, which then receive extensive care over a span of years. Nor did any dinosaur nurse its young via milk-producing mammary glands. It is possible

Predatory dinosaur eggs and nests to same scales (shaded larger scale)

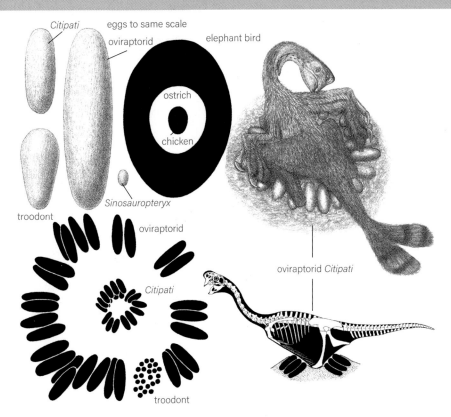

eggs to same scale

Citipati

oviraptorid

elephant bird

ostrich

chicken

Sinosauropteryx

troodont

oviraptorid

Citipati

oviraptorid *Citipati*

troodont

that some dinosaurs produced a "milk"-like substance in the digestive tract that was regurgitated to their young, as pigeons do, but there is no direct evidence of this.

It was long tacitly assumed that, like most reptiles, dinosaurs paid little or no attention to their eggs after burying them. A few lizards do stay with the nest, and pythons actually incubate their eggs with muscle heat. Crocodilians often guard their nests and hatchlings. All birds lavish attention on their eggs. Nearly all incubate the eggs with body heat; the exception is megapode fowl that warm eggs in mounds that generate heat via fermenting vegetation. The fowl carefully regulate the temperature of the nest by adding and removing vegetation to and from the mound. But when megapode chicks hatch, they are so well developed that the precocial juveniles quickly take off and survive on their own. The newly hatched chicks of ratites are also precocial, but they remain under the guardianship of adults that guide them to food sources and protect them from attack. Most bird chicks are altricial: they are so poorly developed when they break out of the egg that they have to be kept warm and fed by adults.

A spate of recent discoveries has revealed that the manner in which dinosaurs deposited eggs and then dealt with them and the offspring varied widely and was both similar to and distinctive from this behavior in living tetrapods.

Laying many eggs each breeding season helped overwhelm the ability of the local predators, including other avepods, to find and eat all the eggs and emerging hatchlings, although a fossil shows a large snake feeding on a just-emerged hatchling. Some dinosaur eggs whose makers have yet to be identified were buried in a manner that implies they were not brooded. Not clear is if any parents stayed close to their nests to guard the eggs; currently there is no evidence that predaceous dinosaurs built heat-generating mounds, although some of the herbivorous variety did. Some prehatchling reptiles in mass nests start vocalizing to better coordinate their synchronous emergence, even though doing so risks attracting egg and hatchling eaters. On the positive side, hatchling chirping can inspire guarding parents to open up the nest and help release its chicks.

Because smaller dinosaurs did not face the problem of accidentally crushing their offspring, they had the potential to be more intensely parental. The best evidence for dinosaur brooding and incubating is provided by the birdlike aveairfoilan theropods, especially oviraptors. The large number of eggs, up to a few dozen in some cases, could not have been produced by a single female, so the nests were probably communal. The big ratites also nest communally, and the resulting brood is therefore that of multiple parents cared for by the locally dominant couple. Oviraptors laid their elongated eggs in two-layered rings with an open center. Laid flat, the eggs were partly buried and partly exposed. Because eggs left open to the elements would die from exposure or predation, eggs were not left exposed unless they were intended to be protected and incubated by adults. A number of oviraptor nests have been found with an adult in classic avian brooding posture atop the eggs, the legs tucked up alongside the hips, the arms spread over the eggs. The egg-free area in the center of the ring allowed the downward-projecting pubis of the deep pelvis to rest between the eggs without crushing them;

flatter-bellied birds do not need this space between their eggs. Presumably the arm and other feathers of oviraptorosaurs completely covered the eggs to protect them from inclement conditions and to retain the incubator's body heat. It is thought that brooding oviraptors were killed in place by sandstorms or more likely dune slides. The giant eggs appear to be of the type laid by oviraptors, and they too are laid in rings, in their case of enormous dimensions (up to 3 m, or 10 ft, across). These are the largest incubated nests known and were apparently brooded by oviraptors weighing a tonne or two—brooding by such big parents was made possible by the body being supported by the pubis between the eggs, rather than the entire body bearing down on the eggs. It may be that these are the biggest nests that are practical for brooding, and that prevented adult oviraptors from becoming even larger. In troodont nests the less-elongated eggs were laid subvertically in a partial spiral ring, again with the center open to accommodate the brooder's pelvis. The size of the adult troodonts found in brooding posture atop their nests is as small as 0.5 kg (1 lb). The half-buried, half-incubated nesting habits of aveairfoilans ideally represent the near-avian arrangement expected in the dinosaurs closest to birds. This scheme was retained in basal birds including enantiornithines, even those that were strongly arboreal. Egg brooding without external heat did not appear until more modern avians, in the Late Cretaceous, and nesting up in vegetation possibly not until the Cenozoic. Megapode fowl have undergone a reversal by incubating eggs in fermenting vegetation mound nests that they carefully maintain at the proper temperature. We humans tend to presume that it was females that did most or all of the brooding, but in birds, including ratites, males often do a lot of the egg and nestling warming, and male cassowaries do all of it.

A problem that all embryos that develop in hard-shelled eggs face is getting out of that shell when the time is right. The effort to do so is all the harder when the egg is large and the shell correspondingly thick. Fortunately, some of the shell is absorbed and used to help build the skeleton of the growing creature. Baby birds use an "egg tooth" to achieve the breakout, and such is likely among the Mesozoic relations. While fossil evidence is not on hand, it is likely that the hatchings of the carnivorous dinosaurs were precocial, being immediately ready to leave the nest and feed themselves like the chicks of ratites and many fowl. Parental care probably ranged from nonexistent as it is for modern megapode fowl chicks to extensive in dinosaurs; in a number of cases it probably exceeded that seen in reptiles or even crocodilians, rivaling that of birds. Upon hatching the babies of adults over half a tonne were probably on their own, the size disparity rendering parental care impractical. Juvenile tyrannosaurids were unusual in having elongated snouts, the opposite of the short faces of juveniles cared for by their parents. This suggests that growing tyrannosaurids hunted independently of the adults, who may have seen the youngsters as potential meals. Suggestions that the gracile juvenile tyrannosaurids hunted prey for their parents are implausible; when food is exchanged between juveniles and adults, it is the latter who feed the former. Collective trackways clearly indicative of dinosaur packs or flocks incorporating potential parents and offspring are not known. Dinosaurs of a few hundred kilograms on down that tended to their nests are the best candidates for parenting. The nonpredatory elaphrosaurs, medium-sized therizinosaurs, oviraptors, and especially the ratite-like ornithomimosaurs were most likely to have practiced the scheme of precocial chicks following their adults as the former feed themselves. If so, it is very possible that males did much or most of the caretaking in some taxa. When broods were large, they would have consisted of juveniles of assorted parents, cared for by a locally dominant couple ready to defend the chicks while leading them to suitable food sources. It is possible that no Mesozoic avepods brought food to helpless altricial nestlings that could not leave their nests.

What no dinosaur did was lavish its offspring with the intensive, often long-term parenting typical of many mammals. And because dinosaurs did not nurse, it is likely that most of them could grow up on their own normally or even if something happened to the grown-ups.

GROWTH

All land reptiles grow slowly. This is true even of giant tortoises and big, energetic (by reptilian standards) monitors. Land reptiles can grow most quickly only in perpetually hot equatorial climates, and even then they are hard pressed to reach a tonne. Aquatic reptiles can grow more rapidly, probably because the low energy cost of swimming allows them the freedom to acquire the large amounts of food needed to put on bulk. But even crocodilians, including the extinct giants that reached nearly 10 tonnes, do not grow as fast as many land mammals. Mature reptiles tend to continue to grow slowly throughout their lives.

Some marsupials and large primates, including humans, grow no faster or only a little faster than the fastest-growing land reptiles. Other mammals, including other marsupials and a number of placentals, grow at a modest pace. Still others grow very rapidly; horses are fully grown in less than two years, and aquatic whales can reach 50 to 100 tonnes in just a few decades. Bull elephants take about 30 years to mature. All living birds grow rapidly; this is especially true of altricial species and the big ratites. No extant bird takes more than a year to grow up, but some of the recently extinct giant island ratites may have taken a few years to complete growth. The secret to fast growth appears

Gigantic *Giganotosaurus*

to be having an aerobic capacity high enough to allow the growing juvenile, or its adult food provider, to gather the large amounts of food needed to sustain rapid growth.

High mortality rates from predation, disease, and accidents make it statistically improbable that unarmored, nonaquatic animals will live very long lives, so they are under pressure to grow rapidly. On the other hand, starting to reproduce while still growing tends to slow down the growth process as energy and nutritional resources are diverted to produce offspring. Few mammals and no living birds begin to breed before they reach adult size. No bird continues to grow once it is mature, nor do most mammals; however, some marsupials and elephants never quite cease growing.

At the microscopic scale the bone matrix is influenced by the speed of growth, and the bone matrix of dinosaurs tended to be more similar to that of birds and mammals, which grows at a faster pace than that of reptiles. Bone ring counts are being used to estimate the growth rate and life span of an increasing number of extinct dinosaurs, but this technique can be problematic because some living birds lay down more than one ring in a year, so ring counts can overestimate age and understate growth rate. There is also the problem of animals that do not lay down growth rings; it is probable that they grow rapidly, but exactly how fast is difficult to pin down. There are additional statistical issues because as animals grow, the innermost growth rings tend to be destroyed, leading to difficulties in estimating the number of missing age markers. Almost all dinosaurs sampled so far appear to have grown at least somewhat faster than land reptiles. A possible exception is a very small, bird-like troodont theropod whose bone rings seem to have been laid down multiple times in a year, perhaps because it was reproducing while growing. Most small dinosaurs fall along the lower end of the mammalian zone of growth, perhaps because they were reproducing while immature. Big dino-predators appear to have grown as fast as similar-sized land mammals, albeit with considerable variation between types. Tyrannosaurids appear to have been on the faster growth rate side of things, with *Tyrannosaurus* reaching final size in about two decades—note that a *Jurassic Park* scenario flaw as glaring as it usually goes unnoticed is the presence of gargantuan artificially bred dinosaurs so soon after the initiation of the paleo theme park project.

The cessation of significant growth of the outer surface of many adult dinosaur bones indicates that most but not all species did not grow throughout life. Medium-sized and large mammals and birds live for only a few years or decades: elephants live about half a century and giant whales can last longer, with the sluggish right whales making it well over 100 years. There is no evidence that dinosaurs lived longer than mammals or birds of similar size. Living in the fast lane, tyrannosaurs combined their rapid growth with rather short life spans of 20 to 30 years. Other giant avepods, such as *Giganotosaurus*, did not grow as extremely fast and lived longer, up to half a century.

ENERGETICS

Vertebrates can utilize two forms of power production. One is aerobiosis, the direct use of oxygen taken in from the lungs to power muscles and other functions. Like air-breathing engines, this system has the advantage of producing power indefinitely but is limited in its maximum power output. An animal that is walking at a modest speed for a long distance, for instance, is exercising aerobically. The other is anaerobiosis, in which chemical reactions that do not immediately require oxygen are used to power muscles. Rather like rockets that do not need to take in air, this system has the advantage of being able to generate about 10 times more power per unit of tissue and time. But it cannot be sustained for an extended period and produces toxins that can lead to serious illness if sustained at too high a rate for too long, which is tens of minutes. Anaerobiosis also builds up an oxygen debt that has to be paid back during a period of recovery. Any fairly fast animal that is running, swimming, or power flying near its top speed is exercising anaerobically.

Most fish and all amphibians and reptiles have low resting bradymetabolic rates and low aerobic capacity. They are therefore bradyenergetic, and even the most energetic reptiles, including the most aerobically capable monitor lizards, are unable to sustain truly high levels of activity for extended periods. Many bradyenergetic animals are, however, able to achieve very high levels of anaerobic burst activity, such as when a monitor lizard or crocodilian suddenly dashes toward and captures prey. Because bradyenergetic animals do not have high metabolic rates, they are largely dependent on external heat sources, primarily the ambient temperature and the sun, for their body heat, so they are ectothermic. As a consequence, bradyenergetic animals tend to experience large fluctuations in body temperature, rendering them heterothermic. The temperature at which reptiles normally operate varies widely depending on their normal habitat. Some are adapted to function optimally at modest temperatures of 12°C (52°F). Those living in hot climates are optimized to function at temperatures of 38°C (100°F) or higher, so it is incorrect to generalize reptiles as "cold blooded." In general, the higher the body temperature is, the more active an animal can be, but even warm reptiles have limited activity potential.

Most mammals and birds have high resting tachymetabolic rates and high aerobic capacity. They are therefore tachyenergetic and are able to sustain high levels of activity for extended periods. The ability to better exploit oxygen for power over time is probably the chief advantage of being tachyenergetic. Tachyenergetic animals also use anaerobic power to briefly achieve the highest levels of athletic performance, but they do not need to rely on this as much as reptiles, are not at risk of serious self-injury, and can recover more quickly. Because tachyenergetic animals have high metabolic rates, they produce most of their body heat internally, so they are endothermic. As a consequence, tachyenergetic animals can achieve more stable body temperatures. Some, like humans, are fully homeothermic, maintaining a nearly constant body temperature at all times when healthy. Many birds and mammals, however, allow their body temperatures to fluctuate to varying degrees, for reasons ranging from going into some degree of torpor to storing excess heat on hot days, on a daily or seasonal basis. So they are semiheterothermic or semihomeothermic depending on the degree of temperature variation. The ability to keep the body at or near its optimal temperature is another advantage of having a high metabolic rate. Normal body temperatures range from 30°C to 44°C (86°F–105°F), with birds always at least at 38°C. High levels of energy production are also necessary to do the cardiac work that creates the high blood pressures needed to be a tall animal.

Typically, mammals and birds have resting metabolic rates and aerobic capacities about 10 times higher than those of reptiles, and differences in energy budgets are even higher. However, there is substantial variation from these norms in tachyenergetic animals. Some mammals, among them monotremes, some marsupials, hedgehogs, armadillos, sloths, and manatees, have modest levels of energy consumption and aerobic performance, in some cases not much higher than those seen in the most energetic reptiles. In general, marsupials are somewhat less energetic than their placental counterparts, so kangaroos are about a third more energy efficient than deer. Among birds, the big ratites are about as energy efficient as similar-sized marsupials. At the other extreme, some small birds share with similarly tiny mammals extremely high levels of oxygen consumption even when their small body size is taken into account.

Widely different energy systems have evolved because they permit a given species to succeed in its particular habitat and lifestyle. Reptiles enjoy the advantage of being energy efficient, allowing them to survive and thrive on limited resources. Tachyenergetic animals are able to sustain much higher levels of activity that can be used to acquire even more energy, which can then be dedicated to the key factor in evolutionary success, reproduction. Tachyenergy has allowed mammals and birds to become the dominant large land animals from the tropics to the poles. But reptiles remain very numerous and successful in the tropics and, to a lesser extent, in the temperate zones.

As diverse as the energy systems of vertebrates are, there appear to be things that they cannot do. All insects have low, reptile-like resting metabolic rates. When flying, larger insects use oxygen at very high rates similar to those of birds and bats. Insects can therefore achieve extremely high maximal/minimal metabolic ratios, allowing them to be both energy efficient and aerobically capable. Insects can do this because they have a dispersed system of tracheae that oxygenate their muscles. No vertebrate has both a very high aerobic capacity and a very low resting metabolism, probably because the centralized respiratory–circulatory system requires that the internal organs work hard even when resting in tachyenergetic vertebrates. An insect-like metabolic arrangement should not, therefore, be applied to dinosaurs. However, it is unlikely that all the energy systems that have evolved in land vertebrates have survived until today, so the possibility that some or all dinosaurs were energetically exotic needs to be considered.

The general assumption until the 1960s was that dinosaur energetics was largely reptilian, but most researchers now agree that their power production and thermoregulation were closer to those of birds and mammals. It is also widely agreed that because dinosaurs were such a large group of diverse forms, there was considerable variation in their energetics, as there is in birds and especially mammals.

Reptiles' nonerect, sprawling legs are suitable for the slow walking speeds of 1–2 km/h (0.5–1 mph) that their low aerobic capacity can power over extended periods. Sprawling limbs also allow reptiles to easily drop onto their belly and rest if they become exhausted. No living bradyenergetic animal has erect legs. Walking is always energy expensive—it is up to a dozen times more costly than swimming the same distance—so only aerobically capable animals can easily walk faster than 3 km/h. The long, erect legs of dinosaurs matched those of birds and mammals and favored the high walking speeds of 3–10 km/h (2–6 mph) that only tachyenergetic animals can sustain for hours at a time. The speed at which an animal of a given size is moving can be approximately estimated from the length of its stride—an animal that is walking slowly steps with shorter strides than it does when it picks up the pace. The trackways of tridactyl dinosaurs show that they normally walked at speeds of 4–7 km/h (average 4 mph), much faster than the slow speeds recorded in the trackways of prehistoric reptiles. Dinosaur legs and the trackways they made both indicate that the dinosaurs' sustained aerobic capacity well exceeded the reptilian maximum. That meant that they could forage over much longer ranges and areas than carnivores with reptilian energetics. They also had the potential to migrate, but big cats, canids, and hyenas tend not to do so, finding it more selectively advantageous to stay in place and go after the local herbivores and any that are passing by on their arduous migrations.

Even the fastest reptiles have slender leg muscles because their low-capacity respirocirculatory systems cannot supply enough oxygen to a larger set of locomotory muscles. Mammals and birds tend to have large leg muscles that propel them at a fast pace over long distances. As a result, mammals and birds have a large pelvis that supports a broad set of thigh muscles. It is interesting that protodinosaurs and first theropods had a short pelvis that could have anchored only a narrow thigh, yet their legs are long

and erect. Such a combination does not exist in any modern animal. This suggests that the small-hipped dinosaurs had an extinct metabolic system, probably intermediate between those of reptiles and mammals. All other avepods had the large hips able to support the large thigh muscles typical of more aerobically capable animals.

That many dinosaurs, therizinosaurs most of all among avepods, could hold their brains far above the level of their hearts indicates that they had the high levels of power production seen in similarly tall birds and mammals.

If we turn to breathing, an intermediate metabolism is compatible with the unsophisticated lungs that protodinosaurs and basal dinopredators appear to have had. The increasingly highly efficient birdlike air-sac-ventilated respiratory complex of avepod theropods is widely understood as being evidence that elevated levels of oxygen consumption were further evolving in these dinosaurs.

Many birds and mammals have large nasal passages that contain respiratory turbinals. These are used to process exhaled air in a manner that helps retain heat and water that would otherwise be lost during the high levels of respiration associated with high metabolic rates. Because they breathe more slowly, reptiles do not need or have respiratory turbinals. Some researchers point to the lack of preserved turbinals in dinosaur nasal passages, and the small dimensions of some of the passages, as evidence that dinosaurs had the low respiration rates of bradyenergetic reptiles. However, some birds and mammals lack well-developed respiratory turbinals, and in a number of birds they are completely cartilaginous and leave no bony traces. Some birds do not even breathe primarily through their nasal passages: California condors, for example, have tiny nostrils. The space available for turbinals has been underestimated in some dinosaurs, including the usually big-snouted dinopredators. Overall, the turbinal evidence does not seem to be definitive.

The presence of a blanket of hollow fibers in a growing array of small dinosaurs is strong evidence of elevated metabolic rates. Such insulation hinders the intake of environmental heat too much to allow ectotherms to quickly warm themselves and is never found adorning bradyenergetic animals. The evolution of insulation early in the group indicates that high metabolic rates also evolved near the beginning of the group or in their ancestors. The uninsulated skin of many dinosaurs is compatible with high metabolic rates, as in mammalian giants, many suids, human children, and even small naked tropical bats. The tropical climate most dinosaurs lived in reduced the need for insulation, and the bulk of large dinosaurs eliminated any need for it.

The low exercise capacity of land reptiles appears to prevent them from being active enough to gather enough food to grow rapidly. In an expression of the principle that it takes money to make money, tachyenergetic animals are able to eat the large amounts of food needed to produce the power needed to gather the additional large

The feathered *Citipati*

fairly recently; the dominant open-area proboscidean used to be one of the biggest land mammals ever, *Palaeoloxodon recki*. A relative of the Asian elephant, it probably had small ears of little use for shedding body heat at any temperature. It is actually small animals that are most in danger of suffering heat exhaustion and heat stroke because their small bodies pick up heat from the environment very quickly. The danger is especially acute in a drought, when water is too scarce to be used for evaporative cooling. Because they have a low surface area/mass ratio, large animals are protected by their bulk against the high heat loads that occur on very hot days, and they can store the heat they generate internally. Large birds and mammals retain the heat they produce during the day by allowing their body temperature to climb a few degrees above normal and then dumping it into the cool night sky, preparing for the cycle of the next day. A basic behavioral means by which flesh-eating dinosaurs would have minimized their heat problem would have been by simply avoiding hunting in the middle of hot, sunny days, instead taking a rest in the shade or a bath in water, if available, instead. Avepods of all sizes have been able to use their air-sac ventilation system to help keep cool while curtailing water loss.

At the other end of the temperature spectrum, the presence of a diverse array of dinosaurs in temperate polar regions and highlands that are known to have experienced freezing conditions during the winter, and were not particularly warm

amounts of food needed to grow rapidly. Tachyenergetic juveniles either gather the food themselves or are fed by their parents. That the predatory dinosaurs, large and small, usually grew at rates faster than those seen in land reptiles of similar size indicates that the former had higher aerobic capacity and energy budgets.

A hot topic has been the concern by many that giant dinosaurs would have overheated in the Mesozoic greenhouse if they had avian- or mammalian-like levels of energy production. However, the largest animals dwelling in the modern tropics, including deserts, are big birds and mammals. And consider that there are no reptiles over a tonne dwelling in the balmy tropics. Further consider that some of the largest elephants—similar in mass to the greatest avepods—live in the Namib Desert of the Skeleton Coast of southwestern Africa, where they often have to tolerate extreme heat and sun without the benefit of shade. It is widely thought that elephants use their ears to keep themselves cool when it is really hot, something dinosaurs could not do. However, elephants flap their ears only when the ambient temperature is below that of their bodies. When the air is as warm as the body, heat can no longer flow out, and flapping the ears actually picks up heat when the air is warmer than the body. Nor was the big-eared African elephant the main savanna elephant until

even in the summer, provides additional evidence that dinosaurs were better able to generate internal heat than reptiles, which were scarce or totally absent in the same habitats. It was not practical for land-walking dinosaurs to migrate far enough toward the equator to escape the cold; it would have cost too much in time and energy, and in some locations, oceans barred movement toward warmer climes. A point that is unknown is whether large polar and high-altitude dinosaurs retained bare skin, in which case they would have needed high internal heat production to ward off frostbite, or whether they were heavily insulated, which also supports tachyendothermy. That the largest known, over a tonne, Early Cretaceous tyrannosauroid living at chilling high altitudes was well feathered hints at possibilities. The largest known polar avepods were Late Cretaceous tyrannosaurids over two tonnes, but the status of their skin remains unknown. The discovery of probable dinosaur burrows in then-polar Australia suggests that some small dinosaurs did hibernate through the winter in a manner similar to bears. While many if not all the burrows were dug by herbivores, the cold-evading flesh eaters may have evicted them on occasion to squat on the property.

Bone isotopes have been used to help assess the metabolism of dinosaurs. These can be used to examine the

Alaskan *Nangsaurus*

Mongolian *Therizinosuarus* vs. *Tarbosaurus*

Winter scenes

temperature fluctuations that a bone experienced during life. If the bones show evidence of strong temperature differences, then the animal was heterothermic on either a daily or seasonal basis. In this case the animal could have been either a bradyenergetic ectotherm or a tachyenergetic endotherm that hibernated in the winter. The results indicate that most dinosaurs, large and small, were more homeothermic, and therefore more tachyenergetic and endothermic, than crocodilians from the same formations.

Bone biomolecules too are being used to restore the metabolic rates of dinosaurs. This effort is in its early stages, and it is not clear that the sample of living and fossil animals of known metabolic levels is yet sufficient to establish the reliability of the method. Also, the sample of dinosaurs is too limited to allow high confidence in the results to date, which is all the more true because the estimates for dinosaurs appear inconsistent in peculiar ways. While the one armored ankylosaur is attributed with a high energy budget that appears excessive for such a relatively slow-moving creature with weak dentition, the sole armored stegosaur is recovered well down in the reptilian range, which looks both too low for an animal with long, erect legs and fairly fast growth and too different from the other armored dinosaur. Also problematic is that reptilian energetics are assigned to the hadrosaur and the ceratopsid examined, not the higher levels expected in animals with such fast food processing and growth and with the large leg muscles and fast-walking pace expected in tachyenergetic endotherms. Also of note is the low metabolism of the giant flying marine pterosaur. The initial biomolecule results indicating that the earliest dinosaurs were endotherms, with avepods big and small remaining so, await further analysis.

Because the most basal and largest of living birds, the ratites, have energy budgets similar to those of marsupials, it is probable that most or all of their Mesozoic relatives did not exceed this limit. This fits with some bone isotope data that seem to indicate that dinosaurs had moderately high levels of food consumption, somewhat lower than seen in most placentals of the same size. Possible exceptions include polar dinosaurs that remained active in the winter and needed to produce lots of warmth. At the opposite end of the spectrum, early dinosaurs, and awkward therizinosaurs, probably had modest energy budgets like those of the less-energetic mammals. It is likely that dinosaurs, like birds, were less prone to controlling their body temperatures as precisely as do many mammals. This is in accord with their tendency to lay down bone rings. Because they lived on a largely hot planet, it is probable that most dinosaurs had high body temperatures of 38°C (100°F) or more to be able to resist overheating. The possible exception was again high-latitude dinosaurs, which may have adopted slightly lower operating temperatures and saved some energy, especially if they were active during the winter. Some researchers have characterized dinosaurs as mesotherms, intermediate between reptiles on the one hand and mammals and birds on the other. But because some mammals and birds themselves are metabolic intermediates, and dinosaurs were probably diverse in their energetics, with some in the avian–mammalian zone, it is not appropriate to tag dinosaurs with a uniform, intermediary label.

Until the 1960s it was widely assumed that high metabolic rates and/or endothermy were an atypical specialization among animals, being limited to mammals and birds, and perhaps to some therapsid ancestors of mammals and the flying pterosaurs. The hypothesis was that being tachyenergetic and endothermic is too energy expensive and inefficient for most creatures and evolved only in special circumstances, such as the presence of live birth and lactation or powered flight. Energy efficiency should be the preferred status of animals, as it reduces their need to gather food in the first place. Since then it has been realized that varying forms of tachyenergy definitely are or probably were present in large flying insects, some tuna and lamnid sharks, some basal Paleozoic reptiles, some marine turtles and the oceangoing plesiosaurs, ichthyosaurs, and mosasaurs, brooding pythons, basal archosaurs, basal crocodilians, pterosaurs, all dinosaurs including birds, some pelycosaurs, therapsids, and mammals. Energy-expensive elevated metabolic rates and body temperatures appear to be a widespread adaptation that has evolved multiple times in animals of the water, land, and air. This should not be surprising in that being highly energetic allows animals to do things that bradyenergetic ectotherms cannot do, and DNA selection acts to exploit available lifestyles that allow reproductive success without a priori caring whether it is done energy efficiently or not. Whatever works, works. So many animals do live on low, energy-efficient budgets, while others follow the scheme of using more energy to acquire yet more energy that can be dedicated to reproducing the species.

A long-term debate asks what specifically it is that leads animals to be tachyenergetic and endothermic. One hypothesis proposes that it is habitat expansion, that animals able to keep their bodies warm when it is cold outside are better or exclusively able to survive in chilly places—near the poles, at high altitudes, in deep waters—or during frosty nights. The other proffers that only tachyenergetic animals with high aerobic capacity can achieve high levels of sustained activity regardless of the ambient temperature, whether at sea level in the tropical daylight or during polar winter nights, and that ability is critical to going high energy. Certainly the first hypothesis is true, but it is also true that all of the many animal groups that feature high energy budgets and warmer-than-ambient body temperatures also thrive in warm and even hot climes, where they beat out the bradyenergetic creatures in activity levels. So both hypotheses are operative, and which is more so depends on the biocircumstances—including being really big on land.

GIGANTISM

Although dinosaurs evolved from small protodinosaurs, and many were small—birds included—dinosaurs are famous for their tendency to develop gigantic forms. The average mammal is the size of a dog, whereas the average fossil dinosaur was bear sized. But those are just averages. Predatory theropods reached as much as 10 plus tonnes, as big as elephants and dwarfing the largest carnivorous mammals by a factor or 10 or more.

Among land animals whose energetics are known, only those that are tachyenergetic have been able to become gigantic on land. The biggest fully terrestrial reptiles, some oversized tortoises and monitors, have never much exceeded a tonne. Land reptiles are probably not able to grow rapidly enough to reach great size in reasonable time. Other factors may also limit their size. It could be that living at 1 g, the normal force of gravity, without the support of water is possible only among animals that can produce high levels of sustained aerobic power. The inability of the low-power, low-pressure reptilian circulatory system to pump blood far above the level of the heart probably helps limit the size of bradyenergetic land animals. That a number of Mesozoic dinosaurs, including those predaceous,

Predatory dinosaur giants compared to big birds
and large carnivorous mammals

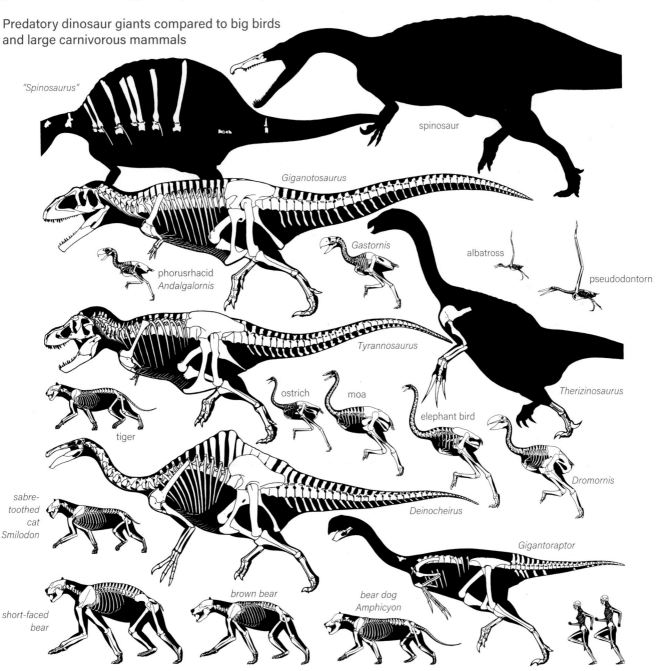

"Spinosaurus"

spinosaur

Giganotosaurus

Gastornis

albatross

pseudodontorn

phorusrhacid
Andalgalornis

Tyrannosaurus

Therizinosaurus

ostrich

moa

elephant bird

tiger

Dromornis

sabre-
toothed
cat
Smilodon

Deinocheirus

Gigantoraptor

short-faced
bear

brown bear

bear dog
Amphicyon

67

exceeded a tonne, as have mammals since then, is compelling evidence that they too had high aerobic power capacity and the correspondingly elevated energy budgets.

The hypothesis that only tachyenergetic animals can grow to enormous dimensions on land is called terramegathermy. An alternative concept, gigantothermy, proposes that the metabolic systems of giant reptiles converge with those of giant mammals, resulting in energy efficiency in all giant animals. In this view, giants rely on their great mass, not high levels of heat production, to achieve thermal stability. This idea reflects a misunderstanding of how animal power systems work. A consistently high body temperature does not provide the motive power needed to sustain high levels of activity; it merely allows a tachyenergetic animal, and only an animal with a high aerobic capacity, to sustain high levels of activity around the clock. A gigantic reptile with a high body temperature would still not be able to remain highly athletic for extended periods. Measurements show that the metabolic rates and aerobic capacity of elephants and whales are as high as expected in mammals of their size and are far higher than those of the biggest crocodilians and turtles, which have the low levels of energy production typical of reptiles. Also pushing animals to be big is improved thermoregulation—the high bulk to relatively low surface area ratio making it easier both to retain internal warmth when it is chilly and to keep external heat out and store heat on hot days.

Another subtle reason that dinosaurs, particularly superavepods, could become so enormous has to do with their mode of reproduction. Because big mammals are slow-breeding K-strategists that lavish attention and care on the small number of calves they produce, there always has to be a large population of adults present to raise the next generation. A healthy herd of elephants has about as many breeding adults as it does juveniles, which cannot survive without parental care. Because there always has to be a lot of grown-ups, the size of the adults has to be limited in order to avoid overexploiting their ecosystem's food resources, which will cause the population to collapse. This constraint appears to limit slow-reproducing mammalian herbivores from exceeding 10–20 tonnes. Flesh eaters live off an even smaller resource base because they prey on the surplus herbivores, and it seems that carnivorous mammals cannot maintain a viable population if they are larger than between 0.5 and 1 tonne.

Because giant dinosaurs were fast-breeding r-strategists that produced large numbers of offspring that could care for themselves, their situation was very different from that of big mammals. A small population of adults, approaching or in the area of the low six figures, was able to produce large numbers of young each year. Even if all adults were killed off on occasion, their eggs and offspring could survive and thrive, keeping the species going over time. Because dinosaurs could get along with smaller populations of adults, the grown-ups were able to grow to enormous dimensions without overexploiting their resource base. This evolutionary scheme allowed plant-eating dinosaurs to grow to over 20 to, perhaps on occasion, 200 tonnes. Because the bulk of the biomass of adult dinoherbivores was tied up in oversized giants, the theropods needed to evolve great size themselves in order to be able to fully access the nutrition tied up in the huge adults—the idea that theropods grew to 6 to 10 tonnes only to "play it safe" by consistently hunting smaller juveniles is not logical—and the fast-breeding and fast-growing, high-energy predators had the ability to reach such tremendous size. The existence of oversized predators in turn may have resulted in a size race in which sauropods evolved great size in part as protection against their enemies, which later encouraged the appearance of supersized theropods that could bring them down.

In the 1800s Edward Cope proposed what has become known as Cope's Rule, the tendency of animal groups to evolve gigantism. The propensity of dinosaurs to take this evolutionary pattern to an extreme means that the Mesozoic saw events on land that are today limited to the oceans. In modern times combat between giants occurs between orcas up to 10 tonnes and whales up to 200. In the dinosaur era it occurred between orca-sized theropods and whale-sized sauropods, hadrosaurs, and ceratopsids.

HUNTING, SCAVENGING, AND DEFENSE

None of the sauropodomorph or ornithischian dinosaurs were archpredators. That does not mean, however, they were purely pacific plant eaters. Ratites are omnivores happy to snatch up small creatures and insects. Even cattle and deer occasionally ingest animal protein and calcium. The dinosaurs least prone to do so would have been those large ornithopod ornithischians with blunt beaks. Most ornithischians, as well as oviraptorosaurs, had sharp beaks, sometimes hooked, and in some cases fangs that would have allowed them to catch and dispatch prey and to scavenge. They would have been suid-like omnivores, including the big-horned ceratopsids that may have competed with tyrannosaurids for access to carcasses. With their long necks, sauropods, prosauropods, therizinosaurs, ornithomimids, and elaphrosaurs would have had no trouble reaching out and up to pick up small creatures, and dine on dead corpses, to supplement their vegetarian diets. While sauropod heads look small, that is relative to the rest of their bodies. In absolute terms their heads could be quite large, the mouth of *Giraffatitan* was a third of a meter broad and could swallow creatures weighing tens of kilograms—the children in *Jurassic Park* would not have been as safe as they seemed.

Among dinosauriforms, only protodinosaurs, herrerasaurs, and theropods—mostly avepods but excluding nonpredaceous elaphrosaurs, ornithomimosaurs, therizinosaurs, oviraptorosaurs, and alvarezsaurs—were full-blown flesh and, to a certain extent, bone-craving predaceous carnivores that made a living by eating other vertebrates for their main sustenance. While doing so the only competition they had to deal with were a few terrestrial crocodilians, none of which were giants.

A big difference between Mesozoic and Cenozoic circumstances is the size factor. In the Age of Mammals, the biggest terrestrial carnivores have been one tonners going after 10 to 20 tonners. In the Age of Dinosaurs, it was 5–10 tonne avepods assaulting sauropods of 20 to 200 tonnes, for reasons that were just discussed.

Another contrast between dinosaur and mammal predator–prey affairs is that therian hunting is limited to adults often involved with raising and feeding their innocuous young and to large juveniles in training, whereas the juveniles of predaceous dinosaurs, especially the larger examples, were deadly hunters competing with similar-sized adults of other species while posing a serious threat to prey dinosaurs. This is a fundamental difference—one driven by radically different reproductive adaptations—between the dinosaur-dominated versus mammal-dominated predator–prey faunas of the Mesozoic versus Cenozoic.

Shared with hunting mammals was that most dinosaur habitats and faunas had more than one big predator in each one: *Allosaurus*, *Ceratosaurus*, and *Torvosaurus* in the Morrison; *Giganotosaurus* and *Ekrixinatosaurus* in the Candeleros; *Carcharodontosaurus* and *Rugops* in the Echkar; *Gorgosaurus* and *Daspletosaurus* in Dinosaur Park; *Tarbosaurus* and *Alioramus* in the Nemegt; and *Tyrannosaurus*, *Nanotyrannus*, and *Stygivenator* in the Hell Creek, Lance, and other formations, with *T. rex* and *T. regina* alive at the same time. This is similar to bears, wolves, and cougars in western North America and lions, leopards, cheetahs, hyaenas, and hunting dogs in eastern and southern Africa. That the Horseshoe Canyon fauna seems to have featured only *Albertosaurus* seems to have been a rarity.

In a given dinosaur habitat each predator species and individual would tend to concentrate on those prey items best suited for the carnivore's characteristics regarding its size, speed, and killing power and techniques. Giant adults would focus on mega prey, while their juvenile and small theropods would be limited to similarly lesser victims on down to insects no grown-up *Tyrannosaurus* would consider dining upon. Even so, there was lots of dietary overlap. Wolves eat caribou and moose as well as the voles and mice coyotes commonly target. Cape hunting dogs snap up rodents, and their packs bring down big ungulates. It is possible that robust *Daspletosaurus* was more prone to taking on the combative parrot-beaked and horned ceratopsids while its more gracile competitor *Gorgosaurus* went after the more vulnerable duck-billed hadrosaurs, but both probably fed on the other's preference on a common basis. While lithe *Stygivenator*, *Nanotyrannus*, and juvenile *Tyrannosaurus* were all targeting different, smaller prey than the massive adults of the latter, the bigger arms, less robust teeth, and more gracile legs of *Stygivenator* and *Nanotyrannus* show they were doing so in a different manner from young *Tyrannosaurus* and quite successfully— they outnumbered the fast-growing *Tyrannosuarus* two to one. In some cases what was eating what is more perplexing. *Allosaurus* was generally larger than contemporary *Ceratosaurus*, but the latter had larger teeth so may have been more prone to attack sauropods and stegosaurs than it first appears.

A main means by which Mesozoic predatory dinosaurs small and large and young and old caught prey was by running it down, all examples having the well-muscled, flexed, long-footed, birdy legs needed to run at good speed. And they all had the elevated aerobic exercise capacity and high body temperatures that allowed them to sustain the high speeds to a greater degree than did bradyaerobic reptiles, this being most true in avepods yet more so in the big-hipped averostrans and beyond. Among the big dinosaurian carnivores, it was long-legged, large-hipped, small-armed abelisaurs and especially the tyrannosaurids of the Late Cretaceous that went the furthest in regard to the speed pursuit factor. The Cenozoic predatory ground birds of South America could have given their Mesozoic predecessors a run for their money. Few mammal hunters could match or exceed the avepods pace, the exception being cheetahs, which are probably the fastest of land animals to have evolved.

In general, dinosaur hunters were faster than their dinosaur prey. Prosauropods were never speedsters, perhaps even less so were derived therizinosaurs, while heavy limbed sauropods and stegosaurs were limited to an amble about a third as fast as their tormentors. Also not speedy were the armored dinosaurs. Ceratopsians small to gigantic could run, trot, and perhaps gallop at a good clip. At least as fast and often more so among ornithischians were lesothosaurs, heterodontosaurs, pachycephalosaurs, and petite to gigantic ornithopods. Among the latter, the big hadrosaurs with their gracile arms should have been able to outpace the clunkier, heavier-armed iguanodonts. The few prey dinosaurs that were really swift were of course the ratite-like ornithomimids, plus caudipterids, avimimids, and alvarezsaurs—but these were all Cretaceous; of them, only the alvarezsaurs are known from south of the equator, where they were not commonplace. Back in the Jurassic the nonpredaceous elaphrosaurs were fairly fleet. This situation of most predators being faster than most of the prey is another big difference with the layout of modern land mammals, in which some of the ungulates possess extremely gracile, long, unguligrade limbs that give them nominally greater velocity than the digitigrade and plantigrade carnivores, with ratites also being faster than the latter. In Australia the bounding kangaroos could outpace the marsupial and super lizard hunters until human invaders did the pouched predator in. The hows and whys of the predator–prey speed contest are interesting. When on the hunt, meat eaters have the advantage of carrying small digestive tracts empty of food that are much lighter than the bulky digestive complexes packed with digesting fodder plant consumers have to carry around all the time, even when fleeing for their lives. Swift quadrupedal mammalian herbivores have been able to get around the gut size disadvantage via the adaptation of legs with very willowy unguligrade feet, while the carnivores have to have flatter-footed limbs robust enough to grapple with prey, so they have trouble keeping up with their toe-tip-running targets. That said, there are videos showing brown

bears successfully chasing down large, apparently fit, and fast elk galloping full tilt across fields. Being bipeds that in most cases did not use their hindlimbs to injure prey, theropods and company were free to maximize their legs' running potential. Meanwhile, bipeds have to have toes that lie flat on the ground to avoid tipping over when not standing, so bipedal herbivorous dinosaurs could not go extreme unguligrade. That the nonpredaceous ornithomimids, caudipterids, avimimids, and alvarezsaurs had exceptional digitigrade running legs may be due to their not being burdened by capacious fodder-fermenting guts; that

Running predatory
dinosaurs

Liliensternus

Andalgornis

the ornithomimids could flee so fast may explain why they were a fairly abundant group in their northerly habitats but not why they did not make it south of the equator. What is not clear is why large-bellied small ornithopods were not able, or did not happen, to evolve similarly speed-capable hindlimbs.

Predators that cannot run fast and/or far are limited to ambushing prey over a short distance and very quickly before their target can escape—that is true of all flesh-eating reptiles. More aerobically capable predators that can run fast and far have the option to chase down intended victims. Those that do sport limb muscles packed with high-aerobic-capacity red muscle tissues lavishly oxygenated by high-capacity respiratory tracts; big-lunged canids and hyenas typify this mode. Smaller-chested and with whiter, more anaerobic muscles, big cats are capable of high burst speeds but cannot sustain them and are short- or modest-range ambush–chase predators; super-fast cheetahs are the extreme of this tactic—racing greyhounds and whippets have been bred to have similar surge performance. Lacking either high-capacity respiratory tracts or leg musculatures, protodinosaurs, herrerasaurs, and nonavepod theropods would have been burst runners of the cat mold. The early avepods with their early grade avian respiration and larger leg muscles should have performed somewhat better. With their near avian respiratory tracts and big pelvic-born muscles, the averostrans were gaining the capacity to be longer-range pursuit predators able to run down prey over distances and/or engage in extended combat with them. This capacity would have been taken to its limits in the abelisaurs and even more the tyrannosaurids, with the caveat that over a tonne or two the power needed to achieve high velocities was so high that doing so would have compelled the limb muscles to emphasize at the cellular level anaerobic burst performance. Thus *Tyrannosaurus*, *Tarbosaurus*, and *Giganotosaurus* should have been more toward the ambush side of the ambush–pursuit spectrum, particularly the latter in view of its modest-sized limbs.

Being primarily visually oriented, dinopredators seeing prey was likely to have been the main means of locating it. Also useful was olfaction, although that only would work well when downwind. Conversely, approaching game from upwind was to be avoided. Hearing was important, which would work over very long distances when herbivores were loudly vocalizing. Again conversely, being quiet while stalking was a must. Finding prey can be a task in of itself. Locations that animals come to drink are likely locations. Herbivore herds are easy to spot.

Having potential prey in sight, the predator must assess its suitability. Keeping the act of eating animals as easy and safe as possible is boosted by selecting vulnerable prey. Preferred victims are those much smaller than the predator, including juveniles; those having limited or minimal weaponry or slow speed; and those that appear disabled by visual indications of age, illness, injury, or malnutrition—although the last has the downside of making a thin meal.

Predators do not normally engage with other predators of similar combat potential, so smaller predators are at risk of being victims. Because of these selective targeting factors, combined with the instinctive and learned stalking and attack skills of the professional predator, the vast majority of attacks do not result in the injury of death of the predator, although some do. That said, predators sometimes attack big, healthy, well-armed adults that can put up a good and sometimes successful fight for their lives or that of their offspring. This type of encounter is most likely to happen when a series of hunting expeditions have failed, and the carnivore is hungry to the point of be willing to take risks. Also, bringing down one large victim and then feasting on it has the advantage of minimizing the number of hunts.

Indeed, hunting success rates are not high, being in the area of one in six to four. Hunting in packs can boost success rates by about half. Carnivorous species tend to be either solitary or pack hunters—closely related species can be very different in this regard as per pride-forming lions versus the other members of its genus *Panthera*, including tigers. It is possible that all predatory dinosaurs were solitary hunters, it is much less probable that they were all pack hunters, but it is plausible that a number were group predators. Assessing this possibility is very difficult. Lesser flesh-eating dinosaurs may have hunted communally to maximize food provision for their offspring, which would not apply to those over a tonne because they would not have cared for their progeny. Whether the dinosaurs had the smarts to be collaborative hunters is up in the air. The available trackway evidence for theropods moving in collective groups is suggestive but ambiguous. Neither are fossil sites featuring multiple remains of the same dino-predator species definitive. Pack-hunting nonavian dinosaurs is credible but by no means certain.

Whether done alone or with assistance, intentionally lethal combat committed for practical purposes is not about confrontation for display. Determined killers generally want to kill with as little effort and danger as possible, all the more so if they do so because they have no alternative to feed themselves, and the best way to achieve successful kills over years and decades is via surprise. Gunners in the old west rarely stood in the middle of streets and quick drew, that being a way to make dying a 50/50 proposition. When possible, shots were fired from behind cover. In fighter versus fighter duels, pilots prefer not to dogfight, that being a frightening and risky bet. The majority of kills are the result of surprise, at which high aces become skilled. In dinosaur movies, and even documentaries that should know better, it is a stock faux drama device to have the terrible theropod stand in view of its terrified prey and issue threat roars before it attacks, the prey well warned of the danger. Such is not seen in reality; documentaries of predators actually attacking prey demonstrate that the last thing the predator wishes to do is alert its desired victims that they need to flee or fight—surprise is always optimal. Thus predatory dinosaurs, when feasible, would work to sneak up on prey, using vegetation and terrain as cover and crouching when needed to make the attack distance as short as possible. It is possible that the normal digitigrades would temporarily adopt a plantigrade foot posture to lower their profile—some trackways seem to show avepods walking on their heels, but most of those tracks were made in deep muds into which the feet had sunk far. The aim was to deliver critical wounds to the victim before it could properly defend itself by running or fighting back. The smaller the hunting dinosaur, the easier to accomplish surprise; likewise arboreal predation among dense leaves. Attack without the prey seeing the approaching predator was not always possible, some habitats being largely open terrain that did not offer much in the way of cover. The hoped for aim was to at best hit the prey target before it could mount a good defense whether that be to bolt and run or to try to fight back. If fighting back does not occur, then the startled and frightened prey running away is the better option for the slayer because the game is less likely to injure the attacker, as it might if it engages in combat. Encouraging flight is all the more important if the prey has formidable defensive armament it can deploy if it does not turn tail.

Upon catching up with the potential meal, the next task is to deliver crippling and ultimately lethal injuries to overcome any defense the prey might mount and convert a living animal into food. The primary—in the case of small, armed ceratosaurs and tyrannosaurids the only—means for arch dinopredators to do so was with long, serrated-blade tooth-lined jaws. The closest living analogs in this particular respect are carnivorous lizards that also sport long jaws bearing recurved blade teeth edged with penetration and slice-enhancing serrations—the quadrupedal lizards are not able to seriously wound their prey with their claws, worn down by chronic contact with the ground. The similarities should not be taken too far: lizard heads for one thing are low and broad, those of dagger-toothed dinosaurs deep and narrow. Whether on dinosaurs or lizards, extended rows of modest-sized teeth are not able to penetrate deeply like the long canines of mammalian carnivores, hyper-elongated saber teeth most of all. It is also of interest that the skulls of most predatory dinosaurs were not particularly strongly constructed, quite different from the solidly built skulls of crocodilians and carnivorous mammals. This lighter construction results from the dinosaurs' multiple large cranial openings. This was most true just in front of the orbit, at the aft end of the snout; because of the enormous preorbital fenestra, and another in the roof of the mouth, the snout was attached to the rest of the skull by struts of modest robustness. Lizard snouts are much more solidly put together. The jaw closing muscles were not extremely powerful, and the leverage over the long jaws was modest. This pattern indicates that most dinopredators, including most of the giants, did not try to deliver hard, crushing bites intended to immediately dispatch the prey or to hold tight onto their targets in order to try to pull them down or drag them under water for drowning, the way water-adapted crocodilians do. Instead, the jaws and teeth delivered slashing wounds to disable limb muscles and cause bleeding, shock, and infection. Neck muscles aided this action by pulling the head and its tooth rows backward along the victim's flesh. This layout was most delicately constructed in small coelophysids, coelurosaurs, and gracile-skulled deinonychosaurs, but it was largely retained in the big to gigantic podokesauroids, megalosaurs, and allosauroids up to 10 tonnes.

Tyrannosaurids that lacked substantial arms and focused their firepower in their heads were significantly different. Their skulls were somewhat more robustly constructed, and the jaw and neck musculature was more powerful at a given size. The tyrannosaurids had deeper, broader, more rounded snout fronts rimmed with D-shaped teeth that formed a cookie cutter arcade. These features indicate that these predators attempted to scoop out crippling chunks of flesh. This scheme was taken further in robust daspletosaurs and tarbosaurs and to an extreme in the ultimate dinosaur predator, *Tyrannosaurus*, which sported larger, in both relative and absolute terms,

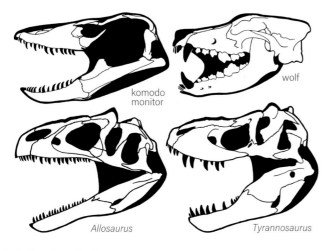

Flesh eater skulls

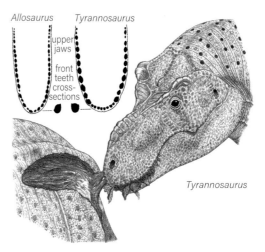

Predatory dinosaur biting

as well as stouter teeth and a set of maximal jaws muscles contained in an extra capacious aft skull. Also lacking arm weapons were abelisaurs, and they too bore atypical heads as primary killing weapons, with shortened snouts powered by deep sets of jaw muscles. Yet their teeth were short, leaving it unclear just how they bit down on their victims. Ceratosaurs, whose arms were on the short side, bore extra-large teeth for slashing their marks. It is interesting that while giant carcharodontosaurids had reduced arms, they retained the standard slashing head form. Among smaller avepods there were some notable divergences from the typical pattern. Some dromaeosaurs such as *Dromaeosaurus*, for instance, had a fairly solidly built skull and large teeth, indicating emphasis on head power.

Also out of the predatory dinosaur norm were spinosaurs, whose crocodilian-style skulls featuring long, slender snouts rimmed with large numbers of stout and/or snaggy teeth indicate that they were used much like those of crocs, including but not exclusively for fishing. Stouterskulled crocodilians also snatch tetrapods from shorelines and when they are swimming. This similarity with crocodilians has led to extended research of the enormous *Spinosaurus*, with it being presented as a short-legged fisher as well adapted for the aquatic predatory lifestyle as modern crocodilians. The restorations of its skeleton are based on sets of partial, disarticulated remains of differing sizes found very long distances from one another and which may not be from a sufficiently narrow time zone to place in one species. It is possible that the tall-spined vertebrae are those of allosauroids. This ambiguity renders attempts to reconstruct the complete skeleton as lying outside rigorous scientific standards and correspondingly problematic. The attention directed to *Spinosaurus* is excessive; it would be better to wait until more complete remains are on hand. And while not as pneumatic as and being denser boned than avepod norms are in line with being more aquatic than usual for the group, that *Spinosaurus* did have some air sacs means it would have been too buoyant to sustain underwater swimming the way crocodilians can. Nor is the great, drag-inducing back sail a feature expected in underwater hunters. And while a tail that may belong to *Spinosaurus* is deep, its slender spines look more suitable for display than for sculling. Spinosaurs were not hunting and killing the same way as standard, deep-snouted, blade-toothed avepods, but they had not become as highly water bound and adept as crocodilians. Splashing along shorelines and shallows picking up swimming items and surface swimming were more their hunting style. Same for the halszkaraptorine dromaeosaurs whose broad, shallow bodies and long necks suggest they spent time floating about while plunging for aquatic creatures.

There is no solid evidence that predaceous dinosaurs used toxins to poison prey as do a few lizards do. Predatory lizards practice good oral hygiene, carefully cleaning their teeth, lips, and gums after a meal with tongue and hands, presumably to minimize oral infections. The idea that their bites are toxically septic appears to be unfounded. Dinosaurs may well have done the same.

While some allosauroids, tyrannosaurs, and abelisaurs had little in the way of arm power, most predaceous dinosaurs did, and they used their forelimbs to help handle game. Not being used for locomotion, the three main fingers' claws, with emphasis on the thumb and central, were sharp-tipped hooks. These were not killing devices; they were not sufficiently powerfully muscled even in the largest armed examples, but they would have allowed the dinosaurian killers to latch onto and manipulate their prey during the assault. That manipulation would have included maintaining an optimal position while dealing the slash bites to the part of the victim being targeted. Not counting those that were using their arms as airfoils, the dinopredators that put the most emphasis on arm power included herrerasaurs, large podokesauroids, *Yangchuanosaurus*, *Allosaurus*, basal tyrannosauroids, dromaeosaurs, and small coelurosaurs—among the last *Sinosauropteryx prima* is notable for its big thumb claw. Of note are the hands of some American tyrannosauroids that had only two fingers yet the hands were fairly large, which seems a functional contradiction. The dryptosaurs' hands seem more useful for hunting and other purposes than the smaller appendages of tyrannosaurids, albeit not by much. But as useful as grappling arms may have been, that they were not critical is shown by their common reduction, often to the point of being of little or no use during predation, in favor of biting power.

The major majority of hunting dinosaurs did not specialize their legs for working with prey. For one thing, the load-bearing toe tips were constantly abraded and trimmed while walking and running in those that did not spend much time climbing in plants. However, these dinosaurs' feet could be of nonlocomotory utility, mainly pinning down relatively small prey items so they can be worked on by jaws and hands, for those who had the latter. The big exception that weaponized the feet were deinonychosaurs. The Late Jurassic arboreal aveairfoilan archaeopterygians modified the second toe to being hyperextendable and bearing an enlarged claw. This would not have been a deadly weapon. It probably initially evolved as a climbing hook and for defense and prey handling and was used as such by the small arboreal and flying microraptorines of the Early Cretaceous. It was the flightless dromaeosaurs that really ran wild with their often big sickle claws, wielded by robust, powerful legs. These claws would have been potent weapons able to severely damage and dispatch prey, somewhat similar to the large rapier second toe claws with which cassowaries can kill animals and people. Prime soft targets would have been digestive tracts and throats. Simulations that claim the super claws would not have been effective are absurd, they being simulations of simulations that are hard pressed to replicate how a living predator would have the muscle knowledge to correctly wield and apply the terrible claws in the manner that maximized their penetration and ripping potency. In the "fighting dinosaurs" fossil, the big claws of the dromaeo-saur are directed into the belly and neck regions of the big beaked protoceratopsid, indicating that they were the primary killing weapons—neither the lightly built head nor gracile arms had such wounding power. Troodont sickle claws were not as large and were borne on more delicate limbs. The living bird with the second toe claw most like deinonychosaurs, the seriema, uses the device to hold prey and then help tear it apart for feeding. The seriema does not use this claw to kill prey because the claw is not very big, is on very slender legs, and the bird is small—not the best model for far much more powerfully built and mus-cled dromaeosaurs as it is for the more gracile deinony-chosaurs. The oversized dromaeosaur claw did not evolve for prey handling and processing: it was a lethal feature.

That giant avepods attacked giant herbivores is supported by forensic evidence. A famous Early Cretaceous Texas trackway records a sauropod approaching two dozen tonnes that is closely paralleled on its immediate left by that of an allosauroid of a few tonnes. That they were moving in synch at about the same speed, and that when the sauropod tracks bend a little to the left so do those of the avepod, indicates the similar paths were not coinciden-tal. When the direction shift occurs, the sauropod appears to have stumbled, while the predator seems to have missed a step. This indicates the latter attacked at the moment, possibly slashing at the tail-based leg retractor muscles in order to ty to cripple its target as part of a killing process. While latched onto its prey the avepod was pulled along and misstepped, while the pained victim was pushed off balance. That the herbivore slowed down after the critical points suggests the attack was successful. Skeptics question the statistical likelihood that such an event would happen to be preserved, but the particular attack was probably one among many as the carnivore worked away at damaging its victim enough to kill it, perhaps over a period of hours and a few kilometers distance.

Doubt has been directed toward terrible tyrannosaurids, as per *Tyrannosau-rus*, risking facing off with dangerously horned and beaked adult ceratopsids such as *Triceratops* in dramatic, direct head-to-head combat like in movies, documentaries, and print. The left brow horn of a *Triceratops* skull was bitten off, as established by bite marks made by giant avepod teeth. The horn bone core then healed for a couple years, so the two mighty dino-saurs battled head-to-head with the herbivore winning in that it survived, the fate of the tyrannosaurid not being known. That such was found on one *Triceratops* out of a sample of a few dozen suggests such encounters were not rare. *Edmontosaurus* tails have healed bite marks of the size expected from a *Tyrannosaurus*.

The psychology of professional hunting animals is not well understood. It would seem that they should always be prudent calculators, not taking any more risks than neces-sary to maximize the Darwinian/Dawkinsian replication of their DNA. But genetics are sloppy, and creatures are not super logical automatons. It is very likely that pred-ators find enjoyment in their jobs, like swallows zipping across fields at speeds while getting their fill of hapless fly-ing bugs. Deriving pleasure from assaulting other creatures could be necessary to help motivate hunting in addition to hunger pains. Also, testing limits is a basic way for an animal to find out what is feasible in its lifestyle. As such predators may on occasion find themselves with their blood up and taking on special challenges, such as tackling very big prey items to see if they can make themselves an extra-large dinner. Not that big predatory dinosaurs would turn down smaller, easier prey to tide them over until the next big banquet, much as wolves will snarf down mice when they are on hand and jackals will snap up swarm-ing flying insects. In any case, dinopredators, including the giants, were high-risk throw-away animals. Fast breeding, fast growing, small brained, and with little parental invest-ment, the DNA replication scheme was to produce more individuals and get them into the ecosystem, with enough reaching reproductive age to keep the operation going. High casualties when hunting were more tolerable than they would be for adult mammal carnivores that put con-siderable effort into nursing and raising their big-brained young. Thus tyrannosaurids typically lived only half as long as slow-breeding rhinos and elephants.

At any given time and place the great majority of predaceous dino-saurs were either small species or juveniles that were not going after big game. Gut contents—including those of the first found *Compsognathus*

misstep by sauropod

missing avepod footprint

Giant avepod attacking sauropod trackway

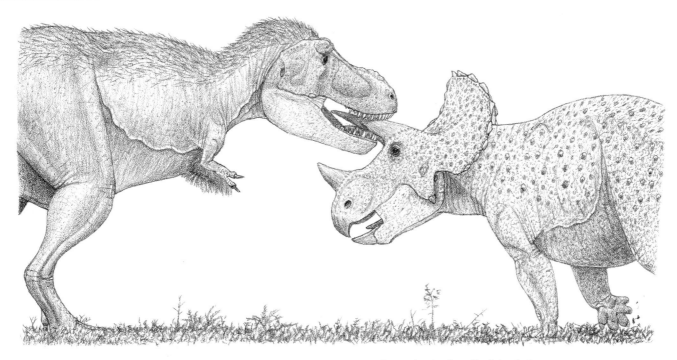

Tyrannosaurus regina biting off the horn of a *Triceratops prorsus*, based on a fossil of the latter

fossil—include fossils of lizards and other small reptiles, small dromaeosaurs, pterosaurs, and fish. That the remains of the prey are sometimes dismembered indicates the predator tore apart the carcass with hands, feet, and jaws working in coordination and swallowed the segments whole like raptors.

While the preponderance of dinosaurs hunted on the ground, a fair number did so among the branches. The winged microraptorine dromaeosaurs were novel in that they had a substantial degree of powered flight. They could have been ultimate ambush predators, dropping down on startled prey walking unawares below them—or perhaps not so unaware. A very atypical feature for dinosaurs are the partly upturned eyes of psittacosaurs of the time that appear well oriented for spotting killers ready to drop on them from branches above. Microraptorines may have also been aerial carnivores, targeting other basal aveairfoilans whose aerial abilities were not sufficient to readily evade the biplane microraptorines. The extra wing of the little dromaeosaurs may have been to enhance aerial agility and improve their chance of catching prey on the wing.

Turning to the perspective of prey on the defense, animals living in predator-dense habitats live lives of constant, chronic fear and dread, knowing that at any moment they may need to disrupt their daily lives to take intense defensive steps, probably having witnessed other herbivores calling out in fear and pain when being assaulted. This dynamic results in "landscapes of fear" in which prey animals are prone to avoiding locations that put them at most risk at a given time when particular predators are most likely to pounce upon them. This behavior can occur within a daily cycle—Yellowstone elk stay out of the woods that

cougars hunt in at night and then shun open flats where wolf packs are most active after sunrise and before sunset. Cryptic colorization helps prey avoid being spotted.

The flying dinosaur *Sinornithosaurus* attacking *Psittacosaurus*

If predators are in the immediate area, detecting their presence and moving away before they initiate the attack is optimal. This detection does not necessarily mean always automatically fleeing predators when seen; if the latter are acting casually, their presence may be tolerated. Being in a group has the disadvantage of the collection being easy for the predator to spot and the advantage of not being the sole target if a strike goes down. Approaching heavily used drinking spots needs to be minimized and, when necessary, on ready to take action. Detecting the distinctive odors of predators is useful but requires the predator to make the mistake of approaching from upwind. Hearing careful stalkers is not likely to work either; visual identification of the shapes of predators imprinted on preys' mind was dinosaurs' most useful defensive sensor system. When evasion by various means fails to prevent assaults, the goal is to survive with little or no injury, perhaps also to protect fellow herd members if they are related and its offspring if it is caring for them. This point leads to the probability that big herbivorous dinosaurs were not as dedicated to protecting fellow species members, including their young, as are mammals. While the latter always raise and protect their own offspring with devotion, post-nestling dinosaur babies were too small to accompany their parents. When herds of large-bodied dinosaurs formed, they were random aggregations in which youngsters would be hoping to enjoy some incidental protection from adults by mere association—they were not parent–offspring groups. So immature individuals were on their own. There would have been no hefty parents trying to fend off a predator from their own progeny, much less bravely determined and organized adults forming defense rings with the hapless juveniles, who probably are not their own, in the center. This lack of care should have aided the ability of theropods to pick off the juveniles without having to contend with battling parents trying to interfere with the kill. With the adult–juvenile size disparity for small- and modest-sized dinoherbivores low enough for the latter to stay with their mothers and/or fathers after leaving the nest, it is possible but not certain that the parents of at least some such species would have put at least some effort into defending their little ones, as is common in birds.

When a hunting predator was spotted, hopefully while still stalking, worse when it was already coming in at velocity from far or near, defensive actions appropriate to the circumstances and to the survival attributes of the dinosaur would be taken. If the prey had the ability to reliably outpace the attacker, then that may have been the best choice, certainly so if not well equipped to fend off the aggressor or, more dire, a pack of predators. Being in a herd or flock could have favored fleeing even if not definitely faster because the predator would probably end up hitting another individual, especially if one or some in the herd were not able to keep up with the others. If a herd one belonged to did run, then being a nonconformist who did not go with the crowd would result in being attacked, so

that is not a good option unless a prey animal needed to stay behind to try to protect its offspring, when the possibility of doing so was viable. Dinosaurs suited for running for their lives included all those with long, flexed running limbs, including protodinosaurs, most small- to medium-sized predaceous avepods when hunted by bigger predaceous avepods, elaphrosaurs, oviraptorosaurs especially avimimids, of course ornithomimids and alvarezsaurs, possibly prosauropods, lesothosaurs, heterodontosaurs, perhaps small ceratopsians, scelidosaurs, ankylosaurs under appropriate circumstances, and all ornithopods. Among those potential runners the elaphrosaurs, avimimids, alvarezsaurs, lesothosaurs, small ornithopods, and hadrosaurs were too lacking in the way of defensive weaponry to have much choice other than to run, hoping for the best. Healed bite marks on the tails of duck-billed hadrosaurs indicate that they survived attacks by pursuing tyrannosaurids. Ornithomimids were better off, being able to hard kick with their long powerful legs—ostriches have been known to fend off and even kill lions, sometimes while defending their chicks. One evasive tactic would be to run into dense brush when on hand.

Another tactic of desperation is to run into a body of water. This brings us to the conceit that was once the conventional wisdom—that dinopredators were hydrophobic to the degree that all an herbivore under pursuit had to do was go for a swim and leave the vexed theropod standing frustrated on the shoreline. The premise was that the narrow toes of theropods left them more prone to get mired or rendered them poorer swimmers. When it was thought that sauropods and duck-bills were predominantly aquatic, this was seen as their go-to predator defense, in part because the hadrosaurs were errantly thought to have webbed feet. This water-escape notion has largely fallen by the wayside with the realization, based partly on the bottom-poling avepod trackways, that dinopredators were of course adept swimmers capable of pursuing their victims into water, all the more so if the pursuing avepod was a semi-aquatic spinosaur. And mammalian carnivores are known to chase down panicked mammalian herbivores that try the river or lake to escape.

But the water trick should not be dismissed out of hand. If a dinosaurian carnivore lived in Late Triassic environs in which watercourses were infested by very large crocodilian like phytosaurs, or Cretaceous supercrocodilians up to 10 tonnes, then dashing into waters graced by such terrestrial-beast-drowning monsters would have been dangerous—and for the prey target, too, the water option may have been leaping from the frying pan into the fire. Fear of crocs may be why carnivorous mammals do not always chase game into tropical waters. And there may be no point for land predators to kill prey in crocodilian-dominated waters because the crocs will happily take over the carcass conveniently floating in the habitat in which they have the advantage. Only if a big dinosaur can quickly carry or drag its victim ashore and out of reach of crocodilian jaws is it advisable to dispatch it in the latter's' territory in the first place. These concerns did not apply to Jurassic avepods because there were no large swimming crocodilians to deal with. If the prey dinosaur was a nonpneumatic prosauropod, or ornithopod that either lived before the age of big crocs or was willing to take the risk posed by Triassic and Cretaceous aquatic archosaurs, resorting to water could be a good option for outfoxing an avepod. The latter would be too buoyant to dive after a nonpneumatic dinosaur if it dove beneath the surface in sufficiently deep water, where the avepod would not even be able to track it.

Running from attack was not the optimal option for all dinosaurs. Elephantine-limbed sauropods and derived stegosaurs were too slow to even try it. Other running dinosaurs were better off fighting for their lives, if they were not fast enough to ensure escape, and had the weaponry to put up a good fight. Sauropod battle qualifications often but not always included sheer size, enormous tails up to 10–15 m (35–50 ft) long and weighing up to 3 to 20 tonnes—often the total mass of the avepods attacking them—and, except for titanosauriforms, big, stout thumb claws. Swinging tails, tipped with long, lashing whips among diplodocoids and titanosaurids, and rearing to deploy hand claws would have posed stout defenses against the biggest avepods. Primary vulnerabilities would be the leg retractors of the tail base and thighs and the slender, delicate neck, at least among sauropods small enough for their neck base to be reached. Sauropods may have been at risk of avepods dashing in, slicing or punch biting out a chunk of flesh, and then darting away before an effective counterattack was made, and doing so on a repeat basis until satisfied. The avepod's aim would not have been to outright kill the land titan, but the resulting wounds may have been debilitating and ultimately lethal, in which case the vast carcass would be a huge smorgasbord for the local avepods until it was consumed or rotted.

The small, slender necks of stegosaurs were vulnerable, although fairly well armored. Their defensive aim was to keep their spike-tipped tails in the faces of their attackers to fend them off. That stegosaur tail spikes are often damaged or even broken and then healed verifies that they were used for combat, presumably against attacking avepods in at least some cases. When the well-armored nodosaurs did not run they could use their shoulder spikes, especially when they were forward directed, to try to damage the legs of big avepods. There is no practical doubt that ankylosaurids took whacks at tyrannosaurids and the like with their tail clubs. Another tactic for hard-pressed ankylosaurs was to lay flat on the ground, letting the armor shield them. Running pachycephalosaurs may have been best off to ram the poorly protected flanks of those that dared to try to convert them into food, if the avepod was not too big to hurt.

The ponderous therizinosaurs would have stood and fought, swinging out with their big, in some cases Freddy Krueger style, saber finger claws to slash up their attackers in the manner of the more recently extinct giant ground sloths. Also dangerous were their big toe claws. As well likely to hold their place and defend with oversized hand claws were the massive deinocheirid ornithomimosaurs; their blunt toe claws were not so useful. Swifter ornithomimosaurs would have done the same when cornered. Many oviraptorosaurs were likely to lash out with large, sharp finger claws.

With their sharp beaks and fangs when they had them, heterodontosaurs and small ceratopsians had the option to engage predatory avepods like nasty pigs and peccaries. Such struggles might not have always been between hunting avepod and ornithischian prey: contests may have occurred over scavenging rights to carcasses—especially when the omnivores outnumbered the predators. Even more spectacular were the titanic battles between multi-tonne tyrannosaurids and horned and beaked ceratopsids. The ideal situation for the former was to spook the latter into running and exposing their vulnerable rears. If that did not work then the tyrannosaurid would have to decide whether to deal with the horns and beak of its potential next repast or abandon the assault in favor of another attempt—the greater the hunger of the predator the less likely it would take the safe route, which may have led to the *Tyrannosaurus* finding itself having to bite off the *Triceratops* horn to try to save the situation. Scavenging disputes may also have led to serious interactions.

For those little protobird dinosaurs with airfoils, taking to the air would have been an option when in danger. Especially when threatened by microraptorines and the like up in trees, parachuting, gliding, or flying away would have been an option. Those that had large and powerful enough wings might have tried to take off for their lives, but against the gravity well of the planet, performance may not have been sufficient to make this an optimal option. When on the ground, maximally displaying and bristling whatever feathers a dinosaur had, or lifting tails, spreading arms wide, and threat gaping, would have been means of trying to impress a predator that looked like it was about to attack in the hope of looking too large, formidable, and prepared to defend oneself to be worth taking on.

Because dinosaurs over half a tonne would not have been able to care for their initially diminutive offspring and lost contact with them when they were eggs or not long after hatching, the small post-nestlings, both herbivorous and predaceous, would have been relatively easy pickings. That would include the offspring of large predatory dinosaurs, cannibalism being common among carnivores. Horned ceratopsids did not form defensive rings around their young like horned musk oxen—the last are the only modern big mammals that do that and only because their charges are so hard to get to adulthood in the Arctic barrens in the first place. Trackways and bonebeds show that growing juvenile dinosaurian herbivores did not join up with adults until they were large enough to keep up with the latter and not get trampled by them, about a tonne for

Yangchuanosaurus shangyouensis vs. unnamed genus *hochuanensis*

giant sauropods. Such youngsters probably joined up with adults that they would not have had familial relationships with to gain the incidental protection of simply being in the presence of great and powerful adults that had no particular interest in the safety of the little ones that choose to accompany them.

A basic survival stratagem for animals subject to high levels of predation is rapid, r-strategy reproduction to replace the losses, which is a reason dinosaurs were usually fast breeders. Do not make the common mistake of thinking that predation is nature's unfortunate but necessary means of keeping the populations of plant eaters in check: the notion of a balance of nature is a romantic conceit promoted to explain away the cruelty of creation that was falsified long ago. Animals that are not subject to serious predation are prone to be slow-breeding K-strategists. The big continental ratites put out a lot of eggs each year because they suffered high losses due to intense predation by carnivorous mammals. The giant island birds laid only an egg or two a year because there were few or no predators to cull the population. Likewise sharks, being predators not subject to frequent attack, are slow breeders. Birds that can readily protect their youngsters in arboreal nests or remote islands are K-strategists. Reproduction is expensive, so animals are prone to reproduce only as fast as needed to sustain and modestly expand the population when not faced by a host of predators. That option was not available to most dinosaurs.

Land animals were not the only vertebrates on many dinopredators' menus. The majority were probably happy to snag fish when they could, as shown by preserved gut contents, and some were adapted for doing so, most especially the snaggle-toothed masiakasaurs and spinosaurs, even though these were not as aquatic as the earlier phytosaurs or contemporary crocodilians.

Being tachyenergetic, flesh-eating dinosaurs needed to eat about as much meat as avian and mammalian predators at a given size, but none of those get over a tonne. Presume that the avepods had energy budgets similar to those of ratites. Mostly not able to deal with bone, they could have consumed about three-quarters of a body. In that case, a 15 kg (33 lb) *Ornitholestes* needed less than a kilogram of flesh a day, and a 10 kg (22 lb) kill would power the avepod for a week or so—if a reptile of the same size ate the same victim, it would last a couple of months. If a pack of twelve 25 kg (55 lb) *Coelophysis* bring down a 50 kg (110 lb) prosauropod, the meal would sustain them for a couple of days. A dozen 70 kg (154 lb) *Deinonychus* could take down a 400 kg (882 lb) tenontosaur. Predators can gorge themselves by about a quarter their body mass, so they chow down on the carcass for two or three days and can go for a dozen days on that. A deceased 20 tonne sauropod is ready to eat: over the next few days, three dozen 1.5 tonne *Allosaurus* would come in to take their portion before the sauropod rots. Each would get a full meal that fueled them for two

Baryonyx walkeri

weeks, given that metabolic rates per unit body mass decline with large body mass so larger avepods could go longer between meals. A 7 tonne *Tyrannosaurus* would need to bolt down 80 kg (176 lb) of flesh per day—one person could be swallowed whole, 500 tonnes over a 30-year life, 10 of that as an adult. That amount is about 1,000 cows, each of which could be consumed in a few bites. More paleo-realistically, that is around 150 *Triceratops* in a lifetime. The adult *Tyrannosaurus* could wolf down the carcass in about a week, crunching and consuming some bone in the process. If half of those are ingested when the avepod is an adult, that is half a dozen or more each year, with each dining lasting over a month and a half. For a pack of four, the carcass would be two weeks of sustenance.

Most portrayals of dinosaurs on the hunt and making kills show them doing so in daylight, but much of the predation would have occurred in twilight and dark hours when it is easier for ambushing predators to hide and the intense combat occurs in cooling air. Needing to pick up small prey items on a regular basis, lesser predatory dinosaurs would have fed many, if not most, hours on a given day. After gorging on a tremendous amount of flesh from a major kill, greater dinopredators would have spent lots of time, days at a stretch, like lions in a pride, lazing about until they felt the need to find further sustenance.

Almost all carnivores scavenge when the opportunity arises—no good reason to turn down a meal that did not need work and risk of injury to acquire, although there may be contention with other scavengers over a body. An exception are cheetahs, as they are too lightly constructed and armed, as well as solitary, to deal with other savanna carnivores. No terrestrial meat eater only scavenges; even the bone-crushing hyenas hunt down much of their food. Spontaneously deceased carcasses are too scarce and too hard to find for land-bound carnivores—only soaring birds able to sight and smell scout out large areas at little energy expense, and then quickly fly at little cost to a body, can be full-term corpse consumers. The situation may have been different in sauropod-dominated habitats where the adult carcasses were colossal food sources, but the big adults may have had low population densities. An excellent sense of smell would have facilitated picking up the odor of rotting meat wafting over long distances, and olfaction is good for hunting too. Storing the calories from carcass flesh as fat would hinder hunting, especially if it was on the long tails because that could make the bipeds tail heavy. And no meat eater with the ability to kill a given animal it encounters will pass up the chance to try to do so when it is hungry and no dead carcasses are on hand. In North America the tyrannosaurids lacked access to massive sauropod bodies. Add to that the forensic evidence that giant avepods, including tyrannosaurids, engaged in full combat with also titanic plant-eating dinosaurs renders the scavenging-only big avepod a myth. For much of the Mesozoic, what blade-toothed dinosaurs did not have to contend with while enjoying a carcass was competition from aerial scavengers. The exception may have been the Early Cretaceous of Eurasia, when the cutting-toothed istiodactylid pterosaurs look to have been drop-in scavengers—the later, straight-beaked, toothless azhdarchids do not appear to have been dedicated scavengers.

Lacking the boosted aerobic exercise capacity and erect limbs that allow and facilitate high cruising speeds, and not needing to eat all that much, large predatory lizards never have home ranges over just a few square kilometers. Cats, canids, and hyenas patrol ranges spanning from around 10 to 1,000 square kilometers (4–400 square miles) in the search for the large amounts of flesh they need in the form of prey or carcasses, in part depending on local prey density. Being similarly aerobically mobile and metabolically hungry, hunting and scavenging dinosaur home ranges would have been comparably expansive, with the larger taxa tending to have greater ranges, perhaps in some cases over 1,000 square kilometers. Being at the apex of the ecological food pyramid, and having the elevated metabolic rates and corresponding high food budgets, predaceous dinosaurs made up only a few percent of the animal biomass in a given locale, being outnumbered by their prey dozens to one, as are carnivorous mammals in modern habitats. Lower energy budgets allow reptiles to have much larger populations relative to their victims.

EATING PLANTS

Theropods and then avepods started out as blade-toothed predators, and the solid majority remained such until the end of the Mesozoic extinction. But herbivores evolved from carnivores, and plant eaters were being spun off of from avepods on a fairly frequent basis. As far as is known, this started in the Middle Jurassic with elaphrosaurs. They were not especially sophisticated vegetarians and presumably consumed softer, more readily broken down plant materials, although gizzard stones helped mash what they swallowed. This little group did not become common and soon disappeared. Elaphrosaurs set an herbivorous-avepod long-running-legs precedence for the Cretaceous successors, the ornithomimosaurs, which were a fairly diverse lot that were often common in their habitats, still going strong right up to the great extinction. Small- and medium-sized ornithomimosaurs, especially ornithomimids, took the fast-running-herbivore motif to its dinosaurian limits, the resulting dinopredator evasion performance probably contributing to their success. As per the ratites they are so similar to, and the earlier elaphrosaurs, the small, simple-headed, narrow-beaked ornithomimids, with limited-capacity guts, were herbivore–omnivores that avoided tough vegetation while picking up a good amount of small animal protein. The massive deinocheirids appear to have retained this basic system, their bellies not being highly expanded. At least some ornithomimosaurs utilized gizzard stones. The long arms and clawed fingers of these bipedal dinoherbivores were likely to have been used to help feed on vegetation.

Omnivorous and herbivorous oviraptorosaurs were a variable lot. Possibly descended from aerial omnivoropterygid like birds whose name indicates their feeding preferences, the similarly configured heads of oviraptorosaurs ranged from fairly small to substantial in size. Oviraptorids developed a distinctive downward-projecting mouth roof that can be easily seen when the mouth is open. Combined with the deep, short, parrot-like jaws, the unusual arrangement appears suited for crushing and perhaps pulping materials, although how and what is not clear. Modest-size bellies suggest that the materials, whatever they were, did not demand extensive breakdown, although gastroliths were sometimes used. A few oviraptorids got fairly large.

The acme of plant-consuming avepods were the therizinosaurs, which may have been the offshoot of the seed-eating jeholornid birds. Therizinosaurs bore blunt, rounded browsing beaks followed by modest dental arrays that could do some initial light mastication before swallowing. Starting with the earliest known examples, therizinosaur abdomens were capacious in the manner that indicates microbial fermentation was used to digest tough plant materials over time. As their evolution proceeded the bellies became bigger due to their increasingly greater breadth in combination with backward expansion due to retroversion of the pubes, until their guts matched the digestive capacity of the highest developed herbivores. It is possible therizinosaurs used a foregut chamber to begin microbial breakdown like ruminant ungulates and hoatzins, although they would not have regurgitated cuds to rechew; the dinosaurs' teeth are not developed well enough to do that. With their up-tilted trunk, therizinosaurs were best adapted for high browsing in tall bushes and short trees, aided by their extra-long arms and elongated great-clawed fingers. Perhaps that is why even those arch herbivores, as far as is known, did not develop the broad, squared-off beaks needed to graze ground cover such as ferns and sedges or the grasses that were beginning to appear.

As much success as a few largely Cretaceous avepod groups had chomping down on plants, they were a small minority of the dinoherbivore population, which was always dominated by a great empire of heavy-limbed sauropodomorphs and beaked and cheeked ornithischians from the Triassic to the K/Pg boundary.

THE EVOLUTION—AND LOSS—OF AVIAN FLIGHT

Powered flight has evolved repeatedly among animals, including numerous times in insects in the late Paleozoic, among gliding fish, and three times in tetrapods—pterosaurs in the Triassic, birds in the Jurassic, and bats in the early Cenozoic. In all cases among vertebrates, flight evolved rapidly in geological terms, so much so that the earliest stages have not yet been found in the fossil record for pterosaurs and bats. The means by which flight evolved in pterosaurs remains correspondingly obscure, although that they have wing membranes attached to their legs favors it starting out as arboreal gliding. The fact that bats evolved from tiny insectivorous mammals, and the recent discovery of an early fossil bat with wings smaller than those of more modern forms, show that mammalian flight evolved in high places.

In terms of fossil material, the origin of birds and their flight is much better understood than it is for pterosaurs and bats. This knowledge extends back to the discovery of Late Jurassic *Archaeopteryx* in the mid-1800s and is rapidly accelerating with the abundant new fossils that have come to light in recent years, especially from the Early Cretaceous and from the middle of the Jurassic before *Archaeopteryx*. We can conclude that dinosaurian flight was not yet developing in the Triassic because while fine-grained sediments from that time preserve an array of flying insects and early pterosaurs, nothing dinoavian has shown up. However, a major fossil gap exists because little is known about what was happening in the Early Jurassic and most of the middle of that era regarding flying creatures. Also hindering greater understanding is that the fine-grained deposits that best preserve small flying creatures are largely limited to a few geographic locations, mainly eastern China in the later portion of the Jurassic and Early Cretaceous and Europe in the Late Jurassic, and it is unknown what was happening in the rest of the world over that time.

When it was assumed that birds did not evolve from dinosaurs, it was correspondingly presumed that their flight evolved among climbers that first glided and then developed powered flight. This theory has the advantage of our knowing that arboreal animals can evolve powered flight with the aid of gravity, as bats almost certainly did and pterosaurs probably too. For that matter the membrane-winged scansoriopterygids look more like capable climbers than adept runners. When it was realized that birds descended from deinonychosaurs that sport long legs, many researchers switched to the hypothesis that running dinosaurs learned to fly from the ground up. This has the disadvantage of uncertainty over whether it is practical for tetrapod flight to evolve among ground runners working against gravity.

The characteristics of birds indicate that they evolved from dinosaurs that had first evolved as bipedal runners and then evolved into long-armed climbers. If the ancestors of birds had been entirely arboreal, then they should have been semiquadrupedal forms whose sprawling legs were integrated into the main airfoil, as in bats. That birds are bipeds whose erect legs are separate from the wings indicates that their ancestors evolved to run. Conversely, how and why ground animals would directly develop the long, strongly muscled arms and wings necessary for powered flight has not been adequately explained. The hypothesis that running theropods developed the ability to fly as a way to enhance their ability to escape up tree trunks—some juvenile birds flap their small wings to do so—itself involves a degree of arboreality. This theory has been challenged by bioaerodynamic analysis: protobirds did not have wing flapping apparatus as well developed as modern baby birds. Opposition to climbing avepods suffers nonintellectual similarities to the once common belief they avoided going into water. More analytically, some have argued that theropod arms and hands were not configured for climbing because they lacked the wide freedom of movement and lower arm rotation often seen in climbing creatures, and the dinosaurs lacked other features common among arborealists. But the grasping hands and feet of small theropods were actually well suited for climbing in the manner of the one modern bird that readily gets about branches using all four limbs, juvenile hoatzins. When endangered, hoatzin nestlings can leap into the water, then scramble back up. After leaving the nest, fledglings continue to get around by climbing with hands and feet until they can fly. Some airfoilan theropods—scansoriopterygids, *Anchiornis*, *Xiaotingia*, *Archaeopteryx*, microraptorine dromaeosaurs—were even better adapted for quadrupedal climbing, having longer hands sporting larger, more strongly hooked finger claws. And their toe claws too were more strongly arced than they are in hoatzins or is normal for nonclimbing avepods. This was not true of all small deinonychosaurs; *Caihong*, as well as *Eosinopteryx* and *Pedopenna*, which may be the same species, had fairly flat claws suggesting more modest arboreal abilities. And the long foot feathers of *Pedopenna* appear heavily worn, as expected if they were constantly being rubbed on gritty ground. In contrast, the even longer foot feathers of some microraptorines not only look like they would have hindered movement on the ground, but their crisp edges indicate they did not get abraded by regular soil contact. Some protobirds lack claw preservation sufficient to fully assess their arboreality. There is evidence that microraptorine legs could sprawl out more to the sides than normal in dinosaurs; if so, that would have facilitated climbing compared with other protobirds whose legs were not as laterally flexible as per the dinoavian norm, including climbing hoatzins. Dinosaurian quadrupedal climbing largely disappeared with the big extinction that left only birds, which usually lack finger claws for climbing, but it works well enough that the scheme has reappeared in the young of at least one group of birds—there may have been more in the Cenozoic.

Avian flight probably evolved among dinopredators that spent time both on the ground and, for those with long arms, in the trees. As climbers they would have had a simple yet significant problem—going down, because they lacked features such as the reversible ankles that make it easy for squirrels to go down tree trunks. The solution to that problem was going down through the air, which had the advantage of getting down more swiftly but needed to be done safely. In scansoriopterygids the initial appearance of arm membranes slowed aerial falls. In early aveairfoilans, expanding limb feathers allowed medium velocity parachute leaps to lower locations. To travel horizontally at speed, leaps between branches could have been lengthened by developing larger, more aerodynamically asymmetrical pennaceous feathers that turned leaps into short glides. As the feathers lengthened, they increased the length of the glides. Large arm membranes allowed scansoriopterygids to glide too, but for unknown reasons they do not appear to have gone beyond this stage, as did pterosaurs and bats, and soon disappeared from the evolutionary scene. As feathered protowings and arm muscles became larger, flapping would have added power, turning the glides into a form of active flight. The same flapping motion subsequently would have aided the rapid climbing of trees via the air rather than scrambling up trunks and branches as arm power and wing area further scaled upward. Arguments that transitioning from gliding to flapping is too aerodynamically difficult were never compelling and have been discredited, all the more so because bats probably did so. Selective pressures promoted increases in arm muscle power and wing size until the performance level seen in *Archaeopteryx* was present. The flying deinonychosaur had an oversized furcula and large pectoral crest on the humerus, which supported an expanded set of muscles for flapping flight. The absence of a large sternum shows that its flight was rather weak by modern standards, but takeoffs from the ground were probably achievable by leaping up or running into the wind. As bird flight further developed, the sternum became a large plate like those seen in microraptorine dromaeosaurs. Fixed on the rib cage with ossified sternal ribs, the plate anchored large wing-depressing muscles. The ribcage was also braced by the ossified uncinate processes you see on intact chicken and turkey breasts, and the outer primary wing feathers were better anchored on somewhat flattened central finger bases. Going further into avian evolution, sterna began to sport a deepening keel that further expanded the flight muscles, especially those used for climbing flight as adaptations at and near the shoulder joint improved the ability of the wing to elevate, helping boost the rate of climb. At the same time, the hand was further flattened and stiffened to better support the primaries, and finger claws were reduced and typically lost. The tail rapidly shortened in most early birds until it was a stub. This means that birds quickly evolved a dynamic form of flight, much more rapidly than pterosaurs, which retained a long tail stabilizer through most of the Jurassic. The above adaptations were appearing in Early Cretaceous birds, and the essentially modern flight system featuring the big, deep-keeled sternum and highly

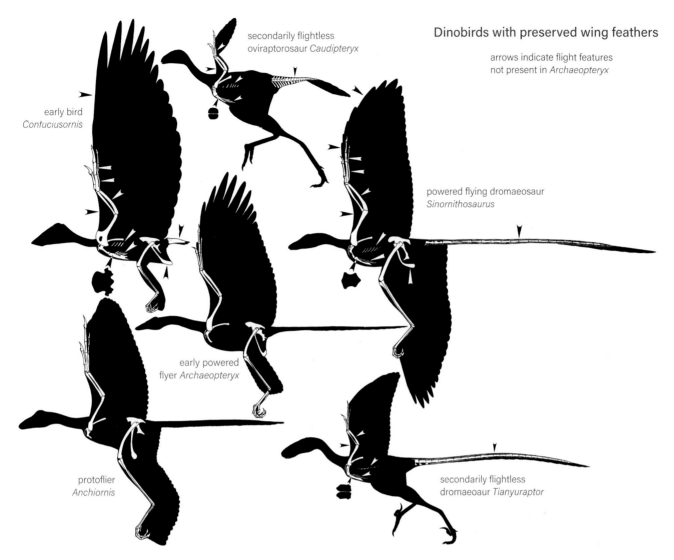

Dinobirds with preserved wing feathers

arrows indicate flight features
not present in *Archaeopteryx*

secondarily flightless
oviraptorosaur *Caudipteryx*

early bird
Contuciusornis

powered flying dromaeosaur
Sinornithosaurus

early powered
flyer *Archaeopteryx*

protoflier
Anchiornis

secondarily flightless
dromaeoaur *Tianyuraptor*

co-ossified, shortened, and clawless hand had evolved by the Late Cretaceous. While dinosaur flight may have evolved just once, multiple origins cannot be ruled out. Borne by sprawling legs, the microraptorines very large tandem hindwings may have been a specialization specific to dromaeosaurs that were an evolutionary sideshow, or they may have been a stage in the development of avian flight later lost, like the long bony tails of basal protobirds.

For obvious reasons, researchers have focused on the development of dinosaur flight as they took to the air, but that has been a mistake regarding sorting out how flight evolved and how protobirds are related to one another. Also critical is reduction or outright loss of flight abilities as this researcher pointed out in *Dinosaurs of the Air* in 2002. For all its advantages, flight has its downsides, including all the mass and energy that is absorbed by the oversized wing tissues, especially the enormous flight muscles. Nor can flying birds be especially large, the limit being around 100 kg (200 lb)—aerial pterosaurs reached a few hundred kilograms. For such reasons, a number of derived birds herbivorous and predatory have lost flight even on the continents, starting in Late Cretaceous with *Gargantuavis* and continuing to today. Among those that lost flight are the phorusrhacid terror bids of South America; the earlier, similarly big-beaked gastornithids; the dromornithid thunder birds of the Neogene of Australia; and assorted fleet-footed ratites in many regions—yet these birds have not been able to do much with their clawless hands.

Biological common sense predicts that late Mesozoic protobirds with only modest flight abilities and clawed hands that could be used for multiple purposes would have been much more prone to losing the ability to take to the air than those that are more sophisticated fliers. Yet the common view of dinoavian evolution includes little if any losses of flight outside birds proper. This scenario is critically flawed because the possibility that weak early fliers rarely if ever lost flight over the nearly 100 million years of the Late Jurassic and Cretaceous is not logically viable: it would have been happening all the time. Early flight loss not being the frequent norm is an extraordinary hypothesis that requires correspondingly extraordinary evidence in the tradition of the Carl Sagan axiom. Continental tetrapods have undergone the massive reversals needed to flip back to flippered marine swimmers numerous times. Losing flight is much easier and simpler to do, all the more so when it occurs among initial, low-ability fliers. Reversals modest and radical are frequent features of Darwinian evolution. The reason for the failure to spot losses of flight that is expected to have happened during the transition from dinosaurs nonavian to avian is a consequence of strong reliance on the cladistics-driven phylogenetics widely used to try to restore dinosaur relationships. Cladistics, which compares anatomical features, is so commonly used because the genetics used to more accurately investigate the relationships of organisms with living representatives cannot be applied to creatures that all went out of existence tens of millions of years ago. Being a delicate molecule that is able to split up repeatedly for reproduction, DNA does not survive outside living organisms for more than a couple of million years, contra *Jurassic Park*, and even then only in chilly deposits. When DNA can be used, genetics overturns cladistic results so often—bird phylogenetics has been radically altered by genomic investigation—that anatomical cladistics is used only when there is not a molecular alternative, in which case the cladistic conclusions have to be taken with a large dose of skepticism. Results should especially be questioned when they lead to illogical outcomes, such as flight not being commonly lost among protobirds. The example of flight loss is all the more true because cladistics are particularly hard pressed to deal with the relatively subtle degrees of anatomical reversals that are expected in matters such as losses of flight. Despite that, many researchers decline to give much consideration to analyses that are not dependent on cladistics, so losses of flight are likely to be frequently missed. Cladistics is not reliable enough to produce the extraordinary evidence needed to discredit what would have been the ordinary norm of frequent flight loss in the early evolution of aerial dinosaurs.

To do a better job of recovering evidence of flight loss starts with analytical functional anatomy. Such evidence involves the presence in basal aveairfoilan nonfliers of flight features that are normally retained in neoflightless birds. These features include large, squarish sternal plates supported by bony sternal ribs; bony uncinate processes on the ribs; strong flexion between a horizontal, strap-shaped scapula blades strongly flexed relative to vertical; elongated coracoids; folding arms; a number of details of the hand, including stiffening and fusing of elements; bird–pterosaur–bat-style propatagium airfoil surfaces stretched from the shoulder to wrist forward of the elbow; and stiffened, slender, pterosaur-like or very short birdlike tails. All these adaptations are undoubtedly involved in dinosaur fight; attempts to explain them as exaptations before the evolution of flight are at best speculative and often a stretch. It has never been logically explained why large, broad sternal plates evolved to help operate enlarged arms for display or predatory purposes when many of the

flesh-eating avepods drastically reduced their arms—that when we know big squarish sternal plates are directly tied to powered flight in both birds and pterosaurs. Why dromaeosaurs evolved pterosaur-like tails for hunting when no other dinosaurs did so, but pterosaurs evolved them for flight, has not been satisfactorily detailed. It is more logical to presume dromaeosaur tails were a flight adaptation that was then repurposed for added predatory agility after flight was lost. Why do some oviraptorosaurs have a propatagium, present only in bats, pterosaurs, and birds flying and flightless? The wing structure has no apparent reason for existing outside of initially appearing for flight needs, and such should be assumed unless shown otherwise.

Many of the flight attributes listed above apply to nonflying dromaeosaurs, whose early examples appear to have been better adapted for flight than *Archaeopteryx*. Small-armed dromaeosaurs were almost certainly neoflightless, like wingless birds. Troodonts like *Anchiornis* and its similarly weak-flight-capable east Asian relations suggest some deinonychosaurs began to lose flight in the Late Jurassic, with troodonts their completely flightless evolutionary offspring. Therizinosaurs and especially oviraptorosaurs show many of the anatomical signs that some level of flight was present early in their evolution. Another means of determining the presence of fight loss is when a group of nonfliers shows strong similarities to a group of fliers such that the former could have spun off from the latter like new television series. It is particularly telling that the early flying jeholornithiform birds and flightless therizinosaurs shared strikingly similar skulls, as did in turn the deeply parrot-beaked omnivoropterygid birds and oviraptorosaurs. Such deep convergence happening once, as the conventional scenario proposes, is not likely, and it happening twice is a real evolutionary stretch. Very close relationships, in both cases involving the loss of flight expected to spawn off from such crude fliers, are the superior bioevolutionary scenario. That basal oviraptorosaurs had retroverted pubes like flying protobirds is further evidence their ancestors once flew Mesozoic skies. The alvarezsaurs are another potential neoflightless clade. That functional and related evidence for the nonextraordinary frequent loss of flight early in the evolution of dinosaur flight overrules the cladistics results means that the latter is being too heavily relied upon and should not be used to automatically exclude more scientifically logical alternatives.

It is possible if not probable that losing flight was going on all the time in dinobirds, with what looks like monophyletic groups actually being the result of multiple groundings. Oviraptorosaurs, for instance, may be such a diverse group because the varying types independently descended from fliers. Also quite possible is that protobirds were losing and then evolving flight on a regular basis, since that would require only reenlarging their recently reduced but still functional arms and feather arrays. Some cladists contend that while dinobird flight evolved a number of times, it was rarely if ever lost, which is not at all logical in that loss is so much easier to achieve than flight acquisition—however many times flight appeared, its loss would have been more frequent, probably much more so. Limited fossil data and analytical procedures prevent detailed understanding of these events now and probably well into the future. There is only so much that can be done with the evidence on hand. In the Late Cretaceous, birds themselves lost flight on occasion, most famously the widely distributed marine hesperornithiform divers, as well as the chicken- to ratite-sized European island birds of uncertain relationships, among them peculiar *Balaur*, known from near the end of the period. Examples of Early Cretaceous neoflightless birds have not turned up but by no means can be ruled out.

MESOZOIC OXYGEN

Oxygen was absent from the atmosphere for much of the history of the planet, until the photosynthesis of single-celled plants built up enough oxygen to overwhelm the processes that tend to bind it to various elements such as iron. Until recently it was assumed that oxygen levels then became stable, making up about a fifth of the air for the last few hundred million years. Instead, it has been calculated that oxygen has fluctuated strongly since the late Paleozoic. The problem is that the results are themselves variable. The results do agree that the oxygen portion of the atmosphere soared to about a never-seen-again third or more during the late Paleozoic, when the great coal forests were forming and, because of the high oxygen levels, often burning. It is notable that this is when many insects achieved enormous dimensions by the standards of the group, including dragonfly relatives with wings over 0.5 m (2 ft) across. Because insects bring oxygen into their bodies by a dispersed set of tracheae, the size of their bodies may be tied to the level of oxygen. And aquatic insect larvae may have been oversized in order to keep oxygen uptake from being too high by minimizing surface area relative to mass.

But in the Mesozoic the situation is less clear. Oxygen levels may have plunged precipitously, sinking to a little over half the current level by the Triassic and Jurassic. In this case, oxygen availability at sea level would have been as poor as it is at high altitudes today. Making matters worse were the high levels of carbon dioxide. Although not high enough to be directly lethal, the combination of low oxygen and high carbon dioxide would have posed a serious respiratory challenge that could have propelled evolution of the efficient air sac respiration of pterosaurs and some dinosaurs. However, other work indicates that oxygen did not plunge so sharply, and in one analysis it never even fell below modern levels, being somewhat higher in most of the Mesozoic, an apparently high rate of fires being evidence for that. Reliably assaying the actual oxygen content of the atmosphere in the dinosaur days remains an important challenge.

PREDATORY DINOSAUR SAFARI

Assume that a practical means of time travel has been invented, and, *The Princeton Field Guide to Predatory Dinosaurs* in hand, you are ready to take a trip to the Mesozoic to see the killer dinosaurs' fantastic world. What would such an expedition be like? Here we ignore the classic time paradox issue that plagues the very concept of time travel: What would happen if a time traveler to the dinosaur era did something that changed the course of events to such a degree that humans never evolved to travel back in time and disrupt the time line in the first place?

One difficulty that might arise could be the lack of modern levels of oxygen and/or extreme greenhouse levels of carbon dioxide (which can be toxic for unprepared animals), especially if the expedition traveled to the Triassic or Jurassic. Acclimation could be necessary, and even then, supplemental oxygen might be needed at least on an occasional basis. Work at high altitudes would be especially difficult. But, as noted earlier, oxygen deprivation may not have been the case. Another problem would be the high levels of heat chronically present in most dinosaur habitats. Relief would be found at high latitudes, as well as on mountains and high plateaus.

If the safari went to one of the classic Mesozoic ecosystems that included gigantic dinosaurs, the biggest problem would be the sheer safety of the expedition members. The bureaucratic protocols developed for a Mesozoic expedition would emphasize safety, with the intent of keeping the chances of losing any participants to a bare minimum. Modern safaris in Africa require the presence of a guard armed with a rifle when visitors are not in vehicles in case of an attack by big cats, hippos, buffalo, rhinos, or elephants. Similar weaponry is often needed in tiger country, in areas with large populations of grizzlies, or in Arctic areas inhabited by polar bears. The potential danger level would be even higher in the presence of flesh-eating dinosaurs as big as rhinos and elephants and able to run down a human, who could potentially be out of breath because of low oxygen. It is possible that carnivorous dinosaurs would not recognize humans as prey, but it is at least as likely that they would, and the latter would have to be assumed. Aside from the desire not to kill members of the indigenous fauna, rifles, even automatic rapid-fire weapons, might not be able to reliably bring down a 5 tonne allosauroid or tyrannosaur, and heavier weapons would be impractical to

Killer dinosaur *Sinosaurus*

carry about. Even medium-sized dinosaurs could pose significant risks. An attack by sickle-clawed dromaeosaurs, for instance, could result in serious casualties.

But there would be another danger that would be as small as it is big. Microbes. Expedition members would be at risk of picking up exotic Mesozoic disease organisms they would not be immune to, and at least as bad would be the danger of contaminating the ancient environment with a host of late Cenozoic viruses, bacteria, and parasites that could seriously disrupt Jurassic and Cretaceous life.

The combined menaces, small and big, would mean that time-traveling dinosaur watchers would probably be banned from directly interacting with the ancient habitats. Instead of walking about under the Mesozoic sun and stars, breathing ancient fresh air, binoculars in hand, they would always have to wear microbe-proof biohazard suits when not in vehicles, and habitats would likewise need to be sealed against microbes getting in or out. Dwelling in dinosaur habitats would be a lot like living on the moon or Mars—a very artificial experience in which paleonauts would be significantly detached from the fascinating world around them, always respiring pretreated air. An advantage of being in biosuits would be temperature control, which would eliminate dealing with the extreme heat prevalent in much of the Mesozoic. Also dealt with would be issues with the composition of the atmosphere. Travel by foot would probably be largely precluded in habitats that included big avepods, sauropods, and ceratopsids. Expedition members would have to move about on the ground in vehicles sufficiently large and strong to be immune from attacks by colossal dinosaurs. Movement away from the vehicles would be possible only when drones showed that the area was safe. Even in places lacking giant dinosaurs, there would be the peril of a biosuit being breached by an assault by a smaller dinosaur—any such penetration from any cause would require some level of medical care, quarantine at least. Defensive weapons might be necessary, although pepper spray guns might suffice. Yet another danger in some Cretaceous habitats would be elephant-sized crocodilians that might snap up and gulp down whole a still-living human unwary enough to go near or in the waters where dinosaurs hung out.

A safe way to observe the prehistoric creatures would be remotely via drones that could observe the winged archosaurs when on the ground and, even better, follow them in the air. Human-carrying ultralights would work too, although they would have their own dangers. Remote static cameras would be as useful for observing Mesozoic wildlife as it is for modern habitats. Also relatively safe would be visiting islands too small to sustain big predators.

A major ethical/scientific issue would be the collection of specimens for investigation. Such would include the comparison of fresh skeleton with the fossils we have collected for purposes of identification. Would killing dinosaurs and other Mesozoic animals be acceptable? Probably not allowed would be bringing organisms of any kind, dead or alive, forward to our times because of the serious biorisks. Artificial intelligence would be an important analysis factor, no doubt.

Whatever the difficulties and dangers, lots of science and results would ensue from Mesozoic expeditions, all the more so if there were many of them dispatched to a wide set of locations over numerous time periods—the more the better. How fast dinosaurs actually were could be directly measured, probably via motion images with radar an option. Were big avepods hunting big plant-eating dinosaurs as well as scavenging them? Were the bigger sauropods immune to attack? How did tyrannosaurids actually battle and bring down ceratopsids? How about the evolution of bird flight, from the trees down or ground up? Their energetics could be measured and thermoregulation observed with sophisticated techniques and laboratory work on site, or after shipping back to the modern world despite the associated issues. How many species of *Tyrannosaurus* there were could be assessed, possibly by superficial anatomy alone if they were visually easy to tell apart or by genetics, which sometimes reveal cryptic species. By genetics and

observations over time the actual intrarelationships of dinosaurs could be sorted out, including how often the protobirds were losing flight and their placement relative to birds. Did theropods have lips? Which had scales versus feathers? How brightly colored were they, or not? Dinosaur scat would be an important source of information just as it is today. How did dinosaur nesting work, including the thermodynamics in their nests? What was the extent of parental care, including among the gigantic examples? Did any of the predaceous dinosaurs, small and large, hunt in packs? Theories would be falsified, others affirmed, and sometimes all would be surprised. While few would actually get to go back on time safaris, there would be endless footage and content for real-life dinosaur documentaries and podcasts set in actual ancient landscapes. Museums, theme parks, and traveling shows would feature full-size replicas static and robotic that faithfully would replicate predaceous dinosaurs, which might displace the dry skeletons currently on display. If any museum specimens remained, they would need to be mounted to be in good accord with the new anatomical knowledge.

A particular issue would be monitoring the events at the terminal Mesozoic, the goal being to see what happened in the wake of the impact and whether the Indian vulcanism played a role. The problem would be the exceptional dangers involved with the shock waves, earthquakes, and pyrosphere experienced everywhere. It would probably be best to set up autonomous devices to experience the effects. For that matter, leaving it all to robots might be the way to go.

A consequence of time travel for paleoartists would be the obsolescence of prior produced life restorations. A possible exception might be the first dinosaur proven to be feathered, little *Sinosauropteryx prima*, its plumage and coloration still being the best understood. Much the same for *Sinornithosaurus zhaoianus*.

IF PREDATORY DINOSAURS HAD SURVIVED

Assume that the K/PG impact is what killed off the dinosaurs, but also assume that the impact did not occur and that nonavian dinosaurs continued into the Cenozoic, perhaps up to today, allowing them to be directly appreciated without time machines—a possibility much more in accord with the physics of our universe. What would the evolution of land animals have been like in that case?

Although much will always be speculative, it is likely that the Age of Dinosaurs would have persisted—indeed the Mesozoic era would have endured—aborting the Cenozoic Age of Mammals. Thirty million years ago western North America probably would have been populated by great dinosaurs rather than the rhino-like titanotheres. Having plateaued out in size for the last half of the Mesozoic, sauropods would probably not have gotten bigger, but the continuation of the ultimate browsers should have inhibited the growth of dense forests. Even so, the flowering

angiosperms would have continued to evolve and to produce a new array of food sources including well-developed fruits that herbivorous dinosaurs would have needed to adapt to in order to exploit.

One thing we can count on is if the big extinction had not wiped out the toothed birds, then the ornithurines would not have been as spectacularly successful as they have been. They would have shared, and might still share, the world with enantiornithines and perhaps toothed euornithines. That would have cut down on the diversity in forms and species of ornithurines, with evolutionary consequences we cannot discern. The oceans might still feature hesperornithiforms rather than penguins. Enantiornithines might have filled roles played by modern birds over much or all of the last 60 million years. Would there be owls, or hawks, or albatross, gulls, pigeons, parrots, crows, and songbirds? We do not know. It is possible that over the eons ornithurines would have proved adaptatively superior and beat out the other birds. Or maybe not.

What is not certain is whether mammals would have remained diminutive or would have begun to compete with dinosaurs for the large-body ecological niches. By the end of the Cretaceous, sophisticated marsupial and placental mammals were appearing, and they may have been able to begin to mount a serious contest with dinosaurs carnivorous and herbivorous as time progressed. Among the herbivorous and omnivorous dinosaurs, therizinosaurs, oviraptors, and ornithomimids should have continued to do well, and it is quite possible that new groups of vegetarian avepods would have come onto the scene. Also continuing would have been the insectivorous alvarezsaurs. Preying on the assorted plant and bug consumers would have been various predaceous avepods. Would the tyrannosaurs have finally expanded their range? How about dramatically new forms of dinopredators evolving? Or new dinofishers? Eventually, southward-migrating Antarctica would have arrived at the South Pole and formed the enormous ice sheets that act as a giant air-conditioning unit for the planet. At the same time, the collision of India and Asia, which closed off the once-great Tethys Ocean and built up the miles-high Tibetan Plateau, also contributed to the great planetary cool-off of the last 20 million years that eventually led to the ongoing ice age despite the rising heat production of the sun. This climate change should have forced the evolution of grazing dinosaurs able to crop the spreading savanna, steppe, and prairie grasslands that thrive in cooler climates. In terms of thermoregulation, dinosaurs should have been able to adapt, although growing winter food shortages may have been a problem for supersauropods. Energetic mammals may have been able to exploit the decreasing temperatures. Perhaps big mammals of strange varieties would have formed a mixed dinosaur–mammal fauna, with the former perhaps including some big birds. If megaavepods continued in force, would mammalian carnivores have become as large? Or would have their live birth system prevented them from exceeding a tonne?

The birdlike dinosaurs evolved brains larger and more complex than those of reptiles toward the end of the Jurassic and beginning of the Cretaceous, but they never exceeded the lower avian range, and they did not exhibit a strong trend toward larger size and intricacy in the Cretaceous similar to the startling increase in neural capacity in Cenozoic mammals. We can only wonder whether predatory dinosaurs would have eventually undergone their own expansion in brain power had they not gone extinct. Perhaps the evolution of large-bodied, big-brained mammals would have compelled dinosaurs to further upgrade thinking performance as well. In such a competition, the better brain efficiency of the bird relatives might have proven an evolutionary advantage, all the more so when combined with full color vision.

The specific species *Homo sapiens* would not have evolved in a timeline that did not include the extinction of dinosaurs, but whether some form of highly intelligent, language- and tool-using animal would have developed is another matter. Modest-sized, bipedal, birdlike predatory theropods with grasping hands might have been able to do so, but the limitation to three fingers may have resulted in inferior manipulation. Or perhaps arboreal theropods with stereo color vision would have become fruit eaters whose evolution paralleled that of the increasingly brainy primates that spawned humans. It is also possible that actual primates would have appeared and evolved above the heads of the great dinosaurs, producing at some point bipedal ground mammals able to create and use tools. On the other hand, the evolution of superintelligent humans may have been a fluke and would not have been repeated in another earthly timeline.

PREDATORY DINOSAUR CONSERVATION

If we take the above scenario to its extreme, assume that some group of smart dinosaurs or mammals managed to survive and thrive in a world of great predatory avepods and became intelligent enough to develop agriculture and civilization as well as an arsenal of lethal weapons. What would have happened to the global fauna?

The fate of large dinosaurs would probably have been grim. We humans may have been the leading factor in the extinction of a large portion of the megafauna that roamed much of the Earth toward the end of the last glacial period, and matters continue to be bad for most wildlife on land and even in the oceans. The desires and practical needs of our imaginary sapients would have compelled them to wipe out the giant avepods, whose low adult populations could have rendered them more susceptible to total loss than the not-as-big mammal carnivores. By the time the sapiens developed industry, the gigantic flesh and plant eaters would probably already have been part of historical lore. If superdinosaurs had instead managed to survive in an industrial world, they would have posed insurmountable problems for zoos. Feeding lions, tigers, and bears is not beyond the means of zoos, but a single tyrannosaur-sized theropod would break the food budget by consuming the equivalent of those thousand cattle-sized animals over a few decades.

And let's not forget the flying dinosaurs, which could include enantiornithines and/or other toothed forms. They would be experiencing the stress common to Holocene birds as it is, with some species in crisis, others under serious pressure, and still others thriving in an industrial world. Vulnerable most of all would be island birds, most of which would be liquidated in short order upon the arrival of seafarers by ancient sail or modern motor.

WHERE PREDATORY DINOSAURS ARE FOUND

Because avepods and the rest of the dinopredators are long gone and time travel probably violates the nature of the universe, we have to be satisfied with finding the remains they left behind. With the possible exception of very high altitudes, dinosaurs lived in all places on all continents, so where they are found is determined by the existence of conditions suitable for preserving their bones and other traces, eggs and footprints especially, as well as by conditions suitable for finding and excavating the fossils. For example, if a dinosaur habitat lacked the conditions that preserved fossils, then that fauna has been totally lost. Or, if the fossils of a given fauna of dinosaurs are currently buried so deep that they are beyond reach, then they are not available for scrutiny. Of dinosaurs, the eaters of meat are the hardest to discover because they were always a minority portion of the general population. Even so, the number of avepods that made up all but a few percent of predatory dinosaurs that lived over time was immense. The populations of giant avepod species living on much of a continent at a given time would have been in the hundreds of thousands, with the majority of that juveniles, most of whom would not make it to full size. With a few megaspecies alive at any time, that would be a global population in the low millions. The smaller avepods would be much more numerous. For example, North American wolves once numbered about two million. The worldly numbers of all avepods at a particular moment would have been in the tens of millions, again most immature. When totaled up over the 170 million years that predatory dinosaurs were around and about, that is well into the trillions—in comparison, the number of humans born so far is circa 100 billion, half getting to at least early adulthood.

All but a very small percentage of the resulting carcasses are destroyed at or soon after death. Many were consumed outright—particularly the juveniles and small adults—or in part by predators and scavengers, and almost all the rested rotted or were weathered away before or after burial at some point. So only a fraction of a percent of those once alive have been preserved. Even so, the number of predatory dinosaur skeletal fossils that still exist on the planet is enormous, in the hundreds of millions or low billions of individuals, although most fragmentary. Of these only a tiny fraction of a percent has been found at or near the small portion of the dinosaur-bearing formations that are exposed on the surface, where the fossils can be accessed, or in the mines that allow some additional remains to be reached. The number of dinosaur fossils that have been scientifically documented to at least some degree is considerable. Some dinosaur bone beds contain the remains of thousands of individuals, and the total number of dinosaur individuals known in that sense is probably in the tens or hundreds of thousands. The question is where to find them.

Much of the surface of the planet at any given time is undergoing erosion. This is especially true of highlands. In erosional areas, sediments that could preserve the bones and other traces are not laid down, so highland faunas are rarely found in the geological record; a notable exception is the Yixian–Jiufotang lake–forest district, deposited in volcanic uplands. Fossilization has the potential to occur in areas in which sediments are being deposited quickly enough, and in large enough quantities, to bury animal remains and traces before they are destroyed by consumption or exposure. Animals can be preserved in deep fissures or caves in highland areas; this is fairly rare but not unknown when dealing with the Mesozoic. Areas undergoing deposition tend to be lowlands downstream of uplifting or volcanic highlands that provide abundant sediment loads, carried in streams, rivers, lakes, or lagoons, that settle out to form beds of silt, sand, or gravel. Therefore, large-scale formation of fossils occurs only in regions experiencing major tectonic and/or volcanic activity. Depositional lowlands can be broad valleys or large basins of varying size in the midst of highlands or can be coastal regions. As a result, most known dinosaur habitats were flatlands, usually riverine floodplains, with little in the way of local topography. In some cases, the eroding neighboring highlands were visible in the distance from the locations where fossilization was occurring; this was especially true in ancient rift valleys and along the margins of large basins. The Yixian–Jiufotang did feature local volcanoes, which generally were not as common as paleoillustrators used to show them. Ashfalls can preserve skeletons, sometimes en masse, but lava flows tend to incinerate and destroy animal remains. In deserts, windblown dunes can preserve bones, as well as dunes that, when wetted by rains, slump. Also suitable for preserving the occasional dinosaur carcass as floating drift are sea and ocean bottoms.

Most sediment deposition occurs during floods, which may also drown animals that are then buried and preserved. The great majority of preserved dinosaurs, however, died before a given flood. The bottoms and edges of bodies of water whether streams, rivers, or lakes are prime preservation locations. In some cases, these watery locations lead to exquisite preservation, including soft tissues. Also good are floodplains, although carcasses not buried before the next seasonal flood results in their degradation by feeding vertebrates and invertebrates and by exposure. An extreme means of fossilization with near-perfect preservation is in tree sap; this can capture only small remains or animal parts. Once burial occurs, the processes that preserve remains are complex and, in many regards, poorly understood. It is being realized, for instance, that bacterial activity is often important in preserving organic bodies. Depending on the circumstances, fossilization can be rapid or very slow, to the point that it never really occurs even after millions of years. The degree of fossilization therefore varies and tends to be more extensive the further back in time the animal was buried. The most extreme fossilization occurs when the original bone is completely replaced by groundwater-borne minerals. Some Australian dinosaur bones have, for instance, been opalized. Most dinosaur bones, however, retain the original calcium structure. The pores have been filled with minerals, converting the bones into rocks much heavier than the living bones. In some locations, such as the Morrison Formation, bacterial activity encouraged the concentration of uranium in many bones—skeletons still below the surface can be detected by their radiation—leading to a radiation risk from stored bones. In other cases, the environment surrounding dinosaur bones has been so stable that little alteration has occurred, leading to the partial retention of some soft tissues near the core of the bones, including a *Tyrannosaurus*.

Although the number of dinosaur bones and trackways that lie in the ground is tremendous, all but a tiny fraction are for practical reasons out of reach. Nearly all are simply buried too deeply. The great majority of fossils that are found are on or within a few feet of the surface. Occasional exceptions include deep excavations, such as construction sites and quarries, or mining operations. Even if deposits loaded with dinosaur fossils occur near the surface, their discovery is difficult if a heavy cover of well-watered vegetation and soil hides the sediments. For example, large tracts of dinosaur-bearing Mesozoic sediments lie on the Eastern Seaboard, running under major cities such as Washington, DC, and Baltimore. But the limited access to the sediments hinders discoveries, which are largely limited to construction sites made available by willing landowners. Of late a dinosaur enthusiast, to the surprise of the local professionals, found abundant prints as well as a squashed baby ankylosaur in suburban Maryland, most in creek beds. Coastal cliffs made up of Mesozoic deposits are another location for dinosaur hunting in forested areas, but they can be dangerous as they erode and collapse.

Prime dinosaur real estate consists of suitable Mesozoic sediments that have been exposed and eroded over large areas that are too arid to support heavy vegetation. This includes shortgrass prairies, badlands, and deserts. There are occasional locations in which dinosaur bone material is so abundant that their remains are easily found with little effort, especially before they have been picked over. Dinosaur Provincial Park in southern Alberta is a well-known example. In some locations countless trackways have been exposed. In most cases dinosaur bones are much less common. Finding dinosaurs has changed little since the 1800s. It normally consists of walking slowly, stooped over, usually under a baking sun, often afflicted by flying insects, looking for telltale traces. If really small remains are being looked for, such as fragmented eggshells, crawling on (padded) knees is necessary. Novices often miss the traces against the background of sediments, but even amateurs soon learn to mentally key in on the characteristics that distinguish fossil remains from rocks. Typically, broken pieces of bones on the surface indicate that a bone or skeleton is eroding out. One hopes that tracing the broken pieces upslope will soon lead to bones that are still in place. In recent years GPS has greatly aided in determining and mapping the position of fossils. Ground-penetrating radar has sometimes been used to better map out the extent of a newly found set of remains, but usually researchers just dig and see what turns up.

Now it becomes a matter of properly excavating and removing the fossil without damaging it while scientifically investigating and recording the nature of the surrounding sediments in order to recover the information they may contain. These basic methods have also not changed much over the years. On occasion thick overburden may be removed by heavy equipment or even explosives. But usually it is a job for jackhammers, sledgehammers, picks, or shovels, depending on the depth and hardness of the sediments and the equipment that can be brought in. When the remains to be recovered are reached, more-careful excavation tools, including trowels, hammers and chisels, picks, and even dental tools and brushes, are used. It is rare to be able to simply brush sand off a well-preserved specimen as in the movies, although this happy circumstance does occur in some ancient dune deposits in Mongolia. Usually sediments are cemented to some degree and require forceful action. At the same time, the bones and other remains are fragile, and care must be taken to avoid damaging them. And their position has to be documented by quarry maps, photography, or laser scanning before removal. Individual bones can be removed, or blocks of sediment including multiple bones or articulated skeletons may have to be taken out intact. Trackways pose their own removal issues—and because there are far more than can be collected, most are left in place where they will be lost to erosion. Again, these operations are usually conducted under conditions that include flying insects, dust, heat, and sun, although tarps can provide shade. In Arctic locations heat is not a problem, but insect swarms are intense during the summer field season. The summer season has been abandoned in some locations because the melted sediments of river cliffs are a serious avalanche danger, so digs are conducted under the extreme Arctic cold.

After exposure, especially fragile bones may be soaked with glue to harden them. On the other hand, the increasingly advanced techniques being applied to bones in the laboratory discourage alteration and contamination of bones. Before removal, most remains are quickly covered with tissue paper that is wetted in place, followed by heavier paper, and over that a thick layer of plaster to form a protective jacket. Wood is usually used to brace the jacket. Once the top is so protected, the remains are undermined and then flipped—a process often requiring considerable exertion and entailing some risk to both excavators and the fossils. Then the other side is papered and plastered, forming a protective cocoon.

Late Triassic (Rhaetian–Norian–Carnian)

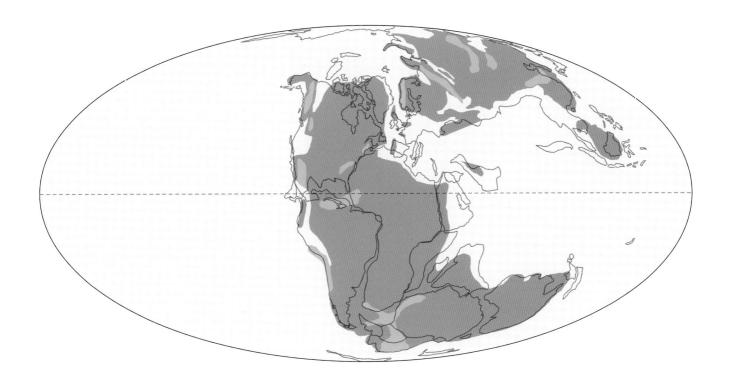

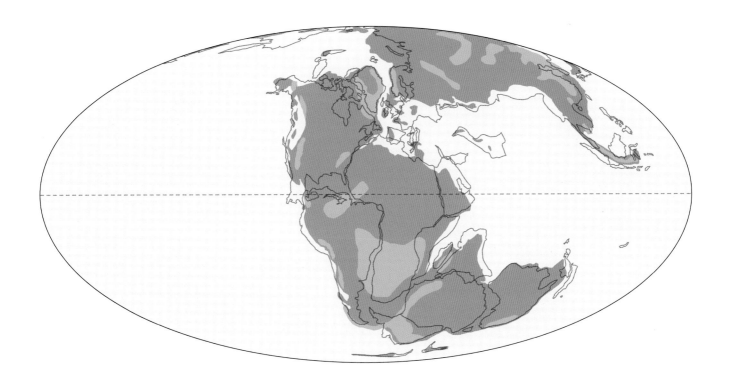

Early Jurassic (Sinemurian)

Middle Jurassic (Callovian)

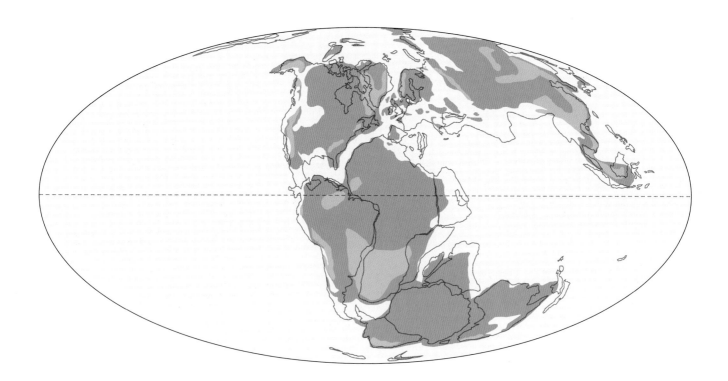

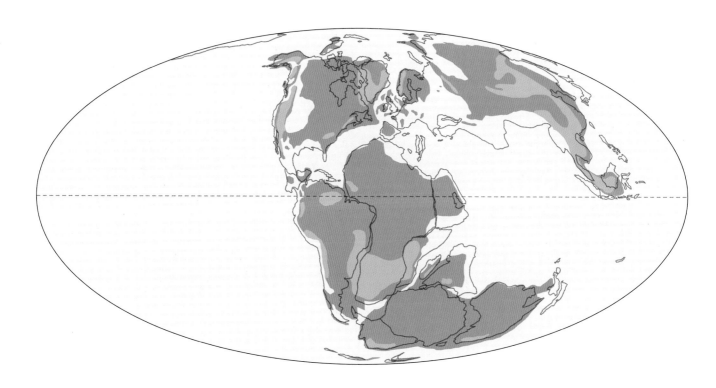

Late Jurassic (Kimmeridgian)

Early Cretaceous (Valanginian–Berriasian)

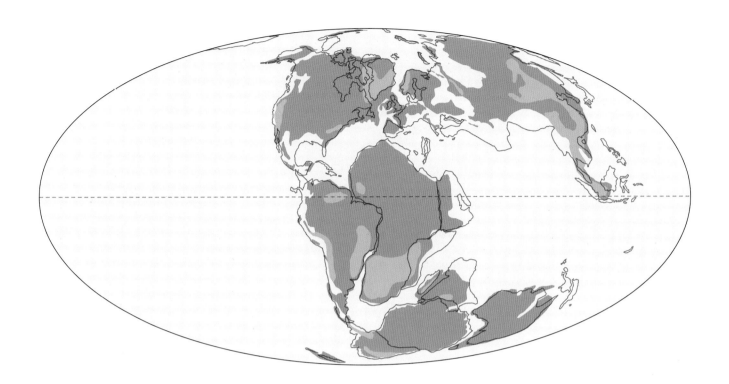

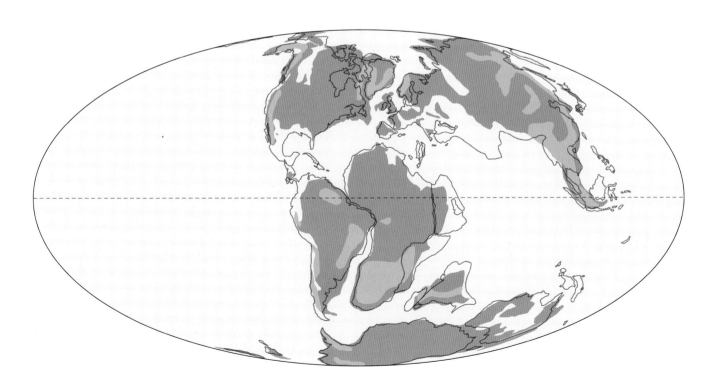

Early Cretaceous (Aptian)

Late Cretaceous (Coniacian)

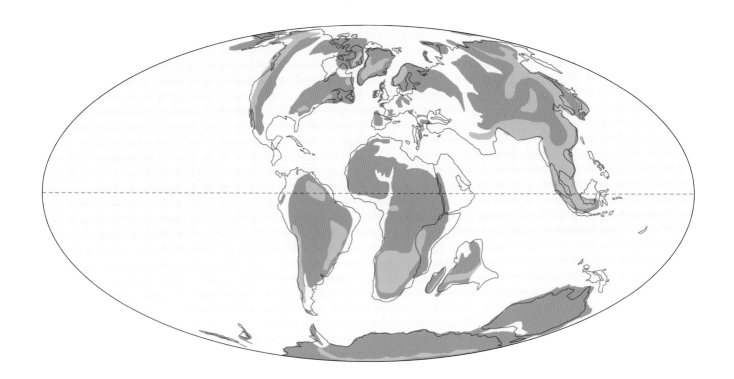

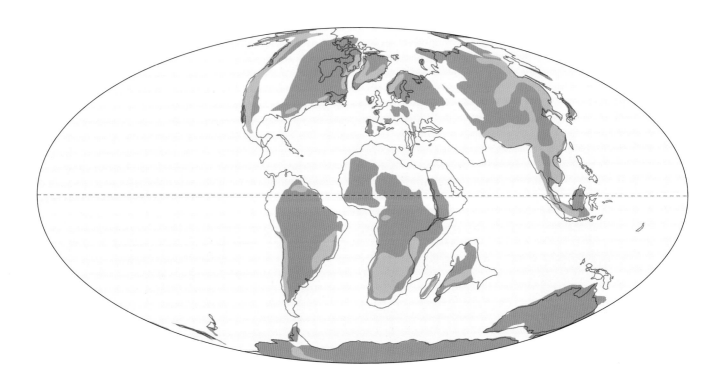

Late Cretaceous (Campanian)

These techniques have changed little in over a century. If the jacketed block is very heavy and not accessible to heavy equipment, a helicopter may be brought in to lift it out. On occasion this requires a heavy-lift helicopter; the US Army is sometimes willing to conduct such operations gratis as part of dissimilar cargo training that allows its crews to learn how to cope with challenging objects rather than standard pallets.

Because dinosaur paleontology is not a high-priority science backed by large financial budgets, and because the number of persons searching for and excavating dinosaurs in the world in a given year is only a few thousand—far more than in the past—the number of dinosaur skeletons that now reside in museums is still just a few thousand. A growing exception is China, where state funding is helping fill warehouses and new museums with material.

In the lab, preparators remove part or all of the jacket and use fine tools to eliminate some or all of the sediment from the bones and any other remains. Most bones are left intact, and only their surface form is documented. In some situations, chemical treatment is required to stabilize bones; this is especially true if the bones are impregnated with pyrite, which gradually swells with moisture. Increasingly, certain bones are opened to reveal their internal structure for various purposes: sectioning is used to examine bone histology and microstructure, to count growth rings, to search for traces of soft tissues, and to sample bone isotopes and proteins. It is becoming the norm to conduct CT scans on skulls and complex bones as a means to determine the three-dimensional structure without invasive preparation, as well as to reduce costs. These scans can be published as conventional hard copies and in digital form. There is increasing reluctance to put original bones in mounted skeletons in display halls because delicate fossils are better conserved when properly curated in the back areas. And mounting heavy, fragile bones poses many difficulties and limitations while removing them from ready study. Instead, the bones are molded, and lightweight casts are used for the display skeleton.

There has never been as much dinosaur-related field activity and research as there is today. At the same time, there is the never-ending shortage of funding and personnel compared with what could otherwise be done. The happy outcome is that there are plenty of opportunities for amateurs to participate in finding and preparing dinosaurs. If nonprofessionals search for fossils on their own, they need to pay attention to laws and to paleontological ethics and practices. In some countries, all dinosaur fossils are regulated by the state—this is true in Canada, for instance. In the United States, fossils found on privately held land are entirely the property of the landowner. Because dinosaur remains in the eastern states usually consist of teeth and other small items, nonhazardous construction sites are often available for exploration on nonwork days. In the West, ranchlands are a primary source of dinosaur remains. Free to dispose of fossils as they wish, ranchers may sell them for profit to private or public collectors. The rising sums of money to be made by selling fossils are making it more difficult for scientific teams to access such lands. That is most true when it concerns the most valuable dinosaur remains, those of dinopredators, especially the big ones—*Tyrannosaurus* most of all. With dinosaur commerce opposed by professional paleontological societies, those considering aiding in such need to take the circumstances into account. Some landowners are not willing to allow dinosaur hunting on their land for various reasons, including theopolitical. Fortunately, dinosaur fossils are a part of western lore and heritage, so many locals favor paleontological activities, which contribute to the tourist trade. All fossils on federal government land are public property and are heavily regulated. Removal can occur only with official permission, which is limited to accredited researchers. Environmental concerns may be involved because dinosaur excavations are in effect small-scale mining operations. Fossils within American Indian reservations may likewise be regulated, and collaboration with resident Native people is indispensable. Dinosaur fossils found by nonprofessionals searching on their own should not be disturbed. Instead, fossils should be reported to qualified experts, who can then properly document and handle the remains. In such cases, the professionals are glad to do so with the assistance of the discoverer.

A number of museums and other institutions offer courses to the public on finding, excavating, and preparing dinosaurs and other fossils. Most expeditions include unpaid volunteers who are trained, often on-site, to provide hands-on assistance to the researchers. Participants are usually expected to pay for their own transportation and general expenses, although food and, in some cases, camping gear as well as equipment may be provided. In order to tap into the growing number of dinosaur enthusiasts, commercial operations led by qualified experts provide a dinosaur-hunting experience for a fee, usually in the western states and Canada. Those searching for and digging up dinosaurs need to take due precautions to protect themselves from sunlight and heat, in terms of UV exposure, dehydration, and hyperthermia, as well as from biting and stinging insects and scorpions. Rattlesnakes are often common in the vicinity of dinosaur fossils. Steep slopes, cliffs, and hidden cavities are potential dangers. In many dinosaur formations, gravel-like caliche deposits that formed in the ancient, semiarid soils create roller-bearing-like surfaces that undermine footing. Flash floods can hit quarries or ill-placed campsites. The use of mechanical and handheld tools when excavating fossils poses risks, as does falling debris from quarry walls. When impact tools are used on hard rock, eye protection may be necessary. Chemicals used while working with fossils require proper handling.

Back in the museums and other facilities, volunteers can be found helping prepare specimens for research and display and cataloging and handling collections. This is important work because, in addition to the constant influx of new specimens with each year's harvest, many dinosaur fossils found as long as a century ago have been sitting on shelves, sometimes still in their original jackets, without being researched.

Landowners who allow researchers onto their land sometimes get a new species found on their property named after them, informally or formally. So do volunteers who find new dinosaurs. Who knows, you may be the next lucky amateur.

USING THE GROUP AND SPECIES DESCRIPTIONS

Over 600 predatory dinosauriform species have been named—about a third of all dinosaurs—but a large portion of these names are invalid. Many are based on inadequate remains, such as teeth or one or a few bones, that are taxonomically indeterminate. Others are junior synonyms for species that have already been named. *Dynamosaurus*, for instance, proved to be the same as the previously named *Tyrannosaurus*, which had been named shortly before, so the former is no longer used. This guide includes mainly those species that are generally considered valid and are based on sufficient remains. A few exceptions are allowed when a species based on a single bone or little more is important in indicating the existence of a distinctive type or group of dinosaurs in a certain time and place. Many of the group and species entries have been changed very little if at all from the first through third editions of the dinosaur guide; corrections and new information have caused a substantial minority to be more heavily revised and, in a few cases, dropped, at the same time that a large array of new species have been added.

The species descriptions are listed hierarchically, starting with major groups and working down the ranking levels to genera and species. Because many researchers have abandoned the traditional Linnaean system of classes, orders, suborders, and families, there is no longer a standard arrangement for the dinosaurs—many dinosaur genera are no longer placed in official families—so none is used here. In general, the taxa are arranged phylogenetically, which presents a number of problems. It is more difficult for the general reader to follow the various groupings. Although there is considerable consensus concerning the broader relationships of some major groups, there is not such for others, and at that and at lower levels the incompleteness of the fossil record hinders a better understanding. The great majority of dinosaur species are not known, many of those that are known are documented by incomplete remains, and it is not possible to examine dinosaur relationships with genetic analysis. Because different cladistic analyses often differ substantially from one another, I have used a degree of personal choice and judgment to arrange the groups and species within the groups. Some of these placements reflect my considered opinion, while others are arbitrary choices between competing research results. Most of the phylogeny and taxonomy offered here is not a formal proposal, but a few new group labels were found necessary and are coined and defined here for future use by others if it proves efficacious. Disputes and alternatives concerning the placement of dinosaur groups and species are often but not always mentioned.

Under the listing for each dinosaur group, the overall geographic distribution and geological time span of its members are noted. This is followed by the anatomical characteristics that apply to the group in general, which are not repeated for each species in the group. The anatomical features usually center on what is recorded in the bones, but other body parts are covered when they have been preserved. The anatomical details are for purposes of general characterization and identification and reflect as much as possible what a dinosaur watcher might use; they are not technical phylogenetic diagnoses. The type of habitat that the group favored is briefly listed, and this varies from specific in some types to very generalized in others. Also outlined are the restored habits that probably characterize the group as a whole. The reliability of these conclusions varies greatly. There is, for example, no doubt that theropods with bladed, serrated teeth consumed flesh rather than plants. There is also little doubt that the sickle-clawed *Velociraptor* regularly attacked the similar-sized herbivore *Protoceratops*—there is even an example of two skeletons still locked in combat. Less certain is exactly how *Velociraptor* used its sickle claws to dispatch prey on a regular basis. It is not known whether *Velociraptor* packs attacked the much larger armored *Pinacosaurus* that lived in the same desert habitat. Some dinosaurs are head-scratchers to varying degrees as far as their habits.

The naming of dinosaur genera and species is often problematic, in part because of a lack of consistency. In some cases, genera are badly oversplit. For example, almost every species of tyrannosaurid receives a generic title despite the high uniformity of the group. On the other hand, the western North American ornithomimids they tried to chase down are in only two genera despite numerous differences. In this work I have attempted to apply more uniform standards to generic and species designations, with the divergences from the conclusions of others noted. Also, a big problem is the inadequate nature of the specimens upon which very many dinosaur taxa are based. Such is true of the original fossils of *Megalosaurus*, *Allosaurus*, *Ornithomimus*, *Struthiomimus*, *Coelophysis*, *Lagosuchus*, *Spinosaurus*, *Nanotyrannus*, *Stygivenator*, *Troodon*, *Stenonychosaurus*, and for that matter *Archaeopteryx*—that starting out as an isolated feather. Appeals to the International Commission on Zoological Nomenclature are helping. But sorting out species, genera, and families is often seriously hindered; these items are noted.

The entry for each species first cites the dimensions and estimated mass of the taxon. The total length (TL) is for the combined skull and skeleton along the line of the vertebrae; any tail feathers that further lengthened the dinosaur are not included. The values are general figures for the size of the largest known adults of the species and do not necessarily apply to the values estimated for specific specimens, most of which can be found via https:// press.princeton.edu/books/hardcover/9780691253169/ the-princeton-field-guide-to-predatory-dinosaurs, which includes the mass estimates for each included specimen. Because the number of specimens for a particular species is a small fraction of those that lived, the largest individuals are not measured; "world record" specimens can be a third or more heavier than is typical—recent estimates that the largest theropod species got to be two-thirds larger are improbable. The sizes of species known only from immature specimens are not estimated. All values are, of course, approximate, and their quality varies depending on the completeness of the remains for a given species. If the species is known from sufficiently complete remains, the dimensions and mass are based on the skeletal restoration. The skeletal restoration is used to estimate the volume of the dinosaur, which can then be used to calculate the mass, with the portion of the volume that was occupied by lungs and any air sacs taken into account. It has been realized of late that animals are generally denser than had been thought. Most land mammals when swimming, for example, are nearly awash and avoid drowning only by swimming, so their density, or specific gravity (SG), is close to 1.0 relative to the density of water. Flying birds with very highly pneumatic skeletons can have low densities, but such values are not applicable to dinosaurs with air sacs because their limb bones are not so air filled, including the early winged forms. The big ratites whose leg bones are hollow float a little higher in water than mammals, suggesting SGs of 0.92. Because pneumatic theropods lacked airy leg bones, their SG was probably between that and 0.96, with large examples being denser. When remains are too incomplete to estimate dimensions and mass directly, these are extrapolated from those of relatives and are considerably more approximate. Both metric and English measurements are included except for metric tonnes, which equal 1.1 English tons; all original calculations are metric, but because they are often imprecise, the conversions from metric to English are often rounded off as well.

The next entry outlines the fossil remains, whether they are skull or skeletal material or both, that can be confidently assigned to the species to date; the number of specimens varies from one to thousands. The accuracy of the list ranges from exact to a generalization. The latter sometimes results from recent reassignment of specimens from one species to another, leaving the precise inventory uncertain. Skeletal and/or skull restorations have been rendered for those species that are known from sufficiently complete remains that have been adequately documented at the time the book was being finalized to execute a reconstruction—or for species that are of such interest that a seriously incomplete restoration is justified (the rather poorly known but oversized oviraptorosaur *Gigantoraptor* being an example). A number of species known from good remains have yet to be made available for research, sometimes decades after their discovery. In some cases, only oblique-view photographs unsuitable for a restoration of reasonably complete skulls and skeletons are obtainable. The pace of discovery is so fast that a few new finds could not be included. A few complete skeletons are so damaged or distorted that a restoration is not feasible. Of special difficulty is *Spinosaurus*, restorations being a questionable amalgam of incomplete and disarticulated fossils from widely differing locations rather than the more complete specimens needed to establish the highly atypical proportions that have been proposed. Only when at least some skull material is on hand is a skeletal included, the exception being a number of the small-headed therizinosaurs, as skulls are scarce in that group. When enough examples of some major dinosaur types are not sufficient, the remains of multiple species have been used to construct a composite representation, such as for the derived alvarezsaurid and therizinosaurid. Despite the absences, this is by far the most extensive predatory skull–skeletal library yet published in print. The core skeletal specimens that have been restored can be found at the above link. The restorations show the bones as solid white set within the solid black restored muscle and keratin profiles; cartilage is not included. In many cases only the skull is complete enough for illustration. When more than one skeleton and/or skull of a given species is included, they are always to the same scale to facilitate size comparisons. In most cases the skull and skeletal restorations are of adults or close to it, but growth series are included for all examples that are achievable, on the premise that young dinosaurs soon left their nests and were part of the general fauna. In some examples the skeletons of the smallest juveniles are reproduced both to scale to the much larger adults and at a larger scale to allow details to be better seen—this guide contains the largest set to date of dinosaur growth stages. The accuracy of the restorations ranges from very good for those that are known from extensive remains that have been described in detail and/or have good photographs on hand, down to approximate if much of the species remains are missing or have not been well illustrated. The restorations have been prepared over four decades to fulfill differing requirements. A number of skeletons and skulls show only those bones that are known—sometimes with replications of one side to the other, ranging from a large fraction to nearly all—whereas others have been filled out to represent a complete skeleton. Reliable information about exactly which bones have and have not been preserved is often not available, so the widely used term "rigorous restoration" of an incomplete skeletal is best avoided in favor of "known bone." Top views of skeletons have been provided when such

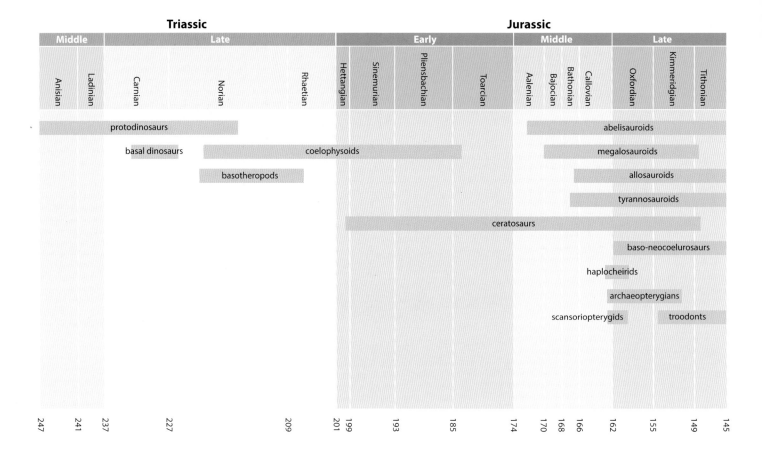

were doable, but in many cases lateral crushing prevents that. A few examples of fore-and-aft views are included. The skeletons are posed in a common basic posture, with the right hind leg pushing off at the end of its propulsive stroke, in order to facilitate cross comparisons. Representative examples of shaded skull restorations have been included with some of the major groups. The same has been done with a sample of muscle studies, whose detailed nature is no less or more realistic than are the particulars in the full-life restorations, which, if anything, involve additional layers of speculation.

The color plates are based on species for which the fully or nearly adult skeletal restorations or skulls are deemed of sufficient quality for a full-life restoration. If it is unlikely that new information will significantly alter the skeletal plan in the near future, a life restoration has been prepared; if the skeletal is not sufficiently reliable, a life restoration has not been done. A few of the earlier life restorations have been replaced because of updates of their skeletal restorations. Most of the skull as well as the skeleton needs to be present to justify a life restoration of an entire animal—although *Majungasaurus* bones are exquisitely preserved, there are too many uncertainties about its unusual proportions to warrant a life drawing. In a few

cases color restorations were executed despite significant questions about the skeletal study because the species is particularly important or interesting for one reason or another, the spectacularly feathered but incomplete *Beipiaosaurus* being an example. In a few cases only the skull is good enough to warrant a life restoration to the exclusion of the overall body. The colors and patterns are entirely speculative except in the few cases in which feather coloration has been restored—there is no consistent effort to coordinate speculative color patterns for a species between the life profiles and full-color scenes. Extremely vibrant color patterns have generally not been used to avoid giving the impression that they are identifying features. Those who wish to use the skeletal, muscle, and life restorations herein as the basis for commercial and other public projects are reminded to first contact the copyright illustrator.

The particular anatomical characteristics that distinguish the species are listed, but these too are for purposes of general identification by putative dinosaur watchers, not for technical species diagnoses. These differ in extent depending on the degree of uniformity versus diversity in a given group as well as the completeness of the available fossil remains. In some cases, the features of the species are not different enough from those of the group to

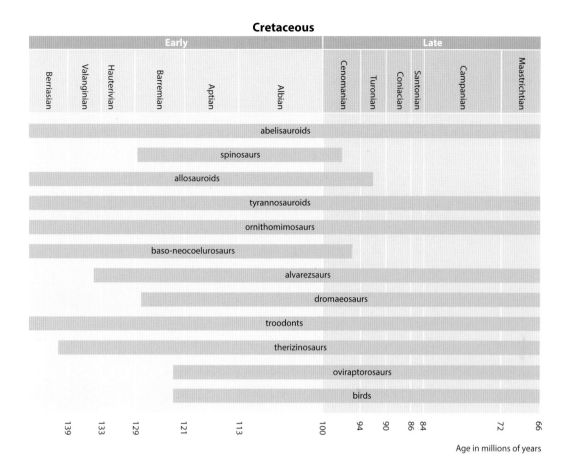

warrant additional description. In other cases, not enough is known to make a separate description possible.

Listed next is the formal geological time period and, when available, the specific stage from which the species is known and the segment of that stage from early to middle to late. As discussed earlier, the age of a given species is known with a precision of within a million years in some cases or as poorly as an entire period in others. The reader can refer to the time scale on the timeline chart (above) to determine the age, or age range, of the species in years (see pp. 98–99). Because most species exist for only a few hundred thousand years before either being replaced by a descendant species or going entirely extinct, when fossils within a single genus are found in sediments that span more than that amount of time, it is presumed they are different species. In some cases, it is not entirely clear whether a species was present in just one time stage or crossed the boundary into the next one. In those cases, the listing includes "and/or," such as late Santonian and/or early Campanian.

Next the geographic location and geological formation from which the species is so far known are listed. The paleomaps of coastlines (see pp. 91–94) can be used to place a species geographically in a world of drifting continents and fast-shifting seaways, with the proviso that no set of maps is extensive enough to show the exact configuration of the ancient lands when each species was extant. Some dinosaur species are known from only a single location and formation, whereas others have been found in an area spread over one or more formations. In the latter situation the first location/formation listed is for the original specimen. I have tended to be conservative in listing the presence of a specific species only in those places and levels where sufficiently complete remains are present. Some formations have not been named, even in areas that are well studied. Many formations were formed over a time span that was longer than that of some or all of the species that lived within them, so when possible the common procedure of simply listing just the overall formation a given species is from is avoided. For example, it has been assumed that one species of *Triceratops* and *Tyrannosaurus* lived during the ~1.5 million years it took to deposit the Hell Creek, Lance, and other formations, but the evidence favors species evolution in both over this period. Because the Morrison Formation was deposited over a span of six million years in the Late Jurassic, there was change over time in the species of allosaurs and ceratosaurs. Because the overlaying Cedar Mountain Formation formed over 40 million years, the fauna underwent radical changes. I have therefore listed the level of the formation that each species comes from, when this information is available. The reader can get an impression of what dinosaur species constituted a given fauna in a particular bed of sediments by using the formation index. Only species present at a given level of a formation could have lived together, but this does not always mean a given set of dinosaurs listed as being from a particular level did so because time separations sometimes exist. In a few cases the sediments a dinosaur is from are not yet named—often but not always because they have not been studied—and the geological group may be named instead, if one is available.

Gallimimus flock

Noted next are the basic characteristics of the dinosaur's habitat in terms of rainfall and vegetation, as well as temperature when it is not generally tropical or subtropical year-round. Environmental information ranges from well studied in heavily researched formations to nonexistent in others. If the habits of the species are thought to include attributes not seen in the group as a whole, then they are outlined. Listed last are special notes about the species when they are called for. Possible ancestor–descendant relationships with close older or younger relatives are sometimes noted, but these are always tentative. This section is also used to note alternative hypotheses and controversies that apply to the groups and species.

Carnotaurus sastrei

PREDATORY DINOSAURIFORMES

SMALL TO GIGANTIC DINOSAURS FROM THE MIDDLE TRIASSIC TO THE MODERN ERA, ALL CONTINENTS.

ANATOMICAL CHARACTERISTICS Heads not massively constructed, skull bones usually somewhat loosely attached to one another, teeth from large serrated blades to absent. Erect leg posture achieved by sub or fully cylindrical femoral head fitting into a deep hip socket and a simple hinge-jointed ankle. Hind-limb dominant in that legs are sole locomotory organs in walking and running and/or are more strongly built than the arms even when are long-armed quadrupeds. Legs flexed at all sizes, long. Hands and feet digitigrade with wrist and ankle held clear of ground. Trackways show that when quadrupedal, hands always at least as far apart from midline as feet or farther; never hopped, and tail almost always held clear of ground. Body scales, when known and present, usually form a nonoverlapping mosaic pattern.

ENERGETICS Thermophysiology intermediate to very high energy.

REPRODUCTION AND ONTOGENY Usually reached sexual maturity while still growing, most were rapid breeders, many laid hard-shelled eggs in pairs, others may have laid soft-shelled eggs, growth rates moderate to rapid.

HABITATS AND HABITS Very diverse. Usually strongly terrestrial from sea level to highlands, from tropics to polar winters, from arid to wet, almost all able to swim, some able to fly. Diets ranged from classic hunting with opportunistic scavenging to omnivorous to full herbivory. Small adults and juveniles with long arm and hook-clawed fingers able to climb. Enormous numbers of trackways laid down along watercourses show that many predatory dinosauriformes of all sizes spent considerable time patrolling shorelines and using them to travel.

PREDATORY BASODINOSAURIFORMES

SMALL PREDATORY DINOSAUROMORPHS, LIMITED TO THE LATE TRIASSIC OF THE AMERICAS, EUROPE, AND AFRICA.

ANATOMICAL CHARACTERISTICS Fairly diverse. Small, lightly built. Teeth blades to leaf like. Necks not elongated. Trunks shallow. Light, midline dorsal armor in at least predatory examples. More erect leg posture achieved by subcylindrical femoral head fitting into a deep hip socket with a small internal opening. Legs long. Predominantly bipedal to quadrupedal, probably able to gallop.

ENERGETICS Thermophysiology probably intermediate, energy levels and food consumption probably low compared with more derived predatory dinosaurs.

HABITS Predatory to herbivorous.

NOTES Basodinosauriformes are dinosauriformes excluding dinosaurs, the protodinosaurs. Existence in Middle Triassic probable but not yet documented by skeletal remains. Only predaceous examples are included below; herbivorous examples include *Asilisaurus*, *Diodorus*, *Eucoelophysis*, *Sacisaurus*, and *Silesaurus*. Absence from other continents may reflect insufficient sampling.

Lagosuchus talampayensis

—— 0.58 m (1.9 ft) TL, 0.2 kg (0.5 lb)

FOSSIL REMAINS Minority of skull and skeletons, some juvenile.

ANATOMICAL CHARACTERISTICS Head medium sized. Arm much shorter than leg, predominantly bipedal.

AGE Late Triassic, early Carnian.

DISTRIBUTION AND FORMATION/S Northwestern Argentina; middle Chanares.

HABITS Small game ambush and pursuit hunter.

NOTES *Marasuchus lilloensis* is probably the adult of this species. Competes with *Ambopteryx, Anchiornis, Caihong, Epidexipteryx, Scansoriopteryx,* and *Serikornis* for title of smallest known nonavian dinosauriform.

Lewisuchus admixtus

—— 1.25 m (4 ft) TL, 2.5 kg (5 lb)

FOSSIL REMAINS Three partial skeletons.

ANATOMICAL CHARACTERISTICS Head large, elongated. Second toe and claw longer than central toe. Arm markedly shorter than leg, predominantly bipedal.

AGE Late Triassic, early Carnian.

DISTRIBUTION AND FORMATION/S Northwestern Argentina; middle Chanares.

HABITS Small-medium game ambush and pursuit hunter.

NOTES Probably includes *Pseudolagosuchus major.* May be a silesaurid, which includes herbivorous examples.

Lewisuchus admixtus

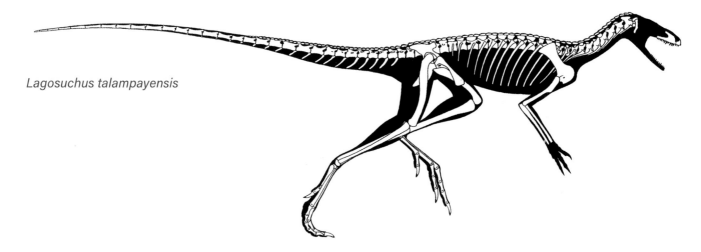

Lagosuchus talampayensis

Lewisuchus admixtus

Tyrannosaurus imperator

PREDATORY DINOSAURS

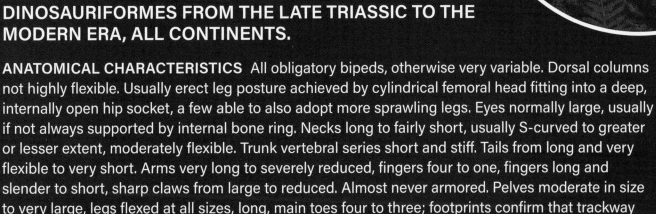

SMALL TO GIGANTIC PREDATORY TO HERBIVOROUS DINOSAURIFORMES FROM THE LATE TRIASSIC TO THE MODERN ERA, ALL CONTINENTS.

ANATOMICAL CHARACTERISTICS All obligatory bipeds, otherwise very variable. Dorsal columns not highly flexible. Usually erect leg posture achieved by cylindrical femoral head fitting into a deep, internally open hip socket, a few able to also adopt more sprawling legs. Eyes normally large, usually if not always supported by internal bone ring. Necks long to fairly short, usually S-curved to greater or lesser extent, moderately flexible. Trunk vertebral series short and stiff. Tails from long and very flexible to very short. Arms very long to severely reduced, fingers four to one, fingers long and slender to short, sharp claws from large to reduced. Almost never armored. Pelves moderate in size to very large, legs flexed at all sizes, long, main toes four to three; footprints confirm that trackway gauge was very narrow. Brains vary from reptilian in size and form to those of birds.

ENERGETICS Thermophysiology moderate to very high energy.

HABITATS Very diverse, from sea level to highlands, from tropics to polar winters, from arid to wet.

HABITS Diets ranged from classic hunting with opportunistic scavenging to full herbivory. Small and juvenile theropods with long arm and hook-clawed fingers were probably able to climb. Enormous numbers of trackways laid down along watercourses show that many theropods of all sizes spent considerable time patrolling shorelines and using them to travel.

NOTES The only dinosaur group that includes archpredators. Already somewhat birdlike at the beginning, generally became increasingly so with time, especially among some advanced groups that include the direct ancestors of birds.

BASAL PREDATORY DINOSAURS
SMALL TO MODERATELY LARGE PREDATORY AND OMNIVOROUS DINOSAURS, LIMITED TO THE LATE TRIASSIC OF THE SOUTHERN HEMISPHERE.

ANATOMICAL CHARACTERISTICS Fairly uniform. Generally lightly built. Heads moderately large, usually long and shallow, subrectangular, fairly robustly constructed, fairly narrow, teeth usually serrated blades, sometimes not serrated and more leaf shaped. Necks moderately long, only gently S-curved. Tails long. Arms and four-fingered hands moderately long, claws well developed. Pelves short but deep, pubes procumbent to a little retroverted. Four load-bearing toes. Beginnings of birdlike respiratory system possibly present. Brains reptilian.
HABITS Pursuit predators, some also omnivorous. Head and arms primary weapons. Jaws and teeth probably delivered slashing wounds to disable muscles and cause bleeding, shock, and infection. Arms used to hold on to and control prey, possibly delivered slashing wounds.

ENERGETICS Thermophysiology probably moderate, energy levels and food consumption probably lower than most or all more derived predatory dinosaurs.

NOTES The most basal dinosaurs, their similarly generalized forms make it difficult to determine relationships with one another and major dinosaur groups, having variously been considered basal theropods, sauropodomorphs, saurischians, ornithoscelids, and paxdinosaurs. These briefly existing early dinosaurs were apparently not able to compete with more sophisticated examples. Absence from at least some continents probably reflects lack of sufficient sampling.

Alwalkeria maleriensis
— 1.5 m (5 ft) TL, 4 kg (8 lb)
FOSSIL REMAINS Minority of skull and skeleton.
ANATOMICAL CHARACTERISTICS Back teeth bladed, front teeth less so, no teeth serrated.
AGE Late Triassic, middle or late Carnian.
DISTRIBUTION AND FORMATION/S Southeastern India; Lower Maleri.
HABITS Probably omnivorous, hunted smaller game and consumed some easily digested plant material.
NOTES May be a basal prosauropod.

Chindesaurus bryansmalli
— 2.5 m (8 ft) TL, 15 kg (30 lb)
FOSSIL REMAINS Fragmentary skeleton, isolated bones.
ANATOMICAL CHARACTERISTICS Insufficient information.
AGE Late Triassic, middle and/or late Norian.
DISTRIBUTION AND FORMATION/S Arizona, New Mexico? Texas?; middle Chinle, Bull Canyon? Tecovas?
HABITAT Well-watered forests, including dense stands of giant conifers.
NOTES It is uncertain whether the remains outside original Chinle specimen belong to one taxon; remains from the Tecovas may belong.

Caseosaurus? crosbyensis
— 2 m (6 ft) TL, 10 kg (20 lb)
FOSSIL REMAINS Fragmentary remains.

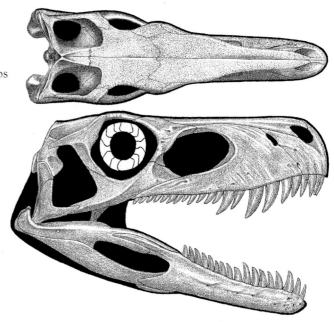

Herrerasaurus shaded skull

ANATOMICAL CHARACTERISTICS Insufficient information.
AGE Late Triassic, Norian.
DISTRIBUTION AND FORMATION/S Arizona, Texas; Tecovas.
NOTES Relationships uncertain.

Sanjuansaurus gordilloi
— 3 m (9 ft) TL, 35 kg (80 lb)
FOSSIL REMAINS Minority of skull and partial skeleton.
ANATOMICAL CHARACTERISTICS Teeth bladed and serrated. Pubis procumbent.
AGE Late Triassic, late Carnian.
DISTRIBUTION AND FORMATION/S Northern Argentina; lower Ischigualasto.
HABITAT Seasonally well-watered forests, including dense stands of giant conifers.

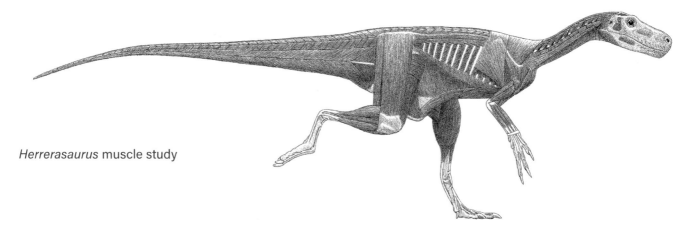

Herrerasaurus muscle study

Staurikosaurus pricei

—— 2.1 m (7 ft) TL, 13 kg (28 lb)

FOSSIL REMAINS Minority of skull and majority of skeleton.

ANATOMICAL CHARACTERISTICS Pubis nearly vertical.

AGE Late Triassic, middle Carnian.

DISTRIBUTION AND FORMATION/S Southeastern Brazil; upper Santa Maria.

NOTES Probably includes *Teyuwasu barberenai*. Along with *Herrerasaurs cabreirai* may be earliest known dinosaurs.

Herrerasaurus (Gnathovorax) cabreirai

—— 3 m (9.5 ft) TL, 60 kg (130 lb)

FOSSIL REMAINS Nearly complete skull and skeleton.

ANATOMICAL CHARACTERISTICS Pubis a little retroverted.

AGE Late Triassic, middle Carnian.

DISTRIBUTION AND FORMATION/S Southeastern Brazil; middle Santa Maria.

HABITS Large game hunter.

NOTES Reliable restoration of incompletely described skeleton not yet possible. May be an ancestor of *H.ischigualastensis*.

Herrerasaurus ischigualastensis

—— 4.5 m (15 ft) TL, 220 kg (450 lb)

FOSSIL REMAINS Two complete skulls and several partial skeletons.

ANATOMICAL CHARACTERISTICS Pubis a little retroverted.

AGE Late Triassic, late Carnian.

DISTRIBUTION AND FORMATION/S Northern Argentina; lower Ischigualasto.

HABITAT Seasonally well-watered forests, including dense stands of giant conifers.

HABITS Large game hunter.

NOTES The classic archaic predaceous dinosaur, includes *Frenguellisaurus ischigualastensis* and *Ischisaurus cattoi*.

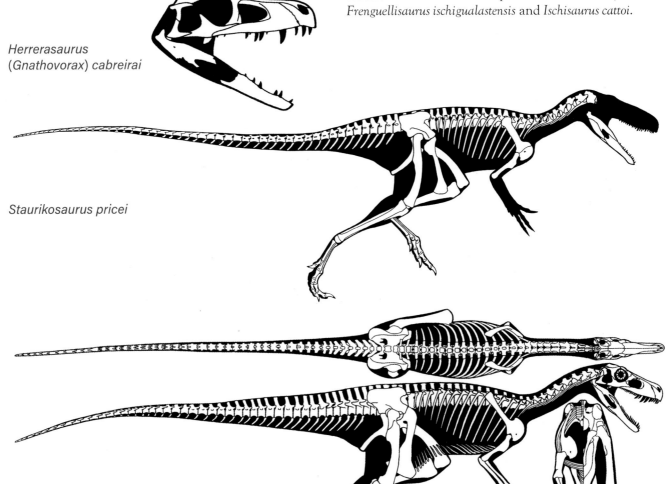

Herrerasaurus
(*Gnathovorax*) *cabreirai*

Staurikosaurus pricei

Herrerasaurus ischigualastensis

Herrerasaurus ischigualastensis

Eodromaeus murphi

— 1.8 m (6 ft) TL, 5.5 kg (12 lb)

FOSSIL REMAINS Majority of skull and skeleton.
ANATOMICAL CHARACTERISTICS Somewhat robust.
Pubis a little procumbent.

AGE Late Triassic, late Carnian and/or earliest Norian.
DISTRIBUTION AND FORMATION/S Northern Argentina;
lower Ischigualasto.
HABITAT Seasonally well-watered forests, including dense
stands of giant conifers.

Eodromaeus murphi

THEROPODS

**SMALL TO GIGANTIC PREDATORY TO HERBIVOROUS DINOSAURS, FROM THE LATE TRIASSIC
TO THE MODERN ERA, ALL CONTINENTS.**

ANATOMICAL CHARACTERISTICS Usually extra joint at middle of lower jaw. Necks long to fairly short, usually
S-curved to greater or lesser extent, moderately flexible. Tails from long and very flexible to very short. Arms very
long to severely reduced, fingers four to one, fingers long and slender to short, sharp claws from large to absent. Pelves
moderate in size to very large, main toes four to three. Brains vary from reptilian in size and form to those of birds.
ENERGETICS Thermophysiology moderate to very high energy.
NOTES Already somewhat birdlike at the beginning, generally became increasingly so with time, especially among
some advanced groups that include the direct ancestors of birds.

BASOTHEROPODS

SMALL TO MODERATELY LARGE THEROPODS, LIMITED TO THE LATE TRIASSIC.

ANATOMICAL CHARACTERISTICS Lightly built. Heads moderately large, fairly narrow, teeth serrated blades. Necks
moderately long, only gently S-curved. Tails long. Arms and four-fingered hands moderately long, claws well
developed. Pelves short but deep, four load-bearing toes. Beginnings of birdlike respiratory system possibly present.
Brains reptilian.
HABITS Pursuit predators. Head and arms primary weapons. Jaws and teeth delivered slashing wounds to disable
muscles and cause bleeding, shock, and infection. Arms used to hold on to and control prey, possibly delivered
slashing wounds.
ENERGETICS Thermophysiology probably moderate, energy levels and food consumption probably lower than most
more derived dinosaurs.
NOTES Basotheropods are basal theropods excluding avepods. May have included some of the above basal dinosaurs.
The most basal theropods, these briefly existing early theropods, were apparently not able to compete with the more
sophisticated avepods. Absence from at least some continents probably reflects lack of sufficient sampling.

Daemonosaurus chauliodus

— Adult size uncertain

FOSSIL REMAINS Nearly complete skull and minority of
skeleton.

ANATOMICAL CHARACTERISTICS Head deep,
subtriangular, snout short, eyes large, upper teeth very
large, those in front procumbent.
AGE Late Triassic, Rhaetian.

DISTRIBUTION AND FORMATION/S New Mexico; probably upper Chinle.
HABITS Probably caught fish as well as land prey.

NOTES Ontogenetic stage of fossil uncertain, so how much unusual form is juvenile or distinctive is also uncertain.

Daemonosaurus chauliodus

Tawa hallae
—— 2.5 m (6 ft) TL, 16 kg (35 lb)
FOSSIL REMAINS Nearly complete skull and majority of a few skeletons, almost completely known.
ANATOMICAL CHARACTERISTICS Snout pointed, upper teeth large. Unfused narrow sternal plates present. Pubis vertical, leg slender.

AGE Late Triassic, middle Norian.
DISTRIBUTION AND FORMATION/S New Mexico; upper Chinle.
HABITAT Seasonally well-watered forests, including dense stands of giant conifers.
HABITS Probably fastest of known basal dinosaurs.

Tawa hallae

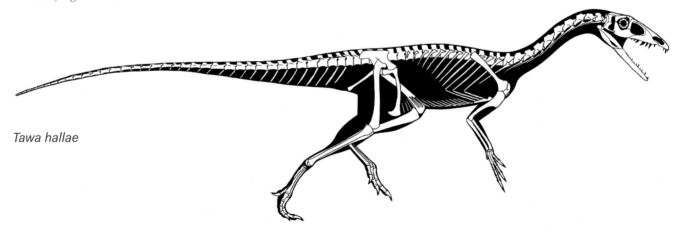

Tawa hallae

AVEPODS

SMALL TO GIGANTIC, THREE-TOED PREDATORY AND HERBIVOROUS THEROPODS, MOST PREDATORS, FROM THE LATE TRIASSIC TO THE MODERN ERA, ALL CONTINENTS.

ANATOMICAL CHARACTERISTICS Highly variable. Head size and shape variable, teeth large and bladed to absent. Necks long to fairly short. Trunks short, stiff. Tails long to very short. Fused furculas often present, arm very long to severely reduced, fingers four to one, usually three, fingers long and slender to short, claws large to reduced. Pelves large, legs long, usually three main toes, inner toe a short hallux, sometimes four load-bearing toes, or two. Skeletons pneumatic, birdlike, air sac–ventilated respiratory system developing. Brains vary from reptilian in size and form to those of birds.
ENERGETICS Energy levels and food consumption probably similar to those of ratites up until other birds.
HABITS Diets ranged from classic hunting in most to full herbivory in some specialized groups. Adept surface swimmers, so able to cross large bodies of water; hard pressed or unable to completely submerge because of pneumatic bodies.
NOTES Theropods lacking contact of metatarsal 1 with ankle so foot is at least nominally tridactyl, or ancestors with same that are in the clade that includes extant birds; has the advantage of encompassing both neotheropods and all taxa more basal than the latter in the clade with the feature. Distinctly birdlike from the start, tridactyl foot most of all.

— PODOKESAUROIDS

SMALL TO LARGE AVEPODS, LIMITED TO THE LATE TRIASSIC AND EARLY JURASSIC, NORTHERN HEMISPHERE AND AFRICA.

ANATOMICAL CHARACTERISTICS Fairly uniform. Generally lightly built. Heads long, snout pointed, narrow, indentation at front of upper jaw often present, lightly constructed paired crests over snout often present, teeth bladed and serrated. Necks long. Trunks not deep. Tails quite long, slender. Arms moderately long, fingers moderately long, claws modest in size. Pelvis moderately large. Brains reptilian. Skeletal pneumaticity partly developed, so birdlike respiratory system developing.

Coelophysis muscle study

ONTOGENY Growth rates moderate.

HABITS Although predominantly fast-pursuit predators, snaggy teeth at tip of kinked upper jaw indicate these were also fishers. Crests when present too delicate for headbutting; for lateral visual display within the species, may or may not have been brightly colored at least during breeding season.

NOTES Podokesauroids are basal avepods excluding averostrans. Group splittable into a number of subdivisions including coelophysoids, which are a smaller group; many divisions and placements of species are uncertain. The most basal avepods, first large examples show avepod theropods reached considerable size as early as the Triassic. All Triassic avepod skeletal remains are podokesauroids, but some footprints suggest more derived groups may have been present by then. Absence from at least some continents probably reflects lack of sufficient sampling.

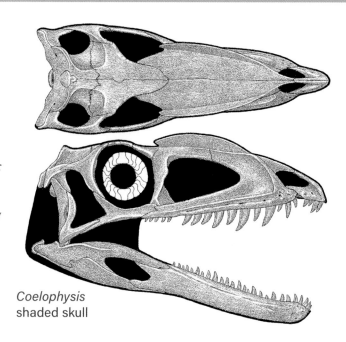

Coelophysis
shaded skull

Dracoraptor hanigani

—— 1.8 m (6 ft) TL, 4 kg (10 lb)

FOSSIL REMAINS Partial skull and minority of skeleton.

ANATOMICAL CHARACTERISTICS Apparently standard for small podokesauroids.

AGE Early Jurassic, earliest Hettangian.

DISTRIBUTION AND FORMATION/S Wales; middle Blue Lias.

NOTES Appears to be the most basal member of the group, probably coelophysoid. Found as drift in nearshore interisland marine sediments. Presence of snout crests uncertain.

Powellvenator podocitus

—— 1.1 m (3.5 ft) TL, 1 kg (2 lb)

FOSSIL REMAINS Fragmentary remains.

ANATOMICAL CHARACTERISTICS Insufficient information.

AGE Late Triassic, middle Norian.

DISTRIBUTION AND FORMATION/S Northern Argentina: upper Los Colorados

Procompsognathus triassicus

—— 1.1 m (3.5 ft) TL, 1 kg (2 lb)

FOSSIL REMAINS Poorly preserved partial skeleton with possible skull.

ANATOMICAL CHARACTERISTICS Apparently standard for small podokesauroids.

AGE Late Triassic, middle Norian.

DISTRIBUTION AND FORMATION/S Southern Germany; middle Lowenstein.

NOTES Remains that belong to this taxon not entirely certain. Smallest known member of the group. Name incorrectly implies an ancestral relationship with the very different Compsognathus. Probable coelophysid. Presence of snout crests uncertain.

Lucianovenator bonoi

—— 4 m (13 ft) TL, 50 kg (100 lb)

FOSSIL REMAINS Minority of four skeletons.

ANATOMICAL CHARACTERISTICS Apparently standard for small podokesauroids.

AGE Late Triassic, late Norian or Rhaetian.

DISTRIBUTION AND FORMATION/S Northwestern Argentina; upper Quebrada del Barro.

HABITAT Semiarid.

NOTES Probable coelophysid. Presence of snout crests uncertain.

Pendraig milnerae

—— Adult size uncertain

FOSSIL REMAINS Fragmentary skeletons, immature.

ANATOMICAL CHARACTERISTICS Insufficient information.

AGE Probably Late Triassic, Rhaetian.

DISTRIBUTION AND FORMATION/S Wales; Magnesian Conglomerate?

HABITAT Forested island.

NOTES Found in ancient fissure fills. May be example of island dwarfism. Presence of snout crests uncertain.

Coelophysis or Camposaurus arizonensis

—— 2 m (6 ft) TL, 12 kg (25 lb)

FOSSIL REMAINS Fragmentary skeleton.

ANATOMICAL CHARACTERISTICS Insufficient information.

AGE Late Triassic, early or middle Norian.

DISTRIBUTION AND FORMATION/S New Mexico; lower Chinle.

NOTES Earliest known avepod. Presence of snout crests uncertain.

Coelophysis bauri

— 3 m (10 ft) TL, 25 kg (50 lb)

FOSSIL REMAINS Hundreds of skulls and skeletons, many complete, juvenile to adult, completely known.

ANATOMICAL CHARACTERISTICS Very lightly built and gracile, overall very long trunked. Head long and shallow, bite not powerful, snout crests absent, teeth numerous and small. Neck long and slender.

AGE Late Triassic, Rhaetian.

DISTRIBUTION AND FORMATION/S New Mexico; uppermost Chinle.

HABITS Predominantly small game hunter but may have occasionally attacked larger prosauropods and herbivorous thecodonts.

NOTES The classic early avepod theropod. In accord with a decision of the committee that handles taxonomic issues, the fossil on which the taxon is based was shifted from inadequate remains in the Chinle to a complete fossil from the famous Ghost Ranch Quarry. A very small element may be fifth metacarpal of hand, or a lateral distal carpal. How hundreds of skeletons came to be concentrated in the quarry remains unsettled.

Coelophysis rhodesiensis

— 2.2 m (7 ft) TL, 14 kg (30 lb)

FOSSIL REMAINS Hundreds of skulls and skeletons, juvenile to adult, completely known.

ANATOMICAL CHARACTERISTICS Same as *C. bauri* except leg longer relative to body.

AGE Early Jurassic, Hettangian.

DISTRIBUTION AND FORMATION/S Zimbabwe; Forest Sandstone.

HABITAT Desert with dunes and oases.

HABITS Same as *C. bauri*, except thecodonts not present.

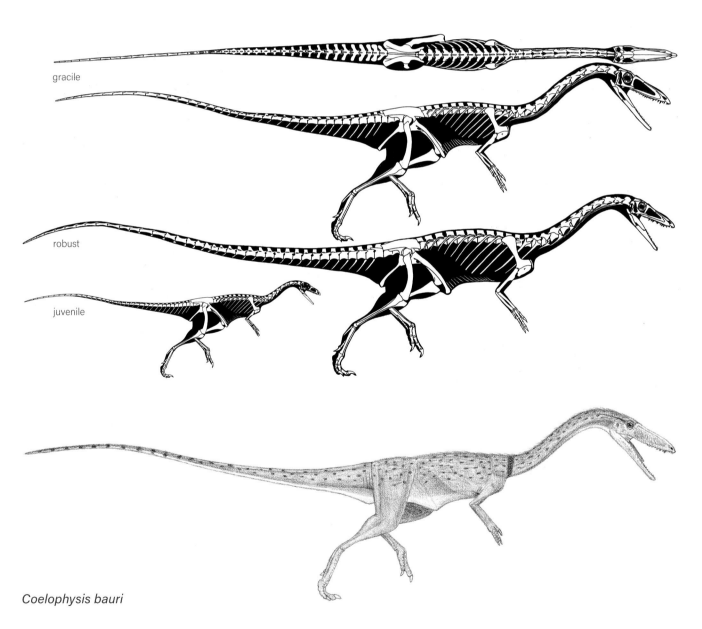

gracile

robust

juvenile

Coelophysis bauri

NOTES Originally *Syntarsus*, that name preoccupied by an insect and replaced by *Megapnosaurus*; species too similar to *C. bauri* to be distinct genus; apparent absence of very small fifth metacarpal may be due to smaller fossil sample. Whether remains from other South African formations belong to this species is uncertain.

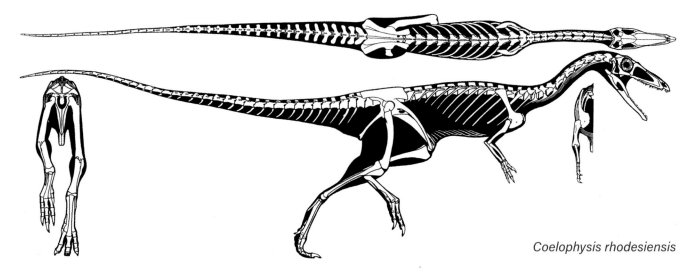

Coelophysis rhodesiensis

Coelophysis kayentakatae
— 2.5 m (9 ft) TL, 30 kg (60 lb)
FOSSIL REMAINS Complete skull and minority of skeleton, other partial remains.
ANATOMICAL CHARACTERISTICS Head fairly deep, snout crests well developed, teeth fairly large and less numerous than in other *Coelophysis*.

AGE Early Jurassic, early Sinemurian.
DISTRIBUTION AND FORMATION/S Arizona; middle Kayenta.
HABITAT Semiarid.
HABITS More robust head and larger teeth indicate this species tended to hunt larger game than other *Coelophysis*.
NOTES Originally in *Syntarsus*.

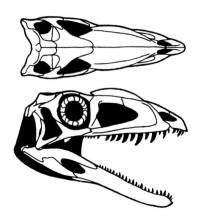

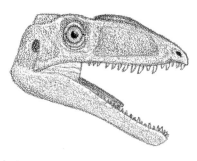

Coelophysis kayentakatae

Podokesaurus (or Coelophysis) holyokensis
— 1 m (3 ft) TL, 1 kg (2 lb)
FOSSIL REMAINS Partial skeleton, possibly juvenile.
ANATOMICAL CHARACTERISTICS Standard for small coelophysoids.
AGE Early Jurassic, Pliensbachian or Toarcian.
DISTRIBUTION AND FORMATION/S Massachusetts; Portland?
HABITAT Semiarid rift valley with lakes.
NOTES Lost in a fire; the original location and age of this fossil are not entirely certain. Presence of snout crests uncertain.

Panguraptor lufengensis
— Adult size uncertain
FOSSIL REMAINS Majority of skull and skeleton, probably a subadult.
ANATOMICAL CHARACTERISTICS Standard for small coelophysoids, snout crests absent.
AGE Early Jurassic, earliest Hettangian.
DISTRIBUTION AND FORMATION/S Southwestern China; lower Lower Lufeng.
NOTES Probable coelophysid.

Panguraptor lufengensis

Segisaurus halli

—— Adult size uncertain

FOSSIL REMAINS Partial skeleton, large juvenile.

ANATOMICAL CHARACTERISTICS Standard for small coelophysoids.

AGE Early Jurassic, Pliensbachian or Toarcian.

DISTRIBUTION AND FORMATION/S Arizona; Navajo Sandstone.

HABITAT Dune desert with oases.

HABITS Largely a small game hunter, probably small prosauropods and ornithischians also.

NOTES Probable coelophysid. Presence of snout crests uncertain.

Notatesseraeraptor frickensis

—— 3 m (10 ft) TL, 25 kg (50 lb)

FOSSIL REMAINS Majority of badly damaged skull and minority of skeleton.

ANATOMICAL CHARACTERISTICS Lightly built. Snout crests absent.

AGE Late Triassic, latest Norian.

DISTRIBUTION AND FORMATION/S Switzerland; uppermost Klettgau.

HABITAT Coastal.

Gojirasaurus quayi

—— 6 m (20 ft) TL, 150 kg (350 lb)

FOSSIL REMAINS Small portion of skeleton.

ANATOMICAL CHARACTERISTICS Insufficient information.

AGE Late Triassic, middle Norian.

DISTRIBUTION AND FORMATION/S New Mexico; Cooper Canyon.

HABITAT Well-watered forests, including dense stands of giant conifers.

NOTES Remains that belong to this taxon uncertain. Probable coelophysoid. Presence of snout crests uncertain.

Lophostropheus airelensis

—— Adult size uncertain

FOSSIL REMAINS Partial skeleton.

ANATOMICAL CHARACTERISTICS Insufficient information.

AGE Latest Triassic and/or Early Jurassic, late Rhaetian and/or early Hettangian.

DISTRIBUTION AND FORMATION/S Northern France; Moon-Airel.

NOTES Presence of snout crests uncertain.

Liliensternus liliensterni

—— 5.2 m (17 ft) TL, 140 kg (300 lb)

FOSSIL REMAINS Majority of skull and two skeletons.

ANATOMICAL CHARACTERISTICS Lightly built like smaller coelophysoids.

AGE Late Triassic, late Norian.

DISTRIBUTION AND FORMATION/S Central Germany; Trossingen.

NOTES Probable coelophysoid. Presence of snout crests uncertain.

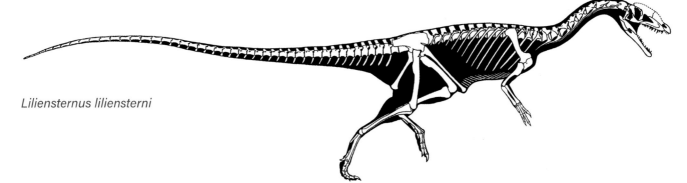

Liliensternus liliensterni

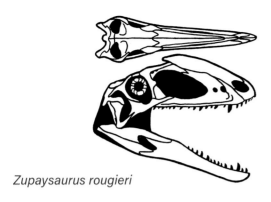

Zupaysaurus rougieri

Zupaysaurus rougieri

—— 6 m (20 ft) TL, 250 kg (550 lb)

FOSSIL REMAINS Almost complete skull and partial skeleton.
ANATOMICAL CHARACTERISTICS Head moderately deep, snout very large, adorned with well-developed paired crests, teeth not large.
AGE Late Triassic, middle Norian.
DISTRIBUTION AND FORMATION/S Northern Argentina; Los Colorados.
HABITAT Seasonally wet woodlands.
HABITS Big game hunter.
NOTES When first described this was considered to be the earliest tetanuran theropod, but other research indicates it is a podokesauroid related to *Dilophosaurus*.

Dracovenator regenti

—— 6 m (20 ft) TL, 250 kg (550 lb)

FOSSIL REMAINS Two partial skulls, juvenile and adult.
ANATOMICAL CHARACTERISTICS Snout crests apparently not large, teeth large.
AGE Early Jurassic, late Hettangian or Sinemurian.
DISTRIBUTION AND FORMATION/S South Africa; upper Elliot.
HABITAT Arid.
HABITS Big game hunter.

Dilophosaurus wetherilli

—— 7 m (22 ft) TL, 285 kg (600 lb)

FOSSIL REMAINS Majority of several skulls and skeletons.
ANATOMICAL CHARACTERISTICS More robustly constructed than smaller coelophysoids. Head large, deep, snout crests large, teeth large.
AGE Early Jurassic, Hettangian or Sinemurian.
DISTRIBUTION AND FORMATION/S Arizona; lower Kayenta.
HABITAT Semiarid.
HABITS Big game hunter.
NOTES Exact shape of crests uncertain. This, *Liliensternus*, *Zupaysaurus*, and *Dracovenator* may form family Dilophosauridae, which may or may not be in coelophysoids; not a big hipped averostran as has been suggested.

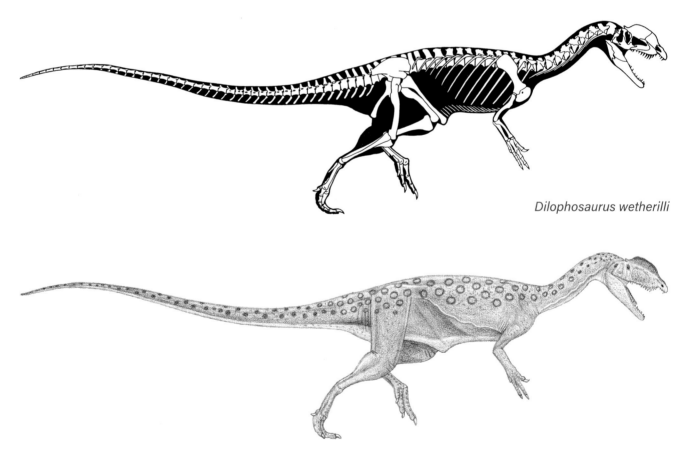

Dilophosaurus wetherilli

Dilophosaurus wetherilli

— AVEROSTRANS

SMALL TO GIGANTIC PREDATORY AND HERBIVOROUS AVEPODS FROM THE EARLY JURASSIC TO THE MODERN ERA, ALL CONTINENTS.

ANATOMICAL CHARACTERISTICS Highly variable. Nasal sinuses better developed. Brains reptilian to avian. Birdlike respiratory system better developed. Pelvic plate large. All known skeletal remains apparently post-Triassic, but some earlier splay-toed footprints may be Triassic averostrans.

— BASAL AVEROSTRAN MISCELLANEA

Berberosaurus liassicus
— 5 m (15 ft) TL, 300 kg (600 lb)
FOSSIL REMAINS Fragmentary skeleton.
ANATOMICAL CHARACTERISTICS Insufficient information.
AGE Early Jurassic, Pliensbachian or Toarcian.
DISTRIBUTION AND FORMATION/S Morocco; Azilial.

Saltriovenator zanellai
— 7 m (23 ft) TL, 1 tonne
FOSSIL REMAINS Fragmentary skeleton.
ANATOMICAL CHARACTERISTICS Insufficient information.
AGE Early Jurassic, early Sinemurian.
DISTRIBUTION AND FORMATION/S Northern Italy; lower Saltrio.

NOTES Found as drift in marine sediments. Largest known early Jurassic predatory dinosaur.

Genyodectes serus
— 6 m (20 ft) TL, 600 kg (1,300 lb)
FOSSIL REMAINS Fragmentary skull.
ANATOMICAL CHARACTERISTICS Insufficient information.
AGE Early Cretaceous, Albian.
DISTRIBUTION AND FORMATION/S Northeastern Argentina, middle Cerro Barcino.
NOTES May be close to ceratosaurids. If so, may indicate survival of that group into Early Cretaceous.

—CERATOSAURIDS

MEDIUM-SIZED TO LARGE PREDATORY AVEROSTRANS FROM THE JURASSIC OF THE AMERICAS, EUROPE, AND AFRICA.

ANATOMICAL CHARACTERISTICS Uniform. Stoutly built. Four fingers. Tails long.

HABITS Ambush and pursuit big game hunters.

NOTES Absence from some continents may reflect lack of sufficient sampling.

Sarcosaurus woodi

—— 3 m (10 ft) TL, 70 kg (150 lb)

FOSSIL REMAINS Minority of skeletons.

ANATOMICAL CHARACTERISTICS Insufficient information.

AGE Early Jurassic, latest Hettengian and/or earliest Sinemurian.

DISTRIBUTION AND FORMATION/S Wales; upper Blue Lias.

NOTES Includes *S. andrewsi*. Relationships with other avepods uncertain.

Ceratosaurus nasicornis

—— 6 m (20 ft) TL, 570 kg (1,250 lb)

FOSSIL REMAINS Two skulls and some skeletons including a juvenile.

ANATOMICAL CHARACTERISTICS Head large, long, rectangular, narrow; large, narrow nasal horn, orbital hornlets large, subtriangular; teeth large. Tail deep and heavy. Arm and hand short. Leg not long. Single row of small, bony scales along back.

AGE Late Jurassic, latest Oxfordian and early Kimmeridgian.

DISTRIBUTION AND FORMATION/S Colorado, Utah; lower and middle Morrison.

HABITAT Short wet season, otherwise semiarid with open floodplain prairies and riverine forests.

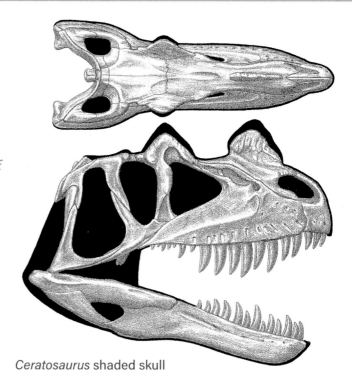

Ceratosaurus shaded skull

ONTOGENY Growth rapid.

HABITS Ambush predators. Large bladed teeth indicate that this hunted large prey by delivering slashing wounds and that the head was a much more important weapon than the small arms. Deep tail used as powerful sculling organ while swimming. Nasal horn for display and possibly headbutting within the species. Armor row for display.

NOTES The species *C. magnicornis* is so similar that it appears to be a member of *C. nasicornis*, or it may represent a descendant of *C. nasicornis*. The only known theropods with any form of body armor. May have been the direct ancestor of *C. dentisulcatus*.

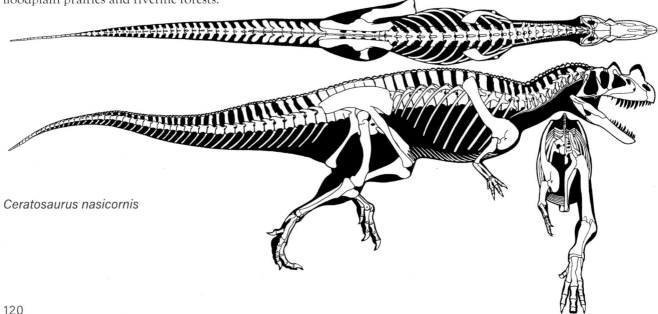

Ceratosaurus nasicornis

Ceratosaurus nasicornis

Ceratosaurus dentisulcatus

—— 7 m (21 ft) TL, 900 kg (2,000 lb)

FOSSIL REMAINS Part of a skull and skeleton.
ANATOMICAL CHARACTERISTICS Head deeper, lower jaw not as curved, and teeth not as proportionally large as in C. *nasicornis*.
AGE Late Jurassic, late middle and/or late Kimmeridgian.
DISTRIBUTION AND FORMATION/S Utah; upper Morrison.
HABITAT Wetter than earlier Morrison, otherwise semiarid with open floodplain prairies and riverine forests.
HABITS Similar to C. *nasicornis*.
NOTES It is uncertain whether C. *dentisulcatus* had a nasal horn or not.

Ceratosaurus? unnamed species

—— 6 m (20 ft) TL, 600 kg (1,300 lb)

FOSSIL REMAINS Minority of skeleton.
ANATOMICAL CHARACTERISTICS Insufficient information.
AGE Late Jurassic, late Kimmeridgian or early Tithonian.
DISTRIBUTION AND FORMATION/S Portugal; lower Lourinhã.
HABITAT Large, seasonally dry island with open woodlands.
NOTES Assignment by some researchers of this fossil to C. *dentisulcatus* is uncertain.

Ceratosaurus nasicornis

— ABELISAUROIDS

SMALL TO GIGANTIC AVEROSTRANS FROM THE EARLY JURASSIC TO THE END OF THE DINOSAUR ERA, LARGELY LIMITED TO THE SOUTHERN HEMISPHERE.

ANATOMICAL CHARACTERISTICS Highly variable. Arms short, four fingers. Vertebrae often flat topped. Tails long. Pelves large. Birdlike respiratory system well developed.
NOTES Abelisaurs show that relatively archaic avepods were able to thrive in the Southern Hemisphere to the end of the dinosaur era as they evolved into specialized forms. Absence from Antarctica and possibly other continents probably reflects lack of sufficient sampling.

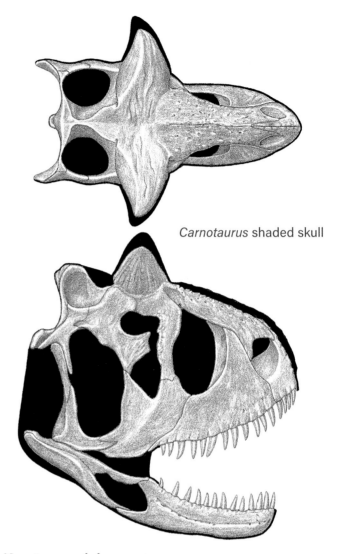

Carnotaurus shaded skull

— ABELISAURIDS

MEDIUM-SIZED TO GIGANTIC ABELISAUROIDS OF THE MIDDLE JURASSIC TO THE END OF THE DINOSAUR ERA, LARGELY LIMITED TO THE SOUTHERN HEMISPHERE.

ANATOMICAL CHARACTERISTICS Fairly uniform. Heads heavily constructed, short and deep, lower jaws slender, teeth short and stout. Arms reduced. Tubercle scales set amid fairly large, flat scales.
HABITATS SEASONALLY DRY TO WELL-WATERED WOODLANDS.
HABITS Ambush and pursuit big game hunters. Reduction of arms indicates that the stout head was the primary weapon, but how the combination of a deep, short skull, slender lower jaw indicating modest musculature, and short teeth functioned is obscure.

Eoabelisaurus mefi
—— 6.5 m (21 ft) TL, 850 kg (1,900 lb)
FOSSIL REMAINS Minority of skull and majority of skeleton.
ANATOMICAL CHARACTERISTICS Robustly built.
AGE Middle Jurassic, Aalenian and/or Bajocian.
DISTRIBUTION AND FORMATION/S Southern Argentina; Cañadón Asfalto.

Spectrovenator ragei
—— Adult size uncertain
FOSSIL REMAINS Complete skull and partial skeleton, possibly juvenile.
ANATOMICAL CHARACTERISTICS Insufficient information.
AGE Early Cretaceous, latest Barremian and/or earliest Aptian.
DISTRIBUTION AND FORMATION/S Southeastern Brazil; lower Quiricó.

Kryptops palaios
—— Adult size uncertain
FOSSIL REMAINS Minority of skull and skeleton, possibly immature.
ANATOMICAL CHARACTERISTICS Insufficient information.
AGE Early Cretaceous, Aptian.
DISTRIBUTION AND FORMATION/S Niger; Elrhaz, level uncertain.
HABITAT Coastal river delta.

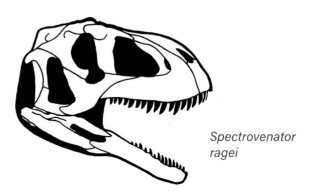

Spectrovenator ragei

Rugops primus

—— 6 m (20 ft) TL, 750 kg (1,600 lb)
FOSSIL REMAINS Partial skull.
ANATOMICAL CHARACTERISTICS Snout very deep and robust, possible low, paired crests on snout.
AGE Late Cretaceous, Cenomanian.
DISTRIBUTION AND FORMATION/S Niger; Echkar.

Rugops primus

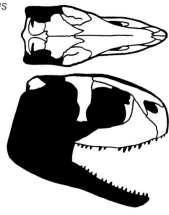

Xenotarsosaurus bonapartei

—— 6 m (20 ft) TL, 750 kg (1,700 lb)
FOSSIL REMAINS Minority of skeleton.
ANATOMICAL CHARACTERISTICS Leg long and gracile.
AGE Late Cretaceous, late Cenomanian or Turonian.
DISTRIBUTION AND FORMATION/S Southern Argentina; lower Bajo Barreal.
HABITAT Seasonally wet, well-forested floodplain.
HABITS Pursuit predator.

Majungasaurus crenatissimus

—— 6 m (20 ft) TL, 730 kg (1,600 lb)
FOSSIL REMAINS Nearly perfect skull and extensive skeletal remains, almost completely known.
ANATOMICAL CHARACTERISTICS Low central horn above orbits. Leg stout, not elongated.
AGE Late Cretaceous, Campanian.
DISTRIBUTION AND FORMATION/S Northwestern Madagascar; Maevarano.
HABITAT Seasonally dry floodplain with coastal swamps and marshes.

ONTOGENY Growth slow.
HABITS Some remains are under preparation. Used horn for display and probably headbutting within the species.

Arcovenator escotae

—— 5 m (16 ft) TL, 500 kg (1,200 lb)
FOSSIL REMAINS Fragmentary skull.
ANATOMICAL CHARACTERISTICS Insufficient information.
AGE Late Cretaceous, late Campanian.
DISTRIBUTION AND FORMATION/S Southeastern France; lower Argiles et Gres a Reptiles.
NOTES If an abelisaur indicates abelisaurs were present in Northern Hemisphere.

Ilokelesia aguadagrandensis

—— 4 m (14 ft) TL, 200 kg (450 lb)
FOSSIL REMAINS Minority of skeleton.
ANATOMICAL CHARACTERISTICS Base of tail exceptionally broad.
AGE Late Cretaceous, late Cenomanian.
DISTRIBUTION AND FORMATION/S Western Argentina; middle Huincul.
HABITAT Short wet season, otherwise semiarid with open floodplains and riverine forests.

Rajasaurus narmadensis

—— 11 m (35 ft) TL, 5 tonnes
FOSSIL REMAINS Majority of skull and partial skeleton, possibly other partial.
ANATOMICAL CHARACTERISTICS Back of head adorned by central crest. Leg stout.

Majungasaurus crenatissimus

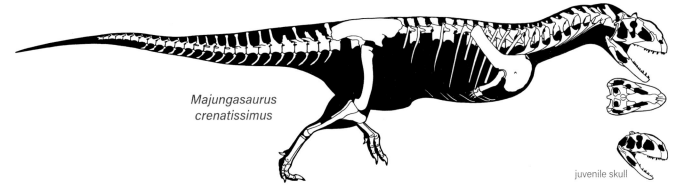

Majungasaurus crenatissimus

juvenile skull

123

AGE Late Cretaceous, late Maastrichtian.
DISTRIBUTION AND FORMATION/S Central India; Lameta.
HABITS Used horn for display and probably headbutting within the species.
NOTES *Coeluroides largus, Ornithomimoides mobilis, Compsosuchus solus,* and *Rahiolisaurus gujaratensis* may be juveniles of this species.

Rajasaurus narmadensis

Ekrixinatosaurus novasi

—— 6.5 m (21 ft) TL, 800 kg (1,800 lb)
FOSSIL REMAINS Minority of skull and partial skeleton.
ANATOMICAL CHARACTERISTICS Insufficient information.
AGE Late Cretaceous, early Cenomanian.
DISTRIBUTION AND FORMATION/S Western Argentina; lower Candeleros.
HABITAT Short wet season, otherwise semiarid with open floodplains and riverine forests.

Skorpiovenator bustingorryi

—— 7.5 m (25 ft) TL, 1.8 tonnes
FOSSIL REMAINS Complete skull and majority of skeleton.
ANATOMICAL CHARACTERISTICS Rugose area around eye socket. Leg long.
AGE Late Cretaceous, middle Cenomanian.
DISTRIBUTION AND FORMATION/S Western Argentina; lower Huincul.
HABITAT Well-watered woodlands with short dry season.

Tralkasaurus cuyi

—— 4 m (13 ft) TL, 300 kg (600 lb)
FOSSIL REMAINS Minority of skull and skeleton.
ANATOMICAL CHARACTERISTICS Insufficient information.
AGE Late Cretaceous, late Cenomanian.
DISTRIBUTION AND FORMATION/S Western Argentina; middle Huincul.
HABITAT Short wet season, otherwise semiarid with open floodplains and riverine forests.

Elemgasem nubilus

—— 4 m (13 ft) TL, 300 kg (600 lb)
FOSSIL REMAINS Minority of skeleton.
ANATOMICAL CHARACTERISTICS Insufficient information.
AGE Late Cretaceous, late Turonian.
DISTRIBUTION AND FORMATION/S Western Argentina; Portezuelo.
HABITAT Well-watered woodlands with short dry season.

Viavenator exxoni

—— 5.5 m (18 ft) TL, 700 kg (1,500 lb)
FOSSIL REMAINS Partial skeleton; at least one immature.
ANATOMICAL CHARACTERISTICS Insufficient information.
AGE Late Cretaceous, Santonian.

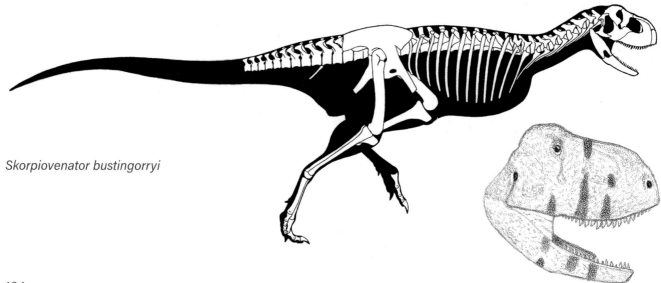

Skorpiovenator bustingorryi

DISTRIBUTION AND FORMATION/S Central Argentina; upper Bajo de la Carpa.
NOTES *Llukalkan aliocranianus* may be juvenile of this species, skull proportions uncertain.

Abelisaurus comahuensis

—— 10 m (30 ft) TL, 4 tonnes
FOSSIL REMAINS Partial skull.
ANATOMICAL CHARACTERISTICS Head unadorned.
AGE Late Cretaceous, late Santonian and/or early Campanian.
DISTRIBUTION AND FORMATION/S Western Argentina; Anacleto.
HABITAT Seasonally semiarid.

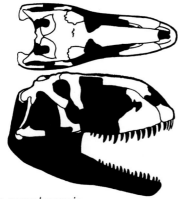

Abelisaurus comahuensis

Abelisaurus (= Aucasaurus) garridoi

—— 5.5 m (18 ft) TL, 740 kg (1,600 lb)
FOSSIL REMAINS Complete skull and nearly complete skeleton.
ANATOMICAL CHARACTERISTICS Head unadorned. Lower arm and hand atrophied. Leg long and gracile, inner toe reduced, toe claws small.
AGE Late Cretaceous, late Santonian and/or early Campanian.

DISTRIBUTION AND FORMATION/S Western Argentina; upper Anacleto.
HABITAT Seasonally semiarid.
HABITS Pursuit predator able to chase prey at high speed.
NOTES Named in a new genus, *Aucasaurus*; the only reason this does not appear to be a juvenile *Abelisaurus comahuensis* is that fusion of skeletal elements indicates it is an adult.

Quilmesaurus curriei

—— 4.5 m (10 ft) TL, 400 kg (1000 lb)
FOSSIL REMAINS Fragmentary skeleton, possibly fragmentary skull and skeleton.
ANATOMICAL CHARACTERISTICS Insufficient information.
AGE Late Cretaceous, late Campanian.
DISTRIBUTION AND FORMATION/S Central Argentina; Allen.
HABITAT Semiarid coastline.
NOTES May include *Niebla antiqua*.

Pycnonemosaurus nevesi

—— 7 m (23 ft) TL, 1.8 tonnes
FOSSIL REMAINS Minority of skeleton.
ANATOMICAL CHARACTERISTICS Insufficient information.
AGE Late Cretaceous, Campanian or Maastrichtian.
DISTRIBUTION AND FORMATION/S Southwestern Brazil; Cachoeira do Born Jardim.

Kurupi itaata

—— 5 m (16 ft) TL, 500 kg (1,200 lb)
FOSSIL REMAINS Fragmentary skeleton.
ANATOMICAL CHARACTERISTICS Insufficient information.
AGE Late Cretaceous, Maastrichtian.
DISTRIBUTION AND FORMATION/S Southeastern Brazil; Marilia.

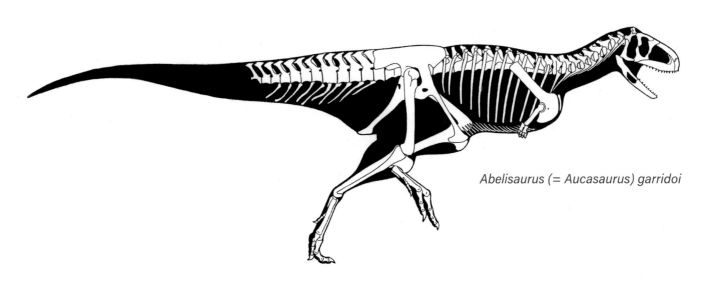

Abelisaurus (= Aucasaurus) garridoi

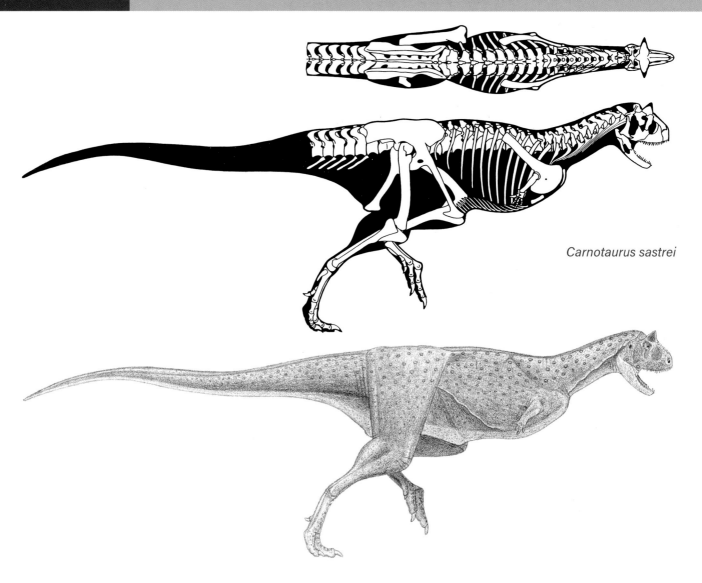

Carnotaurus sastrei

Carnotaurus sastrei

—— 7.5 m (25 ft) TL, 2.2 tonnes

FOSSIL REMAINS Complete skull and majority of skeleton, skin patches.

ANATOMICAL CHARACTERISTICS Head very deep; large, stout brow horns directed sideways. Lower arm and hand atrophied.

AGE Late Cretaceous, Campanian or early Maastrichtian.

DISTRIBUTION AND FORMATION/S Southern Argentina; La Colonia.

HABITS Used horns for display, probably headbutting, and pushing within the species.

NOTES The most specialized known abelisaurid.

Chenanisaurus barbaricus

—— 7.5 m (25 ft) TL, 2 tonnes

FOSSIL REMAINS Fragmentary skull.

ANATOMICAL CHARACTERISTICS Lower jaw deep.

AGE Late Cretaceous, latest Maastrichtian.

DISTRIBUTION AND FORMATION/S Morocco; unnamed.

NOTES Found as drift in marine sediments.

— NOASAURIDS

SMALL- TO MEDIUM-SIZED ABELISAUROIDS OF THE LATE JURASSIC AND CRETACEOUS OF EURASIA, AFRICA, AMERICAS.

ANATOMICAL CHARACTERISTICS Highly variable. Most lightly constructed of the abelisaurs. Heads modest in size, not heavily built.

Carnotaurus sastrei

— ELAPHROSAURINES

MEDIUM-SIZED NOASAURIDS LIMITED TO THE LATE JURASSIC OF ASIA, AFRICA, AND NORTH AMERICA.

ANATOMICAL CHARACTERISTICS Overall build gracile. Heads modest in size, lightly built, at least adults toothless, with blunt beak. Arms short, slender, hand reduced, just two fingers in at least some examples. Pelves usually not highly enlarged, legs gracile.

HABITS Possibly omnivorous, predominantly herbivorous combined with some small animals and insects. Main defense speed, also kicks from legs.

NOTES The first known herbivorous theropods, these Jurassic ostrich mimics evolved feeding and running adaptations broadly similar to those of the even faster and longer-armed Cretaceous ornithomimids. Absence from at least some continents probably reflects lack of sufficient sampling.

Spinostropheus gautieri

—— 4 m (14 ft) TL, 80 kg (170 lb)

FOSSIL REMAINS Minority of skeleton.

ANATOMICAL CHARACTERISTICS Insufficient information.

AGE Late Middle or early Late Jurassic.

DISTRIBUTION AND FORMATION/S Niger; Tiourarén.

HABITAT Well-watered woodlands.

NOTES Originally thought to be from the Early Cretaceous; researchers now place the Tiourarén in the later Jurassic.

Elaphrosaurus bambergi

—— 6 m (20 ft) TL, 260 kg (450 lb)

FOSSIL REMAINS Majority of skeleton.

ANATOMICAL CHARACTERISTICS Trunk elongated and shallow. Tail quite long. Thigh narrow.

AGE Late Jurassic, middle Kimmeridgian.

DISTRIBUTION AND FORMATION/S Tanzania; middle Tendaguru.

HABITAT Coastal, seasonally dry with heavier vegetation inland.

NOTES Remains from much later upper Tendaguru are not same taxon.

Elaphrosaurus bambergi

127

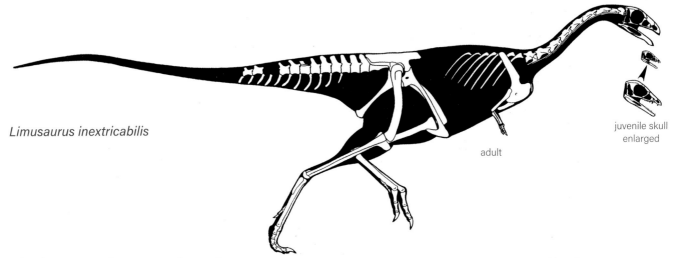

Limusaurus inextricabilis

adult

juvenile skull
enlarged

Elaphrosaurus? unnamed species

—— 4.5 m (15 ft) TL, 100 kg (220 lb)

FOSSIL REMAINS Small portion of skeleton.

ANATOMICAL CHARACTERISTICS Insufficient information.

AGE Late Jurassic, early and/or middle Kimmeridgian.

DISTRIBUTION AND FORMATION/S Colorado; middle Morrison.

HABITAT Short wet season, otherwise semiarid with open floodplains and riverine forests.

NOTES It is uncertain whether these remains are the same genus as *Elaphrosaurus*, probably constitute two species over time.

Limusaurus inextricabilis

—— 2 m (6 ft) TL, 17 kg (35 lb)

FOSSIL REMAINS About a dozen and a half skulls and skeletons, juvenile to adult, gastroliths present.

ANATOMICAL CHARACTERISTICS Head moderately deep. Ossified sternum present. Two functional fingers. Inner toe reduced.

AGE Late Jurassic, early Oxfordian.

DISTRIBUTION AND FORMATION/S Northwestern China; upper Shishugou.

HABITS Presence of well-developed teeth in juveniles indicates shift from carnivory or omnivory to herbivory with growth.

NOTES Juveniles had teeth; juvenile skeleton not yet documented well enough to restore. Parallels tyrannosaurids in having just two fingers.

Berthasaura leopoldinae

—— Adult size not certain

FOSSIL REMAINS Majority of skull and partial skeleton, juvenile.

ANATOMICAL CHARACTERISTICS Insufficient information.

AGE Early Cretaceous, Aptian or Albian.

DISTRIBUTION AND FORMATION/S Southern Brazil; Goio-Erê.

NOTES Juvenile condition and incomplete remains hinder analysis. Lack of teeth suggests may be an elaphrosaurine, large pelvis suggests may be a noasaurine. Juveniles toothless.

—— NOASAURINES

MEDIUM-SIZED NOASAURIDS LIMITED TO THE CRETACEOUS OF AFRICA, SOUTH AMERICA, AND EUROPE.

ANATOMICAL CHARACTERISTICS Teeth present. Arms and hands not strongly reduced.

HABITS Predaceous, fishers.

NOTES Absence from at least some continents probably reflects lack of sufficient sampling.

Dahalokely tokana

—— 4 (13 ft) TL, 100 kg (200 lb)

FOSSIL REMAINS Fragmentary skeleton.

ANATOMICAL CHARACTERISTICS Insufficient information.

AGE Late Cretaceous, Turonian.

DISTRIBUTION AND FORMATION/S Northern Madagascar; Ambolafotsy.

NOTES May be an abelisaurid.

Bahariasaurus ingens and/or Deltadromeus agilis

—— 11 m (35 ft) TL, 4 tonnes

FOSSIL REMAINS Minority of skeletons.

ANATOMICAL CHARACTERISTICS Shoulder girdle massively constructed. Leg long and gracile.

AGE Late Cretaceous, early Cenomanian.

DISTRIBUTION AND FORMATION/S Morocco; Bahariya.

HABITAT Coastal mangroves.

HABITS Fast-running pursuit predator.
NOTES The relationships of *Bahariasaurus* and *Deltadromeus* to other theropods and each other are uncertain; the latter may be a juvenile of the former.

Ligabueino andesi
—— 0.6 m (2 ft) TL, 0.4 kg (1 lb)
FOSSIL REMAINS Minority of skeleton.
ANATOMICAL CHARACTERISTICS Standard for group.
AGE Early Cretaceous, Barremian and/or early Aptian.
DISTRIBUTION AND FORMATION/S Western Argentina; La Amarga.
HABITAT Well-watered coastal woodlands with short dry season.
HABITS Small game hunter.
NOTES If not a juvenile, is one of the smallest theropods outside the birdlike airfoilans.

Velocisaurus unicus
—— 1.5 m (5 ft) TL, 6 kg (15 lb)
FOSSIL REMAINS Two legs.
ANATOMICAL CHARACTERISTICS Leg long, slender, central shaft of upper foot strong with side bones slender.
AGE Late Cretaceous, Santonian.
DISTRIBUTION AND FORMATION/S Central Argentina; Bajo de la Carpa.
HABITS Good runner.

Masiakasaurus knopfleri
—— 2 m (7 ft) TL, 12 kg (25 lb)
FOSSIL REMAINS Minority of skull and skeleton.
ANATOMICAL CHARACTERISTICS Front teeth of lower jaw form a procumbent whorl and are long and weakly serrated; back teeth more conventional.
AGE Late Cretaceous, Campanian.
DISTRIBUTION AND FORMATION/S Northwestern Madagascar; Maevarano.
HABITAT Seasonally dry floodplain with coastal swamps and marshes.
HABITS Hunted small prey, especially fish.

Masiakasaurus knopfleri

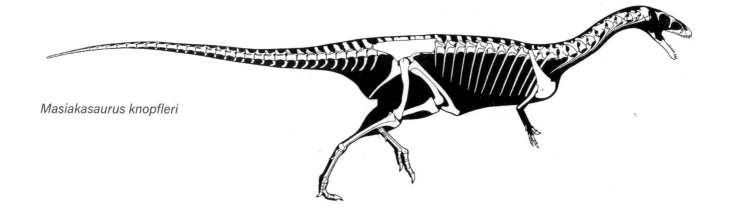

Masiakasaurus knopfleri

Noasaurus leali
—— 1.5 m (5 ft) TL, 6 kg (15 lb)
FOSSIL REMAINS Minority of a skull and skeleton.
ANATOMICAL CHARACTERISTICS Standard for group.
AGE Late Cretaceous, probably early Maastrichtian.
DISTRIBUTION AND FORMATION/S Northern Argentina; Lecho.
HABITS Pursuit predator.
NOTES It was thought that a large claw was a sickle-toe weapon like those of dromaeosaurids, but more likely it belonged to the hand.

Genusaurus sisteronis
—— 3 m (10 ft) TL, 35 kg (70 lb)
FOSSIL REMAINS Minority of skeleton.
ANATOMICAL CHARACTERISTICS Insufficient information.
AGE Early Cretaceous, Albian.
DISTRIBUTION AND FORMATION/S Southeastern France; Bevons Beds.
HABITAT Forested coastline.
NOTES Found as drift in nearshore marine sediments. Placement in noasaurids uncertain, if correct indicates that abelisauroids migrated to the Northern Hemisphere.

Vespersaurus paranaensis
—— 2 m (7 ft) TL, 20 kg (50 lb)
FOSSIL REMAINS Partial remains.
ANATOMICAL CHARACTERISTICS Outer toes somewhat reduced relative to central toe.
AGE Early Late Cretaceous.
DISTRIBUTION AND FORMATION/S Southern Brazil; Rio Paraná.
HABITAT Dune desert.
NOTES May have been semimonodactyl with central toe bearing almost all of animal's mass.

—— TETANURANS
SMALL TO GIGANTIC PREDATORY AND HERBIVOROUS AVEROSTRANS FROM THE MIDDLE JURASSIC TO THE MODERN ERA, ALL CONTINENTS.

ANATOMICAL CHARACTERISTICS Highly variable. Arms very long to very reduced. Birdlike respiratory system usually better developed. Brains reptilian to avian.
ONTOGENY Growth moderately to very rapid.

—— TETANURAN MISCELLANEA

Vectaerovenator inopinatus
—— 4 m (13 ft) TL, 200 kg (400 lb)
FOSSIL REMAINS Fragmentary skeleton.
ANATOMICAL CHARACTERISTICS Insufficient information.
AGE Early Cretaceous; late Aptian.
DISTRIBUTION AND FORMATION/S Southern England; Ferruginous Sands.
NOTES Found as drift in interisland marine sediments.

—— BASO-TETANURANS
SMALL TO LARGE PREDATORY AND HERBIVOROUS TETANURANS FROM THE JURASSIC.

HABITS Crests when present too delicate for headbutting; for visual display, may or may not have been brightly colored at least during breeding season.
NOTES The relationships of the following basal and usually partially known tetanurans are uncertain.

Sinosaurus? sinensis
—— 5.5 m (18 ft) TL, 330 kg (700 lb)
FOSSIL REMAINS Nearly complete skull and skeleton.
ANATOMICAL CHARACTERISTICS Overall build slender. Head adorned by large paired crests. Arm robust. Tail long.
AGE Early Jurassic, early Hettangian.
DISTRIBUTION AND FORMATION/S Southwestern China; lower Lufeng.
HABITS Long crests for lateral display.
NOTES Broadly comparable in time, size, and overall appearance to similarly crested *Dilophosaurus*, it was assumed to be a member of the same genus, but detailed anatomy indicates this is a more derived tetanuran avepod.

Because *S. triassicus?* is problematic, may not be that genus. May be direct ancestor of *S. triassicus?*, which is same genus.

Sinosaurus triassicus?
—— 5.5 m (18 ft) TL, 350 kg (700 lb)
FOSSIL REMAINS A complete and partial skulls.
ANATOMICAL CHARACTERISTICS Head adorned by large paired crests.
AGE Early Jurassic, middle Hettangian.
DISTRIBUTION AND FORMATION/S Southwestern China; upper Lufeng.
HABITS Long crests for lateral display.
NOTES Original fossils very fragmentary and may not be same species as complete skull.

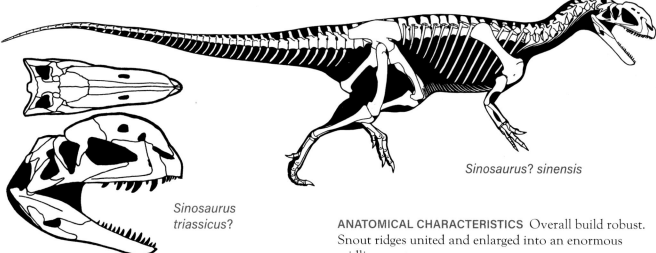

Sinosaurus? sinensis

Sinosaurus triassicus?

Shuangbaisaurus anlongbaoensis

—— 6 m (20 ft) TL, 350 kg (700 lb)

FOSSIL REMAINS Majority of distorted skull.

ANATOMICAL CHARACTERISTICS Head apparently adorned by paired crests.

AGE Early Jurassic, Pliensbachian.

DISTRIBUTION AND FORMATION/S Southwestern China; Fengjiahe.

NOTES Suggested placement in earlier *Sinosaurus* highly problematic.

Monolophosaurus jiangi

—— 5.5 m (18 ft) TL, 540 kg (1,600 lb)

FOSSIL REMAINS Complete skull and majority of skeleton.

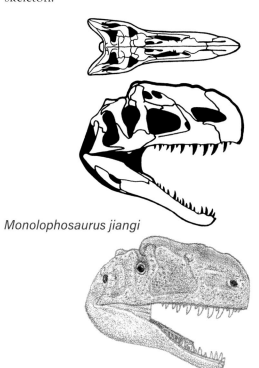

Monolophosaurus jiangi

ANATOMICAL CHARACTERISTICS Overall build robust. Snout ridges united and enlarged into an enormous midline crest.

AGE Middle Jurassic, late Callovian.

DISTRIBUTION AND FORMATION/S Northwestern China; lower Shishugou.

HABITS Crest too delicate for headbutting, for visual display within the species.

Cryolophosaurus ellioti

—— 7 m (23 ft) TL, 775 kg (1,700 lb)

FOSSIL REMAINS Partial skull and two partial skeletons.

ANATOMICAL CHARACTERISTICS Overall build slender. Paired crests very low atop snout, above orbits arc toward middle and join to form large transverse crest. Arm robust.

AGE Early Jurassic, Sinemurian or Pliensbachian.

DISTRIBUTION AND FORMATION/S Central Antarctica; lower Hanson.

HABITAT Polar forests with cold, dark winters.

Cryolophosaurus ellioti

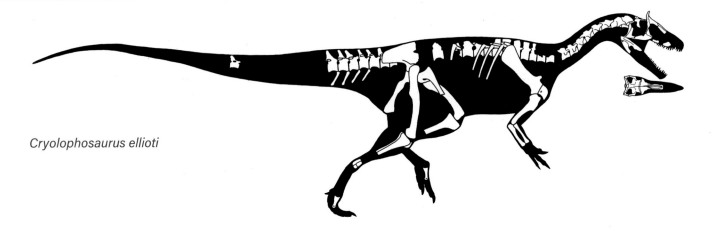

Cryolophosaurus ellioti

HABITS Broad, tall crest for frontal display.
NOTES Largest known Early Jurassic avepod. The only theropod known from continental Antarctica, this reflects lack of more extensive exposed deposits and difficult research conditions.

Shidaisaurus jinae
—— 6 m (20 ft) TL, 700 kg (1,600 lb)
FOSSIL REMAINS Minority of skull and partial skeleton.
ANATOMICAL CHARACTERISTICS Neural spines form shallow sail over trunk and base of tail.
AGE Early Middle Jurassic.
DISTRIBUTION AND FORMATION/S Southwestern China; Chuanjie.

Xuanhanosaurus qilixiaensis
—— 4.5 m (15 ft) TL, 250 kg (500 lb)
FOSSIL REMAINS Minority of skeleton, other possible remains.
ANATOMICAL CHARACTERISTICS Stoutly built. Arm and hand well developed.
AGE Late Jurassic, probably Bathonian.
DISTRIBUTION AND FORMATION/S Central China; lower Xiashaximiao.

HABITAT Heavily forested.
HABITS Arms probably important in handling prey.
NOTES May include *Kaijiangosaurus lini* and *Gasosaurus constructus*.

Chilesaurus diegosuarezi
—— 2.5 m (8 ft) TL, 17 kg (35 lb)
FOSSIL REMAINS Partial skull and majority of skeletons.
ANATOMICAL CHARACTERISTICS Overall build slender. Head small, teeth leaf shaped, lightly serrated. Neck fairly long. Trunk elongated, pubis retroverted. Arm robust, third digit absent. Inner toes well developed, so foot almost tetradactyl.
AGE Late Jurassic, middle Tithonian.
DISTRIBUTION AND FORMATION/S Chile; Toqui.
HABITS Herbivorous or omnivorous.
NOTES Exact relationships with other avepods uncertain. Another example of avepods evolving herbivory and of reenlarging the inner toe—almost as much as therizinosaurs and similar to *Balaur*; and another example of dinosaurs retroverting the pubis, in this case to enlarge the gut for herbivory. Fragmentary Asian Early Jurassic *Eshanosaurus deguchiianus* may be a relative.

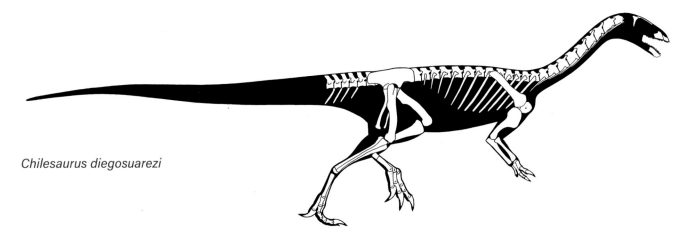

Chilesaurus diegosuarezi

—MEGALOSAUROIDS

LARGE TO VERY LARGE PREDATORY TETANURANS OF THE JURASSIC ON MOST CONTINENTS.

ANATOMICAL CHARACTERISTICS Heads large, low. Tails long.
NOTES The relationships of the following megalosauroids are uncertain. Absence from Australia and Antarctica probably reflects lack of sufficient sampling.

Eustreptospondylus oxoniensis
— 4.5 m (15 ft) TL, 240 kg (500 lb)
FOSSIL REMAINS Majority of skull and skeleton.
ANATOMICAL CHARACTERISTICS Lightly built. Teeth widely spaced.
AGE Middle Jurassic, late Callovian.
DISTRIBUTION AND FORMATION/S Southern England; middle Oxford Clay.

Marshosaurus bicentesimus
— 4.5 m (15 ft) TL, 250 kg (500 lb)
FOSSIL REMAINS Minority of skeletons.
ANATOMICAL CHARACTERISTICS Insufficient information.
AGE Late Jurassic, early and/or middle Kimmeridgian.
DISTRIBUTION AND FORMATION/S Utah; middle Morrison.
HABITAT Wetter than earlier Morrison, otherwise semiarid with open floodplain prairies and riverine forests.

Piatnitzkysaurus floresi
— 4.5 m (15 ft) TL, 300 kg (650 lb)
FOSSIL REMAINS Minority of skull and majority of skeleton.
ANATOMICAL CHARACTERISTICS Lightly built.
AGE Early Jurassic.
DISTRIBUTION AND FORMATION/S Southern Argentina; Cañadón Asfalto.

Condorraptor currumili
— 4.5 m (15 ft) TL, 250 kg (500 lb)
FOSSIL REMAINS Minority of skeleton.
ANATOMICAL CHARACTERISTICS Insufficient information.
AGE Middle Jurassic.
DISTRIBUTION AND FORMATION/S Southern Argentina; Cañadón Asfalto.

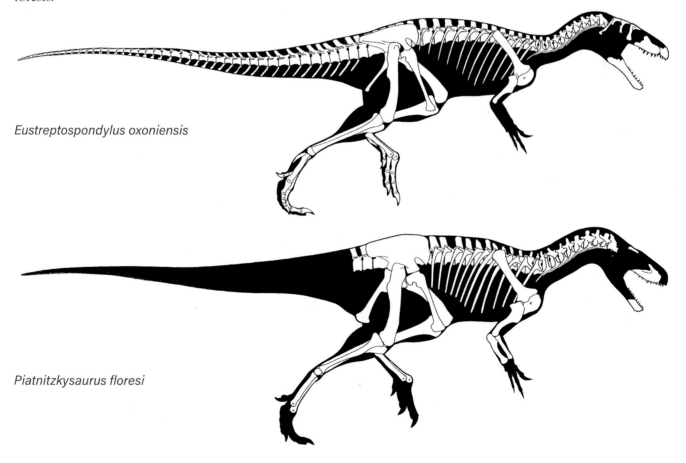

Eustreptospondylus oxoniensis

Piatnitzkysaurus floresi

Afrovenator abakensis

—— 8 m (25 ft) TL, 1.5 tonne
FOSSIL REMAINS Majority of skull and partial skeleton.
ANATOMICAL CHARACTERISTICS Lightly built. Orbital hornlet modest, teeth large. Leg long.
AGE Late Middle or early Late Jurassic.
DISTRIBUTION AND FORMATION/S Niger; Tiourarén.
HABITAT Well-watered woodlands.
HABITS Pursuit predator.
NOTES Originally thought to be from the Early Cretaceous; Tiourarén now placed in the later Jurassic. This, *Dubreuillosaurus*, and *Magnosaurus* may form subfamily Afrovenatorinae.

Afrovenator abakensis

Dubreuillosaurus valesdunensis

—— 5 m (16 ft) TL, 350 kg (750 lb)
FOSSIL REMAINS Majority of skull and partial skeleton.
ANATOMICAL CHARACTERISTICS Teeth large.
AGE Middle Jurassic, middle Bathonian.
DISTRIBUTION AND FORMATION/S Northwestern France; Calcaire de Caen.
HABITAT Coastal mangroves.
NOTES Not a species of *Poekilopleuron*, as originally thought.

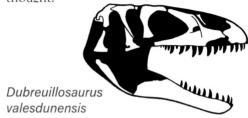

Dubreuillosaurus valesdunensis

Magnosaurus nethercombensis

—— 4.5 m (15 ft) TL, 250 kg (500 lb)
FOSSIL REMAINS Minority of skull and skeleton.
ANATOMICAL CHARACTERISTICS Insufficient information.
AGE Middle Jurassic, Aalenian or Bajocian.
DISTRIBUTION AND FORMATION/S Southwestern England; Inferior Oolite.

—— MEGALOSAURIDS

LARGE TO VERY LARGE PREDATORY MEGALOSAURIDS LIMITED TO THE MIDDLE AND LATE JURASSIC OF EUROPE AND NORTH AMERICA.

ANATOMICAL CHARACTERISTICS Fairly uniform. Massively constructed. Teeth stout. Lower arms short and stout. Pelves broad and shallow. Brains reptilian.
HABITAT Seasonally dry to well-watered woodlands.
HABITS Ambush predators.

Duriavenator hesperis

—— 7 m (23 ft) TL, 1 tonne
FOSSIL REMAINS Partial skull.
ANATOMICAL CHARACTERISTICS Teeth in lower jaw widely spaced.
AGE Middle Jurassic, late Bajocian.
DISTRIBUTION AND FORMATION/S Southern England; upper Inferior Oolite.

Megalosaurus bucklandii

—— 6 m (20 ft) TL, 700 kg (1,600 lb)
FOSSIL REMAINS Lower jaw and possibly skeletal parts.
ANATOMICAL CHARACTERISTICS Standard for group.
AGE Middle Jurassic, middle Bathonian.
DISTRIBUTION AND FORMATION/S Central England; Stonesfield Slate.
NOTES Over the years *Megalosaurus* became a taxonomic grab bag into which a large number of remains from many places and times were placed. The genus and species may be limited to the original fossil, the full extent of which is not entirely certain.

Poekilopleuron? bucklandii

—— 7 m (23 ft) TL, 1 tonne
FOSSIL REMAINS Partial skeleton.
ANATOMICAL CHARACTERISTICS Appears to be standard for group.
AGE Middle Jurassic, middle Bathonian.
DISTRIBUTION AND FORMATION/S Northwestern France; Calcaire de Caen.
NOTES Because at least some bones of this and contemporary and nearby *Megalosaurus bucklandii* are very similar, it is possible that this is the same genus and even species as the latter, or some of the original remains placed in the British megalosaurs may belong to this theropod. Original remains destroyed by allied bombardment in World War II.

composite megalosaurid

Yunyangosaurus puanensis
— 5 m (15 ft) TL, 500 kg (1000 lb)
FOSSIL REMAINS Fragmentary skeleton.
ANATOMICAL CHARACTERISTICS Insufficient information.
AGE Early Middle Jurassic.
DISTRIBUTION AND FORMATION/S Central China; upper Xintiangou.

Wiehenvenator albati
— 9 m (30 ft) TL, 2 tonnes
FOSSIL REMAINS Partial skull and minority of skeleton.
ANATOMICAL CHARACTERISTICS Appears to be standard for group.
AGE Middle Jurassic, middle Callovian.
DISTRIBUTION AND FORMATION/S Northwestern Germany; upper Ornatenton.
HABITAT Interisland coastal/marine.

Torvosaurus tanneri
— 9 m (30 ft) TL, 2 tonnes
FOSSIL REMAINS Majority of a skull and partial skeletons.

ANATOMICAL CHARACTERISTICS Standard for group.
AGE Late Jurassic, early and/or early middle Kimmeridgian.
DISTRIBUTION AND FORMATION/S Colorado, Wyoming, Utah; middle Morrison.
HABITAT Short wet season, otherwise semiarid with open floodplain prairies and riverine forests.
NOTES Remains imply this genus or close relations were present in the lower Morrison and/or the later Portuguese lower and middle Lourinhã Formation; in the latter a nest with eggs and embryos may belong to that taxon.

Torvosaurus tanneri

— SPINOSAURIDS

LARGE TO GIGANTIC FISHING AND PREDATORY TETANURANS OF THE CRETACEOUS OF EURASIA, AFRICA, AND SOUTH AMERICA.

ANATOMICAL CHARACTERISTICS Fairly uniform. Long trunked. Heads very long and shallow; snouts elongated, narrow, and tips hooked; tips of lower jaws expanded, teeth conical; low central crest above orbits, lower jaws could bow outward. Tails long. Arms well developed, three fingers, claws large hooks. Legs short. Brains reptilian. Skeletons less pneumatic than usual for tetanurans.
HABITAT Large watercourses or coastlines.
HABITS Trophically flexible, being able to prey on large animals and also small game hunters with specializations for fishing using crocodilian-like heads and teeth, outward-bowing pelican-like mandibles, and hooked hand claws. Reduced pneumatic spaces indicate more adapted for swimming than other theropods but were still too buoyant to deep swim and fish like crocodilians. Head crests for display within the species. Being less pneumatic than most avepods diving ability improved, but not as deep diving as heavy-boned, armored crocodilians.
NOTES Very large Moroccan snout represents largest known theropod at about 15 m (50 ft) long and 13 tonnes. Absence from at least some continents may reflect lack of sufficient sampling.

Ichthyovenator laosensis
— 8.5 m (27 ft) TL, 2 tonnes
FOSSIL REMAINS Partial skeleton.
ANATOMICAL CHARACTERISTICS Vertebral spines

moderately tall and broad.
AGE Early Cretaceous, late Barremian or early Cenomanian.
DISTRIBUTION AND FORMATION/S Laos; Grès supérieurs.

Baryonyx or Camarillasaurus cirugedae
— Adult size uncertain
FOSSIL REMAINS Fragmentary skeleton, possibly immature.
ANATOMICAL CHARACTERISTICS Insufficient information.
AGE Early Cretaceous, early Barremian.
DISTRIBUTION AND FORMATION/S Northeastern Spain; Camarillas.
NOTES Remains of this and *Iberospinus* too incomplete to tell whether or not they belong to *Baryonyx*.

Baryonyx or Iberospinus natarioi
— 8 m (27 ft) TL, 1.5 tonnes
FOSSIL REMAINS Fragmentary skull and skeleton.
ANATOMICAL CHARACTERISTICS Insufficient information.
AGE Early Cretaceous, early Barremian.
DISTRIBUTION AND FORMATION/S Portugal; Papo Seco.

Baryonyx walkeri
— 7.5 m (25 ft) TL, 1.3 tonnes
FOSSIL REMAINS Partial skull and skeleton.
ANATOMICAL CHARACTERISTICS Small central crest over orbits.
AGE Early Cretaceous, Barremian.
DISTRIBUTION AND FORMATION/S Southeastern England; Weald Clay.

Baryonyx (= Suchomimus) tenerensis
— 9.5 m (30 ft) TL, 3.1 tonnes
FOSSIL REMAINS Partial skull and skeleton.
ANATOMICAL CHARACTERISTICS Small central crest over orbits. Vertebral spines moderately tall.
AGE Early Cretaceous, late Aptian.
DISTRIBUTION AND FORMATION/S Niger; upper Elrhaz.
HABITAT Coastal river delta.
NOTES Probably includes *Cristatusaurus lapparenti*.

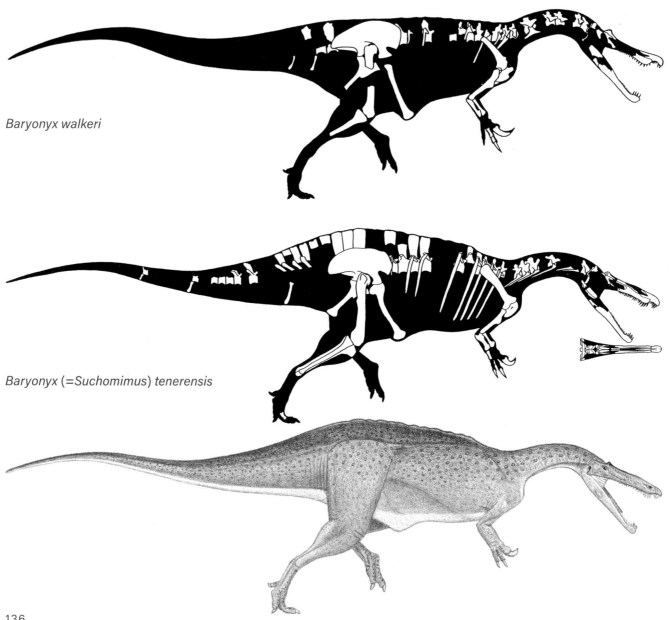

Baryonyx walkeri

Baryonyx (=Suchomimus) tenerensis

Baryonyx (=Suchomimus) tenerensis

Riojavenatrix (or *Baryonyx*) *lacustris*
—— 8 m (27 ft) TL, 1.5 tonnes
FOSSIL REMAINS Minority of skeleton.
ANATOMICAL CHARACTERISTICS Insufficient
information.
AGE Early Cretaceous, latest Barremian, or more likely
early Aptian.
DISTRIBUTION AND FORMATION/S Northeastern Spain;
Encisco.

Irritator challengeri
—— 7.5 m (25 ft) TL, 1 tonne
FOSSIL REMAINS Majority of skull.
ANATOMICAL CHARACTERISTICS Long, low midline
crest over back of head, back of head deep.
AGE Early Cretaceous, probably Albian.
DISTRIBUTION AND FORMATION/S Eastern Brazil;
Santana.

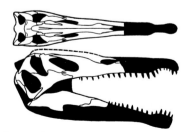

Irritator challengeri

NOTES Found as drift in nearshore marine sediments, a
snout tip labeled *Angaturama limai* from same formation
may belong to this species or even same fossil. There is
evidence of predation on a pterosaur.

Oxalaia quilombensis
—— 12 m (37 ft) TL, 7 tonnes
FOSSIL REMAINS Fragmentary skull.
ANATOMICAL CHARACTERISTICS Insufficient
information.
AGE Early Cretaceous, Cenomanian.
DISTRIBUTION AND FORMATION/S Northeastern Brazil;
Alcantara.
HABITAT Coastal.

Vallibonavenatrix cani
—— 9 m (30 ft) TL, 2 tonnes
FOSSIL REMAINS Partial skeleton.
ANATOMICAL CHARACTERISTICS Trunk vertebral spines
moderately tall.
AGE Early Cretaceous, late Barremian.
DISTRIBUTION AND FORMATION/S Eastern Spain;
Arcillas de Morella.
HABITAT Coastal.
NOTES May include fragmentary *Protathlitis cinctorrensis*.

Spinosaurus aegyptiacus

—— 11 m (35 ft) TL, 6 tonnes
FOSSIL REMAINS Uncertain.
ANATOMICAL CHARACTERISTICS Uncertain.
AGE Late Cretaceous, early Cenomanian.
DISTRIBUTION AND FORMATION/S Egypt; Bahariya.
HABITAT Coastal mangroves.
NOTES Destroyed by Allied bombing during World War II, not certain if very incomplete original remains including the very tall vertebral spines represent one or two types of theropods. Skull material is like those of other spinosaurids, tall spined vertebrae may be allosauroid. If that is so then the fossil is not all the holotype of *S. aegyptiacus*, and only the jaws can be used to diagnose Spinosauridae as currently configured. Assignment of assorted disarticulated fossils from distant regions of Africa that may not be from same very narrow time period is too problematic to meet standards for producing a reliable restoration, so extensive assessments of function and habits based on attempted reconstructions appear premature. Reconstructions with very short legs and tail sails are problematic.

—— AVETHEROPODS

SMALL TO GIGANTIC PREDATORY AND HERBIVOROUS TETANURANS FROM THE MIDDLE JURASSIC TO THE MODERN ERA, ALL CONTINENTS.

ANATOMICAL CHARACTERISTICS Highly variable. Extra joint in lower jaws usually better developed. Arms very long to very reduced, enlarged semilunate inner carpal block for increased wrist flexion usually at least incipient. Birdlike respiratory system highly developed. Brains reptilian to avian.
NOTES Absence from Mesozoic Antarctica probably reflects lack of sufficient sampling.

—— AVETHEROPOD MISCELLANEA

Lourinhanosaurus antunesi

—— Adult size uncertain
FOSSIL REMAINS Fragmentary skeleton, possibly juvenile, possible eggs.
ANATOMICAL CHARACTERISTICS Insufficient information.
AGE Late Jurassic, late Kimmeridgian or Tithonian.
DISTRIBUTION AND FORMATION/S Portugal; Amoreira-Porto Novo.
HABITAT Large, seasonally dry island with open woodlands.

Datanglong guangxiensis

—— 8 m (25 ft) TL, 1.2 tonnes
FOSSIL REMAINS Minority of skeleton.
ANATOMICAL CHARACTERISTICS Insufficient information.
AGE Early Cretaceous.
DISTRIBUTION AND FORMATION/S Central China; Xinlong.

Metriacanthosaurus parkeri

—— 7 m (23 ft) TL, 1 tonne
FOSSIL REMAINS Minority of skeleton.
ANATOMICAL CHARACTERISTICS Trunk vertebral spines moderately tall.
AGE Late Jurassic, early Oxfordian.
DISTRIBUTION AND FORMATION/S Southeast England; upper Oxford Clay.
NOTES May be a yangchuanosaurid.

Siamotyrannus isanensis

—— 6 m (20 ft) TL, 500 kg (1000 lb)
FOSSIL REMAINS Minority of skeleton.
ANATOMICAL CHARACTERISTICS Insufficient information.
AGE Early Cretaceous, Valanginian or Hauterivian.
DISTRIBUTION AND FORMATION/S Northeastern Thailand; Sao Khua.
NOTES May be an allosauroid.

Gualicho shinyae

—— 6 m (20 ft) TL, 500 kg (1000 lb)
FOSSIL REMAINS Partial skeleton(s).
ANATOMICAL CHARACTERISTICS Not heavily built. Arm reduced in size, outer fingers severely reduced to only two developed fingers.
AGE Late Cretaceous, late Cenomanian.
DISTRIBUTION AND FORMATION/S Western Argentina; middle Huincul.
HABITAT Short wet season, otherwise semiarid with open floodplains and riverine forests.
NOTES May include *Aoniraptor libertatem*. May be close to megaraptors. Reduction to two fingers apparently convergent with tyrannosaurids.

— ALLOSAUROIDS OR CARNOSAURS

LARGE TO GIGANTIC PREDATORY AVETHEROPODS FROM THE MIDDLE JURASSIC TO THE EARLY LATE CRETACEOUS, ON MOST CONTINENTS.

ANATOMICAL CHARACTERISTICS Moderately variable. Conventional avetheropod form, heads and trunks moderately robustly built and muscled. Heads generally subtriangular in side view, not very broad, preorbital hornlets modest to well developed, bladed teeth not very large. Tails long. Arm length medium to short. Legs moderately long. Brains reptilian.

HABITS Ambush and pursuit big game hunters. Heads and arms used as weapons. Extreme size of some species indicates that adults hunted adult as well as younger sauropods, armored ornithischians, and large ornithopods by means of heads and long tooth rows powered by powerful neck muscles to dispatch victims with slashing bites intended to cripple prey so it could be safely consumed. Arms used to handle and control prey when necessary.

NOTES Fragmentary remains imply presence in Australia. If megaraptorids are allosauroids rather than tyrannosaurids, as some research indicates, then this group survived until the end of the dinosaur era. The standard big theropod type. Title of group uncertain.

— ALLOSAUROID MISCELLANEA

Asfaltovenator vialidadi
—— 7 m (23 ft) TL, 1 tonne
FOSSIL REMAINS Skull and partial skeleton.
ANATOMICAL CHARACTERISTICS Snout large and upper profile gently arced, snout ridges not well developed, rounded brow hornlet not prominent. Thumb robust. Pubic boot may not be large.
AGE Late Middle or early Late Jurassic.

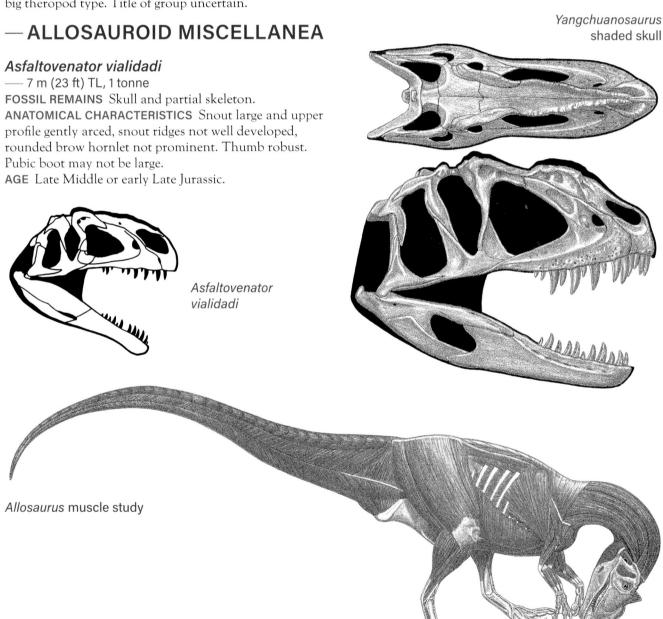

Yangchuanosaurus
shaded skull

Asfaltovenator vialidadi

Allosaurus muscle study

DISTRIBUTION AND FORMATION/S Southern Argentina; Cañadón Asfalto, level uncertain.
HABITAT Short wet season, otherwise semiarid with riverine forests and open floodplains.
NOTES May be most basal known allosauroid.

Erectopus superbus
—— Adult size uncertain
FOSSIL REMAINS Fragmentary skeleton(s), possibly immature.

ANATOMICAL CHARACTERISTICS Insufficient information
AGE Early Cretaceous.
DISTRIBUTION AND FORMATION/S Northwestern France; Penthievre beds.
NOTES Found as drift in interisland marine sediments.

—— METRIACANTHOSAURIDS OR SINRAPTORIDS
LARGE TO GIGANTIC ALLOSAUROIDS LIMITED TO THE MIDDLE AND LATE JURASSIC OF EURASIA.

ANATOMICAL CHARACTERISTICS Uniform. Brow hornlets not prominent. Remnant of fourth finger present.
NOTES Relationships within group and naming of group are uncertain, in part because placement of *Metriacanthosaurus* is uncertain.

Metriacanthosaurus parkeri
—— 7 m (23 ft) TL, 1 tonne
FOSSIL REMAINS Minority of skeleton.
ANATOMICAL CHARACTERISTICS Trunk vertebral spines moderately tall.
AGE Late Jurassic, early Oxfordian.
DISTRIBUTION AND FORMATION/S Southeast England; upper Oxford Clay.

Yangchuanosaurus? zigongensis
—— Adult size uncertain
FOSSIL REMAINS Minority of two skeletons.
ANATOMICAL CHARACTERISTICS Insufficient information.
AGE Middle Jurassic, probably Bathonian.

DISTRIBUTION AND FORMATION/S Central China; lower Xiashaximiao.
HABITAT Heavily forested.
NOTES Originally placed in *Szechuanosaurus campi*, which is based on inadequate remains; the identity of these remains is uncertain.

Yangchuanosaurus (= Sinraptor) dongi
—— 8 m (26 ft) TL, 1.5 tonnes
FOSSIL REMAINS Complete skulls and majority of a skeleton.
ANATOMICAL CHARACTERISTICS Snout ridges not well developed.
AGE Late Jurassic, early Oxfordian.

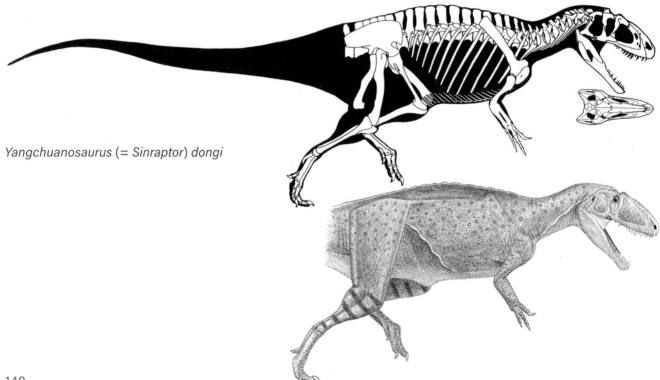

Yangchuanosaurus (= Sinraptor) dongi

DISTRIBUTION AND FORMATION/S Northwestern China; upper Shishugou.
NOTES This species barely differs from *Y. shangyouensis*.

Yangchuanosaurus shangyouensis
—— 11 m (35 ft) TL, 2.9 tonnes
FOSSIL REMAINS A few complete skulls and the majority of some skeletons, completely known.
ANATOMICAL CHARACTERISTICS Snout ridges well developed.
AGE Late Jurassic, probably Oxfordian.

DISTRIBUTION AND FORMATION/S Central China; lower Shangshaxiamiao.
HABITAT Heavily forested.
NOTES From the same formation, very similar and progressively larger in size, *Y. hepingensis*, *Y. shangyouensis*, and *Y. magnus* appear to form a progressive growth series within a single species. It is possible that the modest-sized, fairly complete but poorly described *Leshansaurus qianweiensis* from this formation may be a juvenile of this species, or it could be an afrovenator.

adult

Yangchuanosaurus shangyouensis

immature

Yangchuanosaurus shangyouensis

Yangchuanosaurus shangyouensis

— ALLOSAURIDS

LARGE TO GIGANTIC ALLOSAUROIDS LIMITED TO THE LATE JURASSIC OF THE AMERICAS, EUROPE, AND AFRICA.

ANATOMICAL CHARACTERISTICS Uniform. Heads not especially large, back of head more rigidly braced; large, triangular, sharp-tipped brow hornlets present. Tails long. Boots of pubes large. Fourth finger entirely lost.
ONTOGENY Growth rates moderately rapid, adult size reached in about two decades; life spans normally not exceeding three decades.

Allosaurus jimmadseni
—— Adult size uncertain
FOSSIL REMAINS Two skulls and skeletons, immature.
ANATOMICAL CHARACTERISTICS Head fairly elongated, narrow, temporal region not deep because edge of back of upper jaws is on same line as forward edge; nasal opening shallow, brow hornlet large. Arm and hand not large and/or slender.
AGE Late Jurassic, late Oxfordian.
DISTRIBUTION AND FORMATION/S Colorado, Utah; lower Morrison.
HABITAT Short wet season, otherwise semiarid with open floodplain prairies and riverine forests.
HABITS Not as powerful a predator as later, more robust *A. fragilis.*

Allosaurus **unnamed species?**
—— 8.5 m (28 ft) TL, 1.7 tonnes
FOSSIL REMAINS Majority of two skulls.
ANATOMICAL CHARACTERISTICS Head fairly elongated, narrow, temporal region deep because edge of back of upper jaws is set down relative to forward edge; nasal opening deep, brow hornlet large. Arm and hand not large and/or slender.
AGE Late Jurassic, early Kimmeridgian.
DISTRIBUTION AND FORMATION/S Wyoming; lower middle Morrison.
HABITAT Short wet season, otherwise semiarid with open floodplain prairies and riverine forests.
NOTES May be a different species than later *A. fragilis.* Long versus short heads, found in same quarry, are probably genders or contemporary species.

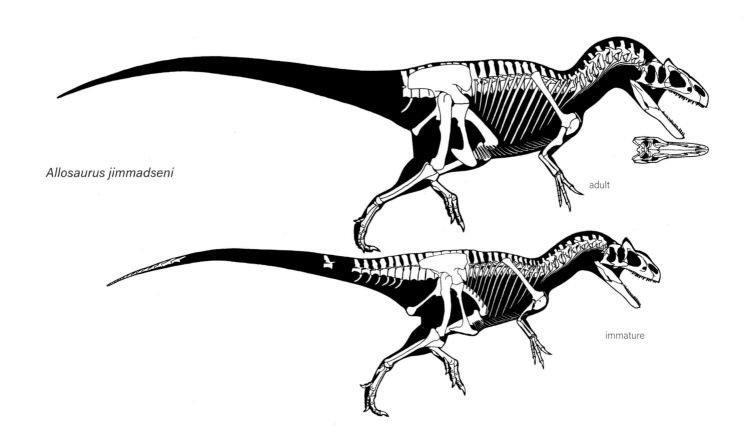

Allosaurus jimmadseni

adult

immature

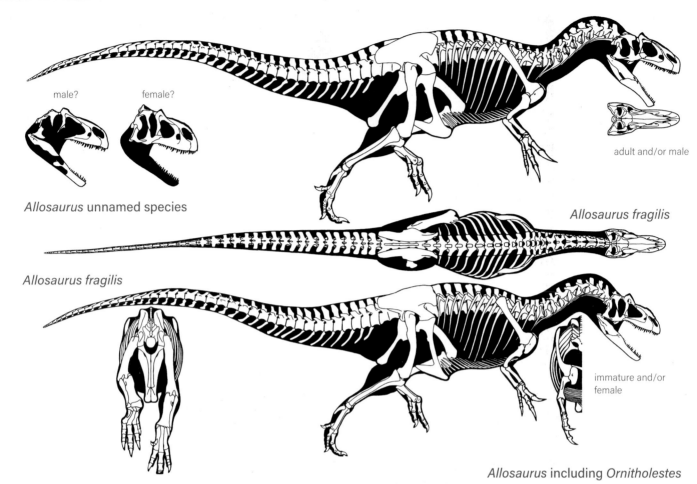

male? female?

Allosaurus unnamed species

Allosaurus fragilis

adult and/or male

Allosaurus fragilis

immature and/or female

Allosaurus including *Ornitholestes*

Allosaurus fragilis

—— 8.5 m (28 ft) TL, 1.7 tonnes

FOSSIL REMAINS A large number of complete and partial skulls and skeletons.

ANATOMICAL CHARACTERISTICS Head length/depth varies from moderately deep to deep, fairly broad, temporal region deep because edge of back of upper jaws is set down relative to forward edge; nasal opening shallow, brow hornlet large. Arm and hand large and stout.

AGE Late Jurassic, middle Kimmeridgian.

DISTRIBUTION AND FORMATION/S Utah, Wyoming, Colorado; upper middle Morrison.

HABITAT Short wet season, otherwise semiarid with open floodplain prairies and riverine forests.

NOTES The classic non-tyrannosaur large theropod. Heads are not similar in proportions, long versus short heads are probably genders or contemporary species; includes *Antrodemus valens* and *Allosaurus lucasi* based on inadequate remains.

Saurophaganax (or Allosaurus) maximus

—— 10.5 m (35 ft) TL, 3 tonnes

FOSSIL REMAINS Minority of the skeleton.

ANATOMICAL CHARACTERISTICS Insufficient information.

AGE Late Jurassic, late middle and/or late Kimmeridgian.

DISTRIBUTION AND FORMATION/S Oklahoma; upper Morrison.

HABITAT Wetter than earlier Morrison, otherwise semiarid with open floodplain prairies and riverine forests.

HABITS Able to hunt larger sauropods.

Allosaurus? europaeus

—— Adult size uncertain

FOSSIL REMAINS Partial skull and skeletal remains, most or all immature and juvenile, possibly dozens of eggs with embryos.

ANATOMICAL CHARACTERISTICS Temporal region of skull deep because edge of back of upper jaws is set down relative to forward edge, nasal opening deep, brow hornlet large.

AGE Late Jurassic, early or middle Tithonian.

DISTRIBUTION AND FORMATION/S Portugal, middle Lourinhã.

HABITAT Large, seasonally dry island with open woodlands.

NOTES May or may not be same genus as earlier *Allosaurus* or *Lourinhanosaurus*.

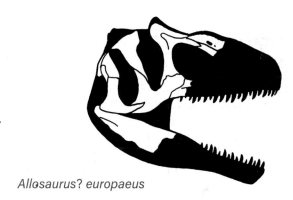

Allosaurus? europaeus

—— CARCHARODONTOSAURIDS

LARGE TO GIGANTIC ALLOSAUROIDS OF THE LATE JURASSIC AND CRETACEOUS OF THE WESTERN HEMISPHERE, EURASIA, AND AFRICA.

ANATOMICAL CHARACTERISTICS Fairly variable. Boots of pubes further enlarged. Arms reduced.

HABITS Arms used less when hunting than in other allosauroids.

ONTOGENY Growth rates moderately rapid, adult size of larger examples reached in about three decades; life spans up to half a century.

NOTES Absence from additional continents may reflect lack of sufficient sampling.

Lusovenator santosi

—— Adult size uncertain

FOSSIL REMAINS Minority of juvenile skeleton.

ANATOMICAL CHARACTERISTICS Insufficient information.

AGE Late Jurassic, Kimmeridgian.

DISTRIBUTION AND FORMATION/S Portugal; Praia da Amoreira-Porto Novo.

Concavenator corcovatus

—— 5 m (16 ft) TL, 330 kg (700 lb)

FOSSIL REMAINS Nearly complete skull and skeleton.

ANATOMICAL CHARACTERISTICS Brow hornlet well developed. Vertebral spines immediately in front of and behind pelvis tall, forming double sail back, especially in front of hips. Bumps on trailing edge of upper arm indicate large quills.

AGE Early Cretaceous, late Barremian.

DISTRIBUTION AND FORMATION/S Spain; Calizas de la Huérguina.

NOTES Possible quill nodes imply presence of feathery structures in allosauroids.

Concavenator corcovatus

Siamraptor suwati

—— 6 m (20 ft) TL, 500 kg (1000 lb)
FOSSIL REMAINS Fragmentary skull and skeleton.
ANATOMICAL CHARACTERISTICS Insufficient
information.
AGE Early Cretaceous, middle Aptian.
DISTRIBUTION AND FORMATION/S Northeastern
Thailand; Khok Kraut.
HABITAT Seasonally arid.

Acrocanthosaurus atokensis

—— 11 m (35 ft) TL, 4.9 tonnes
FOSSIL REMAINS Complete skull and majority of
skeletons.
ANATOMICAL CHARACTERISTICS Upper profile of snout
gently arced, aft head markedly broader than snout,
rounded brow hornlet not prominent, back of lower jaw
deep. Tall vertebral spines from neck to tail form a low
sail. Arm and hand moderately reduced.
AGE Early Cretaceous, Aptian or middle Albian.
DISTRIBUTION AND FORMATION/S Oklahoma, Texas;
Antlers, level uncertain, Twin Mountains.
HABITAT Coastal floodplains with swamps and marshes.
NOTES Placement of all remains from many locations
and uncertain spread of ages in same taxon is
problematic. May be an allosaurid, a
carcharodontosaurid, or its own group.

Lajasvenator ascheriae

—— 3 m (10 ft) TL, 100 kg (200 lb)
FOSSIL REMAINS Minority of skull and skeleton.
ANATOMICAL CHARACTERISTICS Insufficient
information.
AGE Early Cretaceous, late Valanginian.
DISTRIBUTION AND FORMATION/S Western Argentina;
middle Mulichinco.

Eocarcharia dinops

—— Adult size uncertain
FOSSIL REMAINS Minority of skull, possibly large
juvenile.
ANATOMICAL CHARACTERISTICS Insufficient
information.
AGE Early Cretaceous, Aptian.
DISTRIBUTION AND FORMATION/S Niger; Elrhaz, level
uncertain.

Kelmayisaurus petrolicus

—— 10 m (34 ft) TL, 4 tonnes
FOSSIL REMAINS Fragmentary skull.
ANATOMICAL CHARACTERISTICS Insufficient
information.
AGE Early Cretaceous.
DISTRIBUTION AND FORMATION/S Northwestern
China; Tugulu.

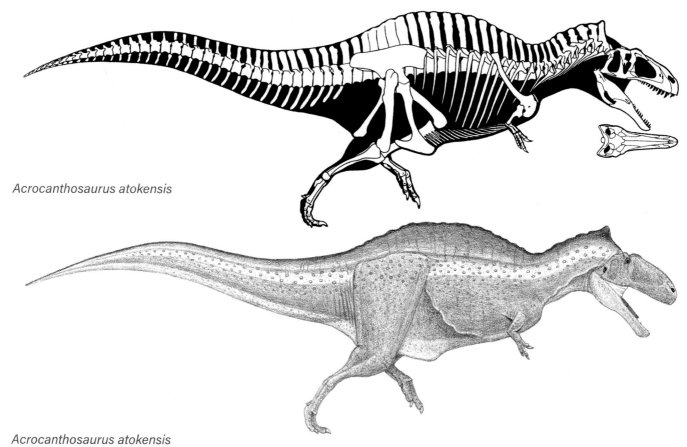

Acrocanthosaurus atokensis

Acrocanthosaurus atokensis

Tyrannotitan chubutensis

—— 13 m (42 ft) TL, 9 tonnes

FOSSIL REMAINS Minority of skull and skeleton.
ANATOMICAL CHARACTERISTICS Vertebral spines over tail rather tall.
AGE Early Cretaceous, Aptian.
DISTRIBUTION AND FORMATION/S Southern Argentina; middle Cerro Barcino.
NOTES Among the largest avepods.

Carcharodontosaurus saharicus

—— 12 m (40 ft) TL, 7 tonnes

FOSSIL REMAINS Partial skull and parts of skeletons.
ANATOMICAL CHARACTERISTICS Brow hornlet small.
AGE Late Cretaceous, early Cenomanian.
DISTRIBUTION AND FORMATION/S Egypt, possibly Morocco and other parts of North Africa; Bahariya, upper Kem Kem Beds, etc.
HABITAT Coastal mangroves.
NOTES Whether remains from a large number of formations actually belong to this species is problematic.

Carcharodontosaurus saharicus

Carcharodontosaurus iguidensis

—— 10 m (34 ft) TL, 4 tonnes

FOSSIL REMAINS Minority of several skulls and small portion of skeleton.
ANATOMICAL CHARACTERISTICS Standard for group.
AGE Late Cretaceous, early Cenomanian.
DISTRIBUTION AND FORMATION/S Niger; Echkar.
NOTES Was placed in *C. saharicus*.

Giganotosaurus (or Carcharodontosaurus) carolinii

—— 12.7–13.7 m (42–45 ft) TL, 8–10 tonnes

FOSSIL REMAINS Partial skull and majority of skeleton, minority of skull.
ANATOMICAL CHARACTERISTICS Brow hornlet not prominent. Front tip of lower jaw squared off.
AGE Late Cretaceous, early Cenomanian.
DISTRIBUTION AND FORMATION/S Western Argentina; lower Candeleros.
HABITAT Short wet season, otherwise semiarid with open floodplains and riverine forests.
NOTES Partial carcharodontosaur skulls have been restored with too great a length. Arm in skeletal based on *Meraxes*. Size estimates complicated by incompleteness of largest known remains.

Meraxes gigas

—— 10 m (34 ft) TL, 4 tonnes

FOSSIL REMAINS Majority of skull and partial skeleton.
ANATOMICAL CHARACTERISTICS Brow hornlet not prominent. Arm and hand moderately reduced. Inner main toe claw enlarged.
AGE Late Cretaceous, middle Cenomanian.
DISTRIBUTION AND FORMATION/S Western Argentina; lower Huincul.
HABITAT Short wet season, otherwise semiarid with open floodplains and riverine forests.

Meraxes gigas

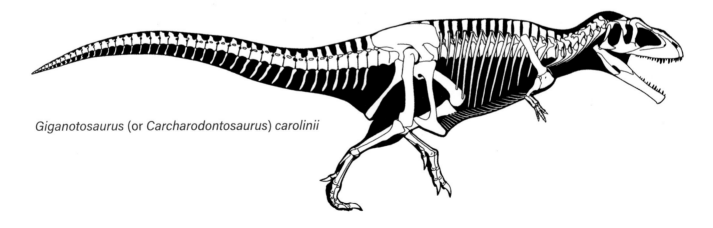

Giganotosaurus (or Carcharodontosaurus) carolinii

Mapusaurus roseae
—— 11.5 m (38 ft) TL, 6 tonnes
FOSSIL REMAINS Large number of skull and skeletal bones.
ANATOMICAL CHARACTERISTICS Standard for group.
AGE Late Cretaceous, late Cenomanian.
DISTRIBUTION AND FORMATION/S Western Argentina; middle Huincul.
HABITAT Short wet season, otherwise semiarid with open floodplains and riverine forests.

Shaochilong maortuensis
—— 5 m (16 ft) TL, 500 kg (1,000 lb)
FOSSIL REMAINS Partial skull and skeleton.
ANATOMICAL CHARACTERISTICS Insufficient information.
AGE Late Cretaceous, Turonian.
DISTRIBUTION AND FORMATION/S Northern China; Ulansuhai.
NOTES Fossil may be a subadult.

—— NEOVENATORIDS
LARGE TO GIGANTIC ALLOSAUROIDS OF THE CRETACEOUS OF EURASIA.

ANATOMICAL CHARACTERISTICS Fairly variable. Arms well developed.

Neovenator salerii
—— 7 m (23 ft) TL, 1.3 tonnes
FOSSIL REMAINS Minority of skull and skeleton.
ANATOMICAL CHARACTERISTICS Lightly built, leg long. Head narrow.
AGE Early Cretaceous, early Barremian.
DISTRIBUTION AND FORMATION/S Southern England; lower Wessex.

Chilantaisaurus tashuikouensis
—— 11 m (35 ft) TL, 5 tonnes
FOSSIL REMAINS Partial skeleton.
ANATOMICAL CHARACTERISTICS Skeleton heavily constructed. Arm well developed.
AGE Late Cretaceous, Turonian.
DISTRIBUTION AND FORMATION/S Northern China; Ulansuhai.
NOTES The last of the known allosauroids, if this taxon belongs to this group.

—— COELUROSAURS
SMALL TO GIGANTIC PREDATORY AND HERBIVOROUS AVETHEROPODS OF THE MIDDLE JURASSIC TO THE MODERN ERA, ALL CONTINENTS.

ANATOMICAL CHARACTERISTICS Highly variable. Tails long to very short. Arms from longer than leg to severely reduced. Legs extremely gracile to robust, toes four to three. Feathers often preserved.
HABITS Extremely variable, from big game predators to fully herbivorous.

—— TYRANNOSAUROIDS
SMALL TO GIGANTIC PREDATORY AVETHEROPODS OF THE MIDDLE JURASSIC TO THE END OF THE DINOSAUR ERA, ALL CONTINENTS EXCEPT ANTARCTICA.

ANATOMICAL CHARACTERISTICS Front teeth of upper jaw D-shaped in cross section. Arms long to severely reduced. Legs long. Brains reptilian.
HABITS Pursuit and ambush predators. Dispatched victims with powerful, deep, punch-like bites rather than slashing, wounding bites.
NOTES Proving to be long-lived and widely distributed. Although tyrannosauroids are not descended directly from allosauroids, the two groups may share a close common ancestor.

—— BASO-TYRANNOSAUROIDS
SMALL TO GIGANTIC TYRANNOSAUROIDS OF THE MIDDLE JURASSIC TO THE END OF THE DINOSAUR ERA.

ANATOMICAL CHARACTERISTICS Fairly variable. Arms not severely reduced, usually three fingers. Legs not as gracile as those of tyrannosaurids.
HABITS Arms as well as head used to handle and wound prey.

Proceratosaurus bradleyi
— 3.5 m (11 ft) TL, 110 kg (250 lb)
FOSSIL REMAINS Majority of skull.
ANATOMICAL CHARACTERISTICS Head subrectangular, snout fairly deep and adorned with nasal midline hornlet or crest, back of head rigidly built, teeth fairly large.
AGE Middle Jurassic, middle Bathonian.
DISTRIBUTION AND FORMATION/S Central England; Forest Marble.
HABITS Crest too delicate for headbutting, for display within the species.
NOTES Name incorrectly implies an ancestral relationship with the very different *Ceratosaurus*. This, *Kileskus*, *Guanlong*, *Dilong*, and *Sinotyrannus* may form family Proceratosauridae.

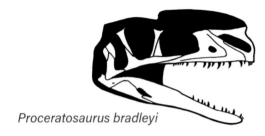

Proceratosaurus bradleyi

Kileskus aristotocus
— 5 m (15 ft) TL, 200 kg (450 lb)
FOSSIL REMAINS Majority of skull and minority of skeleton.
ANATOMICAL CHARACTERISTICS Snout adorned with nasal midline hornlet or crest.

AGE Middle Jurassic, Bathonian.
DISTRIBUTION AND FORMATION/S Central Russia; upper Itat.
HABITS Crest too delicate for headbutting, for display within the species.

Guanlong wucaii
— 3.5 m (11 ft) TL, 110 kg (250 lb)
FOSSIL REMAINS Nearly complete skull and partial skeleton.
ANATOMICAL CHARACTERISTICS Snout ridges united and enlarged into an enormous midline crest with a backward projection.
AGE Middle Jurassic, late Callovian.
DISTRIBUTION AND FORMATION/S Northwestern China; lower Shishugou.
HABITS Crest too delicate for headbutting, for display within the species.
NOTES Probably not as suggested a juvenile of *Monolophosaurus*.

Dilong paradoxus
— Adult size uncertain
FOSSIL REMAINS A few nearly complete skulls and partial skeletons, external fibers, possibly juvenile.
ANATOMICAL CHARACTERISTICS Head long and low, low Y-shaped crest on snout. Hand fairly long. Leg very long. Full extent of protofeather covering uncertain.
AGE Early Cretaceous, latest Barremian and earliest Aptian?

Guanlong wucaii

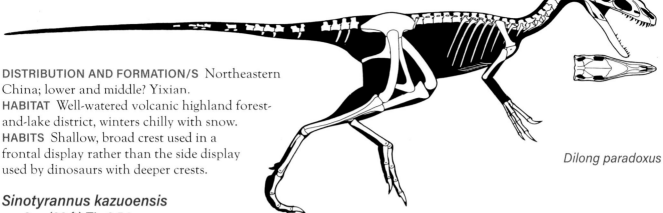

Dilong paradoxus

DISTRIBUTION AND FORMATION/S Northeastern China; lower and middle? Yixian.
HABITAT Well-watered volcanic highland forest-and-lake district, winters chilly with snow.
HABITS Shallow, broad crest used in a frontal display rather than the side display used by dinosaurs with deeper crests.

Sinotyrannus kazuoensis

—— 9 m (30 ft) TL, 2.5 tonnes
FOSSIL REMAINS Partial skull and minority of skeleton.
ANATOMICAL CHARACTERISTICS Insufficient information.
AGE Early Cretaceous, late Barremian.
DISTRIBUTION AND FORMATION/S Northeastern China; lower Jiufotang.
HABITAT Well-watered volcanic highland forest-and-lake district, winters chilly with snow.

Aviatyrannis jurassica

—— 1 m (3 ft) TL, 4 kg (10 lb)
FOSSIL REMAINS Small portion of skeleton, possibly juvenile.
ANATOMICAL CHARACTERISTICS Insufficient information.
AGE Late Jurassic, late Kimmeridgian.
DISTRIBUTION AND FORMATION/S Portugal; Camadas de Alcobaça.
HABITAT Large, seasonally dry island with open woodlands.

Juratyrant langhami

—— 5 m (16 ft) TL, 300 kg (600 lb)
FOSSIL REMAINS Partial skeleton.
ANATOMICAL CHARACTERISTICS Lightly built.
AGE Late Jurassic, early Tithonian.
DISTRIBUTION AND FORMATION/S Southern England; Kimmeridge Clay.
NOTES Originally placed in the smaller baso-tyrannosaurid *Stokesosaurus*, the original species of which, *S. clevelandi* from the Late Jurassic of western North America, is based on one bone.

Eotyrannus lengi

—— 3 m (10 ft) TL, 70 kg (150 lb)
FOSSIL REMAINS Minority of skull and skeleton.
ANATOMICAL CHARACTERISTICS Head strongly built, front of head deep. Skeleton lightly built. Arm long. Leg long and gracile.

AGE Early Cretaceous, early Barremian.
DISTRIBUTION AND FORMATION/S Southern England; lower Wessex.

Yutyrannus huali

—— 7.5 m (25 ft) TL, 1.3 tonnes
FOSSIL REMAINS Three largely complete skulls and skeletons, two juvenile, external fibers.

Yutyrannus huali

Yutyrannus huali

ANATOMICAL CHARACTERISTICS Fairly robustly built. Snout fairly deep, large midline prominence atop back of head. Arm long. External fibers apparently covered much of body including upper foot.
AGE Early Cretaceous, latest Barremian and/or earliest Aptian.
DISTRIBUTION AND FORMATION/S Northeastern China; Yixian, level uncertain.
HABITAT Well-watered volcanic highland forest-and-lake district, winters chilly with snow.

Suskityrannus hazelae
—— 3 m (10 ft) TL, 70 kg (150 lb)
FOSSIL REMAINS Two fragmentary skulls and skeletons.
ANATOMICAL CHARACTERISTICS Insufficient information.
AGE Late Cretaceous, middle Turonian.
DISTRIBUTION AND FORMATION/S New Mexico; lower Moreno Hill.
HABITAT Coastal swamps and marshes.

Xiongguanlong baimoensis
—— 5 m (15 ft) TL, 200 kg (450 lb)
FOSSIL REMAINS Majority of a distorted skull and minority of skeleton.
ANATOMICAL CHARACTERISTICS Head and especially snout long, low.
AGE Early Cretaceous, probably Aptian or Albian.

DISTRIBUTION AND FORMATION/S Central China; upper Xiagou.
NOTES Shows that some tyrannosauroids were fairly large in the mid-Cretaceous.

Jinbeisaurus wangi
—— 5 m (15 ft) TL, 200 kg (450 lb)
FOSSIL REMAINS Minority of skull and fragmentary skeleton.
ANATOMICAL CHARACTERISTICS Insufficient information.
AGE Late Cretaceous.
DISTRIBUTION AND FORMATION/S Eastern China; upper Huiquanpu.

Timurlengia euotica
—— 3 m (10 ft) TL, 70 kg (150 lb)
FOSSIL REMAINS Minority of skull and skeletons.
ANATOMICAL CHARACTERISTICS Insufficient information.
AGE Late Cretaceous, Turonian.
DISTRIBUTION AND FORMATION/S Uzbekistan; Bissekty.

Moros intrepidus
—— 3 m (10 ft) TL, 70 kg (150 lb)
FOSSIL REMAINS Minority of skeleton.
ANATOMICAL CHARACTERISTICS Insufficient information.

AGE Late Cretaceous, early Cenomanian.
DISTRIBUTION AND FORMATION/S Utah; uppermost Cedar Mountain.
HABITAT Short wet season, otherwise semiarid with floodplain prairies, open woodlands, and riverine forests.

Megaraptor namunhuaiquii
—— 8 m (25 ft) TL, 2 tonnes
FOSSIL REMAINS Minority of a few skeletons.
ANATOMICAL CHARACTERISTICS Hand claws slender.
AGE Late Cretaceous, late Turonian.
DISTRIBUTION AND FORMATION/S Western Argentina; Portezuelo.
HABITAT Well-watered woodlands with short dry season.
NOTES Incorrectly thought to be the biggest dromaeosaurid; others consider this a spinosaur or an allosauroid. This, *Fukuiraptor, Australovenator, Tratayenia, Murusraptor, Maip, Phuwiangvenator, Aerosteon,* and *Orkoraptor* may form family Megaraptoridae.

Fukuiraptor kitadaniensis
—— 5 m (16 ft) TL, 300 kg (600 lb)
FOSSIL REMAINS Partial skeleton.
ANATOMICAL CHARACTERISTICS Lightly built, leg long.
AGE Early Cretaceous, Albian.
DISTRIBUTION AND FORMATION/S Central Japan; Kitadani.

Australovenator wintonensis
—— 6 m (20 ft) TL, 500 kg (1,000 lb)
FOSSIL REMAINS Minority of skull and skeleton.
ANATOMICAL CHARACTERISTICS Lightly built, leg long.
AGE Early Cretaceous, late Cenomanian and/or early Turonian.
DISTRIBUTION AND FORMATION/S Northeastern Australia; upper Winton.
HABITAT Well-watered, cold winters with heavy snows.

Tratayenia rosalesi
—— 11 m (35 ft) TL, 5 tonnes
FOSSIL REMAINS Fragmentary skeleton.
ANATOMICAL CHARACTERISTICS Insufficient information.
AGE Late Cretaceous, Santonian.
DISTRIBUTION AND FORMATION/S Western Argentina; Bajo de la Carpa.
HABITAT Semiarid.

Murusraptor barrosaensis
—— 8 m (25 ft) TL, 2 tonnes
FOSSIL REMAINS Partial skull and minority of skeleton.
ANATOMICAL CHARACTERISTICS Insufficient information.
AGE Late Cretaceous, Coniacian.
DISTRIBUTION AND FORMATION/S Western Argentina; middle Sierra Barrosa.

Maip macrothorax
—— 8 m (25 ft) TL, 2 tonnes
FOSSIL REMAINS Minority of skeleton.
ANATOMICAL CHARACTERISTICS Insufficient information.
AGE Late Cretaceous, middle Maastrichtian.
DISTRIBUTION AND FORMATION/S Southern Argentina; Chorrillo.

Phuwiangvenator yaemniyomi
—— 6 m (20 ft) TL, 500 kg (1,000 lb)
FOSSIL REMAINS Two fragmentary skeletons, one juvenile.
ANATOMICAL CHARACTERISTICS Insufficient information.
AGE Early Cretaceous, upper? Barremian.
DISTRIBUTION AND FORMATION/S Northeastern Thailand; Sao Khua.
NOTES *Vayuraptor nongbualamphuensis* may be juvenile of this species.

Aerosteon riocoloradensis
—— 6 m (20 ft) TL, 500 kg (1,000 lb)
FOSSIL REMAINS Minority of skull and partial skeleton.
ANATOMICAL CHARACTERISTICS Lightly built, leg long.
AGE Late Cretaceous, late Santonian and/or early Campanian.
DISTRIBUTION AND FORMATION/S Western Argentina; Anacleto.
HABITAT Seasonally semiarid.

Orkoraptor burkei
—— 6 m (20 ft) TL, 500 kg (1,000 lb)
FOSSIL REMAINS Minority of skull and skeleton.
ANATOMICAL CHARACTERISTICS Insufficient information.
AGE Late Cretaceous, early Maastrichtian.
DISTRIBUTION AND FORMATION/S Southern Argentina; Pari Aike.
NOTES If megaraptorids were allosauroids, as some researchers indicate, then *Orkoraptor* indicates that allosauroids lasted until close to and probably to the end of the dinosaur era.

Santanaraptor placidus
—— 1.5 m (5 ft) TL, 15 kg (30 lb)
FOSSIL REMAINS Minority of skeleton.
ANATOMICAL CHARACTERISTICS Insufficient information.
AGE Early Cretaceous, probably Albian.
DISTRIBUTION AND FORMATION/S Eastern Brazil; Santana.
NOTES Found as drift in nearshore marine sediments.

Labocania anomala
—— 7 m (23 ft) TL, 1.5 tonnes
FOSSIL REMAINS Small portion of skull and skeleton.
ANATOMICAL CHARACTERISTICS Massively constructed.
AGE Late Cretaceous, probably Campanian.
DISTRIBUTION AND FORMATION/S Baja California, Mexico; La Bocana Roja.
HABITS Ambush big game hunter.

Appalachiosaurus? montgomeriensis
—— Adult size uncertain
FOSSIL REMAINS Partial skull, skeleton.
ANATOMICAL CHARACTERISTICS Typically gracile for smaller tyrannosaurids.
AGE Late Cretaceous, early Campanian.
DISTRIBUTION AND FORMATION/S Alabama; Demopolis.
Notes Found as drift in nearshore marine sediments. May have had fairly large arms. May be same genus as *Dryptosaurus*.

Dryptosaurus aquilunguis
—— 7.5 m (25 ft) TL, 1.5 tonnes
FOSSIL REMAINS Minority of skull and skeleton.
ANATOMICAL CHARACTERISTICS Arm and especially hand large, probably two fingers.
AGE Late Cretaceous, late Campanian or early Maastrichtian.
DISTRIBUTION AND FORMATION/S New Jersey; New Egypt.
HABITS Arms used as weapons.
NOTES Found as drift in nearshore marine sediments.

Stygivenator? unnamed species
—— Adult size uncertain
FOSSIL REMAINS Complete subadult skull and skeleton (Bloody Mary), hand (Jodi).
ANATOMICAL CHARACTERISTICS Medium sized, very gracile. Small lacrimal hornlet present; very long snout very shallow and sharply triangular as is front end of maxillary depression, dentary tip upturned, is very elongated and has long groove on side, teeth modest-sized blades. Lower arm and especially elongated two-fingered hand longer than femur. Lower leg very long.
DISTRIBUTION AND FORMATION/S Montana; lower? Hell Creek.
HABITAT Well-watered coastal woodlands, climate warmer than earlier in Maastrichtian.
HABITS Pursued similar-sized dinosaurs including ornithomimids. Arms used as weapons.
NOTES Complete remains not yet described in detail. Cannot be, as often believed, a juvenile *Tyrannosaurus* because hand is larger than that of latter's adults, and shrinkage of appendages with maturity does not occur in other tyrannosauroids, any dinosaurs, or amniote vertebrates. Also has too many teeth to be juvenile *Tyrannosaurus* because number of teeth does not decrease with growth in other tyrannosauroids, dinosaurs, or reptiles. Instead the large lower arms means is not a tyrannosaurid and indicates that tyrannosaurs' reduction to two fingers evolved at same time arms were decreasing in size. May or may not be same genus as incomplete *S.molnari*, of which this may be an ancestor.

Stygivenator molnari
—— Adult size uncertain
FOSSIL REMAINS Minority of skull (Jordan).
ANATOMICAL CHARACTERISTICS Medium sized. Snout fairly deep, sharply triangular as is front end of maxillary depression, dentary tip upturned, lacks groove on side, upper teeth are large blades.
DISTRIBUTION AND FORMATION/S Montana; upper? Hell Creek.
HABITAT Well-watered coastal woodlands, climate warmer than earlier in Maastrichtian.
NOTES Also not a juvenile *Tyrannosaurus* or tyrannosaurid; probably a close relative of apparently earlier *Stygivenator*? unnamed species.

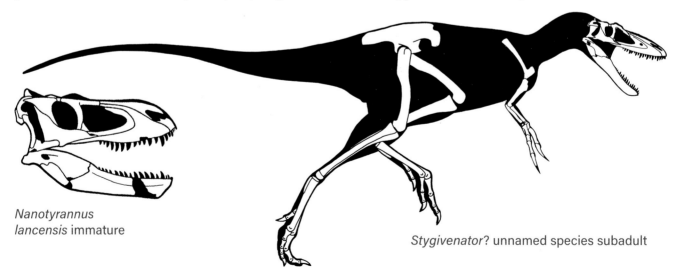

Nanotyrannus lancensis immature

Stygivenator? unnamed species subadult

Nanotyrannus lancensis

—— Adult size uncertain

FOSSIL REMAINS Subadult skull (Nano).

ANATOMICAL CHARACTERISTICS Medium sized. Snout moderately deep, back of head may be especially broad; eyes face more strongly forward, increasing overlap of fields of vision; dentary tip not upturned, is not very long and has long groove on side, 15 maxillary and 16 dentary teeth, are modest-sized blades.

DISTRIBUTION AND FORMATION/S Montana; lower? Hell Creek.

HABITAT Well-watered coastal woodlands, climate warmer than earlier in Maastrichtian.

NOTES Often believed to be a juvenile *Tyrannosaurus* but has too many teeth and former lacks dentary groove. Also very different from *Stygivenator*, and distinctive from next taxon.

Nanotyrannus or unnamed genus and unnamed species

—— Adult size uncertain

FOSSIL REMAINS Majority of skull and skeleton, subadult (Jane), other possible incomplete remains.

ANATOMICAL CHARACTERISTICS Medium sized, very gracile. Small lacrimal hornlet present; snout fairly shallow, back of head may be especially broad, 15 maxillary and 17 dentary teeth are modest-sized blades. Upper arm not elongated, but lower arm and especially hand may have been longer than those of tyrannosaurids. Lower leg very long.

DISTRIBUTION AND FORMATION/S Montana; upper? Hell Creek.

HABITAT Well-watered coastal woodlands, climate warmer than earlier in Maastrichtian.

HABITS Pursued similar-sized dinosaurs, including ornithomimids.

NOTES Often believed to be a juvenile *Tyrannosaurus* but has too many teeth, former lacks the dentary groove, and growth pattern preserved in bones show this is not a young juvenile. Placement in *Nanotyrannus* also problematic; may be part of an array of modest-sized, large-handed tyrannosauroids present in *Tyrannosaurus* habitat, where they outnumbered the juvenile of the latter two to one. The small hands, more conical teeth, and shorter legs inherent to juveniles that were growing into mammoth *Tyrannosaurus* may have hindered their ability to compete with the nontyrannosaurids that had evolved specifically as specialized, medium-sized predators and had recently invaded what had been solely tyrannosaurid western territories from the east.

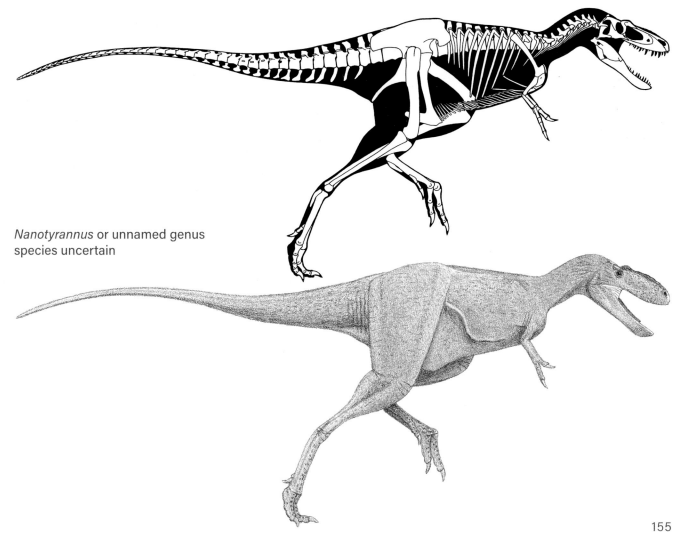

Nanotyrannus or unnamed genus species uncertain

— TYRANNOSAURIDS

LARGE TO GIGANTIC TYRANNOSAUROIDS, LIMITED TO THE LATER LATE CRETACEOUS OF NORTH AMERICA AND ASIA.

ANATOMICAL CHARACTERISTICS Fairly uniform, but juveniles and smaller species gracile, large adults more robust. Heads large and long, robustly constructed, stout bars in the temporal region invade the side openings and further strengthen the skull, skulls of juveniles and smaller species somewhat long, shallow, and graceful, those of adults deeper and shorter snouted, midline ridge on snout rugose, probably bore long, possibly low ridge boss, small brow hornlets or bosses over orbits, bosses usually subtle discs that do not project well above skull top rim, usually small midline prominence atop back of head, back half of head a broad box accommodating exceptionally powerful jaw muscles, eyes face partly forward, and some degree of stereo vision possible, front of snout broader and more rounded than usual, supporting a U-shaped arc of teeth that are D-shaped in cross section, teeth stouter and more conical than in general, lower jaw deep, especially back half. Necks strongly constructed, powerfully muscled. Trunks short and deep. Tails shorter and lighter than standard in other large theropods. Arms severely reduced in size to about two-thirds femur length, outer fingers severely reduced to only two developed digits, yet hands still functional. Pelves very large and leg very long, so leg muscles exceptionally well developed, feet very long and strongly compressed from side to side. Reduction of tails and arms in favor of enlarged and elongated leg indicates greater speed potential than in other giant theropods. Scales small and pebbly, may have been mixed with fibers that, if present, may have been limited to top surfaces. Skeletons of juveniles very gracile, becoming increasingly robust as size increases, but basic characteristics unaltered. Brains larger than usual in large theropods, olfactory bulbs especially large.

ONTOGENY Growth rates moderately rapid, adult size reached in about two decades; life span normally not exceeding three decades. Some small species that have been named are the juveniles of giant taxa; whether any species were small as adults is uncertain.

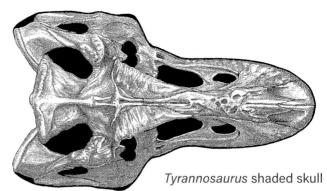

Tyrannosaurus shaded skull

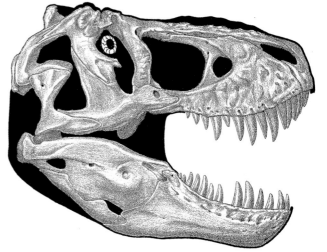

Tyrannosaurus muscle study

HABITS Pursuit and perhaps ambush predators; able to chase running prey at unusually high speeds. Compact form also enhanced maneuverability. Head the primary if not sole weapon. Long snouts of juveniles imply they were independent hunters. Giant adults used massive heads and strong teeth to dispatch victims with powerful, deep, punch-like bites rather than slashing, wounding bites aimed with forward vision, powered by very strong jaw and neck muscles and intended to cripple prey so it could be safely consumed. Function of arms if any poorly understood; they appear too short and small to be useful in handling prey, may have provided grip for males while mating. Head bosses presumably for headbutting during intraspecific contests.

NOTES Overall the most advanced and sophisticated of large theropods, better adapted for gigantism than other tyrannosauroids. Large numbers of hunting juveniles and the smaller species may have swamped their habitats, suppressing the populations of smaller predaceous avepods such as dromaeosaurs and troodonts.

Alectrosaurus olseni
—— Adult size uncertain
FOSSIL REMAINS Partial skull, skeleton, possibly immature.
ANATOMICAL CHARACTERISTICS Typically gracile for smaller tyrannosaurids.
AGE Late Cretaceous.
DISTRIBUTION AND FORMATION/S China; Iren Dabasu.
HABITAT Seasonally wet-dry woodlands.
HABITS Assuming the known fossils are adults, pursued similar-sized dinosaurs including ornithomimids.

Alioramus remotus
—— Adult size uncertain
FOSSIL REMAINS Skull, some parts of skeleton, possibly immature.
ANATOMICAL CHARACTERISTICS Typically gracile for smaller tyrannosaurids. Crenulated midline crest on snout.
AGE Late Late Cretaceous.
DISTRIBUTION AND FORMATION/S Mongolia; Nogoon Tsav.
HABITS Presuming the known fossils are adults, pursued similar-sized dinosaurs including ornithomimids.

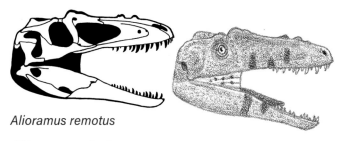

Alioramus remotus

Alioramus altai
—— Adult size uncertain
FOSSIL REMAINS Immature skulls, some parts of skeleton.
ANATOMICAL CHARACTERISTICS Head unusually long and low even for a tyrannosaur of this size. Crenulated midline crest on snout. Supraorbital boss slender.
AGE Late Cretaceous, late Campanian and/or early Maastrichtian.
DISTRIBUTION AND FORMATION/S Mongolia; Nemegt, level not available.

HABITAT Temperate, well-watered, dense woodlands with seasonal rains and winter snow.
HABITS Assuming the known fossils are adults, pursued similar-sized dinosaurs including ornithomimids.
NOTES Thought to be somewhat different in time from *A. remotus*; if not, may be the same species.

Alioramus (or Qianzhousaurus) sinensis
—— Adult size uncertain
FOSSIL REMAINS Nearly complete skull and minority of skeleton.
ANATOMICAL CHARACTERISTICS Snout unusually slender even for a tyrannosaur of this size. Supraorbital boss slender.
AGE Late Cretaceous, early Maastrichtian.
DISTRIBUTION AND FORMATION/S Northern China; Yuanpu.

Alioramus (or *Qianzhousaurus*) *sinensis* skull

Albertosaurus (Gorgosaurus) libratus
—— 8 m (27 ft) TL, 2.7 tonnes
FOSSIL REMAINS A number of skulls and skeletons, adult and juvenile, small skin patches, completely known.
ANATOMICAL CHARACTERISTICS Skeleton not heavily built. Brow hornlet fairly prominent.
AGE Late Cretaceous, late middle? and/or early late Campanian.
DISTRIBUTION AND FORMATION/S Alberta, Montana?; at least middle Dinosaur Park, upper Judith River?, upper Two Medicine?
HABITAT Well-watered, forested floodplain with coastal swamps and marshes, cool winters, uplands drier.
HABITS Relatively gracile build indicates adults specialized in hunting unarmed hadrosaurs, although ceratopsians and ankylosaurs were probably occasional victims.

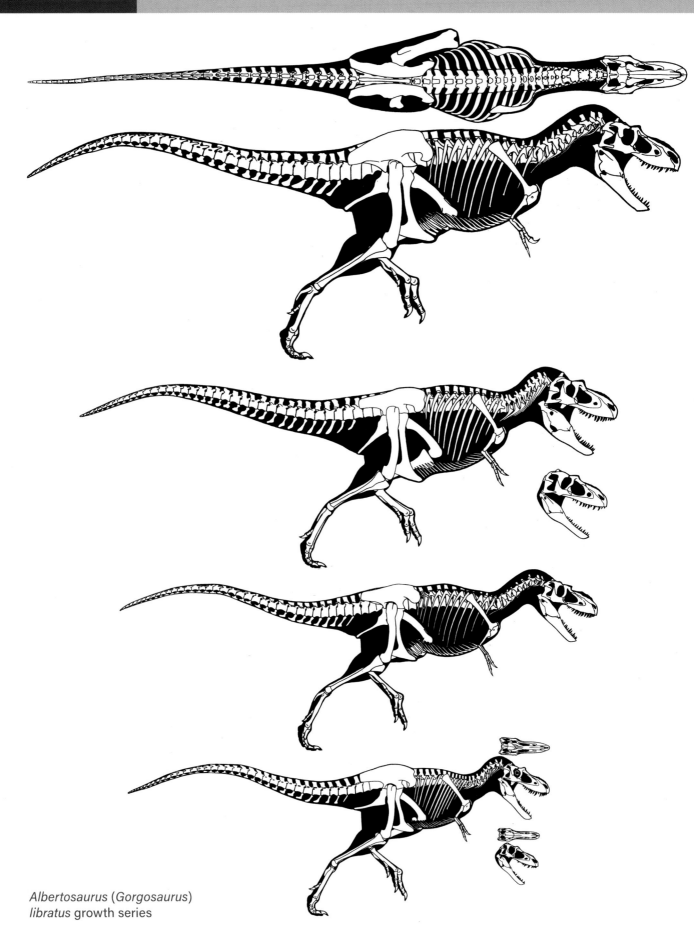

Albertosaurus (*Gorgosaurus*)
libratus growth series

NOTES Often a separate genus, very similar to *Albertosaurus sarcophagus*. Differences including robustness of head may indicate different genders and/or species, remains from Montana may be latter, whether species lived through the entire time span of Dinosaur Park Formation also uncertain. May be direct ancestor of *A. sarcophagus*.

Albertosaurus (Albertosaurus) sarcophagus
—— 8 m (27 ft) TL, 2.9 tonnes
FOSSIL REMAINS Some skulls and partial skeletons, adult and juvenile, well known.
ANATOMICAL CHARACTERISTICS Very similar to *A. libratus*. Leg may have been somewhat longer.
AGE Late Cretaceous, early Maastrichtian.
DISTRIBUTION AND FORMATION/S Alberta, Montana; lower to middle Horseshoe Canyon.
HABITAT Well-watered, forested floodplain with coastal swamps and marshes, cool winters.
HABITS Relatively gracile build indicates this species also preyed mainly on hadrosaurs.
NOTES Probably includes *A. arctunguis*; fossils from upper Horseshoe Canyon may be a distinct taxon. Apparently a fairly rare example of only one large predatory avepod species present in a dinosaur habitat.

Last of the rhino-sized western North American tyrannosaurids, others such as *Tyrannosaurus? mcraeensis* were already larger.

Dynamoterror dynastes
—— 9 m (30 ft) TL, 2.5 tonnes
FOSSIL REMAINS Fragmentary skull and skeleton.
ANATOMICAL CHARACTERISTICS Insufficient information.
AGE Late Cretaceous, early Campanian.
DISTRIBUTION AND FORMATION/S New Mexico; upper Menefee.
NOTES New remains may further detail this species.

Lythronax argestes
—— 6.5 m (20 ft) TL, 1.5 tonnes
FOSSIL REMAINS Majority of skull and minority of skeleton.

Lythronax argestes

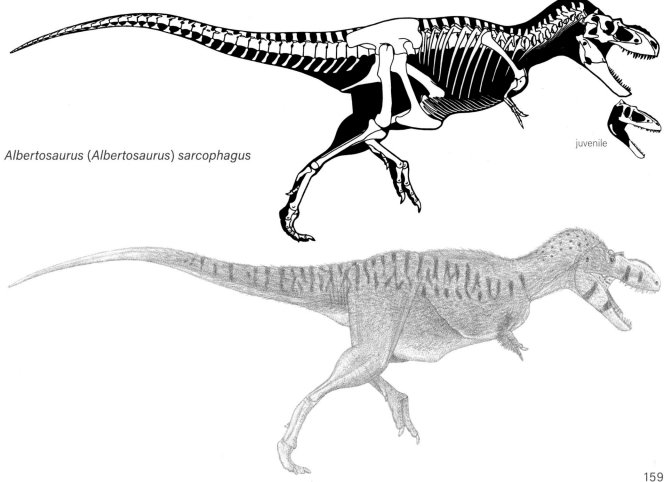

Albertosaurus (Albertosaurus) sarcophagus

juvenile

159

ANATOMICAL CHARACTERISTICS Insufficient information.
AGE Late Cretaceous, middle Campanian.
DISTRIBUTION AND FORMATION/S Southern Utah; Wahweap.
NOTES Restoration of back of incomplete skull as exceptionally broad is problematic.

Teratophoneus curriei

—— 8 m (27 ft) TL, 2.5 tonnes
FOSSIL REMAINS Majority of skull and minority of skeleton.
ANATOMICAL CHARACTERISTICS Small triangular hornlet above orbit, modest supraorbital boss half-moon shaped.
AGE Late Cretaceous, late Campanian.
DISTRIBUTION AND FORMATION/S Utah; middle Kaiparowits.
HABITS Tremendous strength of head indicates specialized for hunting horned ceratopsids.
NOTES There is some uncertainty about adult size. Extreme broadening of back of head similar to *Tyrannosaurus*, may be ancestral to latter.

Teratophoneus curriei

Bistahieversor sealeyi

—— 8 m (27 ft) TL, 2.5 tonnes
FOSSIL REMAINS Nearly complete skull and skeleton.
ANATOMICAL CHARACTERISTICS Snout fairly deep, quite large midline prominence atop back of head.
AGE Late Cretaceous, late Campanian.
DISTRIBUTION AND FORMATION/S New Mexico; lower Kirtland.
NOTES Information on skeleton insufficient for restoration.

Bistahieversor sealeyi

Daspletosaurus torosus

—— 9 m (30 ft) TL, 2.8 tonnes
FOSSIL REMAINS Complete skulls and majority of skeleton, other remains including juveniles.
ANATOMICAL CHARACTERISTICS Head broad, strongly constructed, supraorbital boss knob shaped, not highly prominent, teeth robust. Skeleton robustly built. Leg shorter than usual for group.
AGE Late Cretaceous, middle middle Campanian.

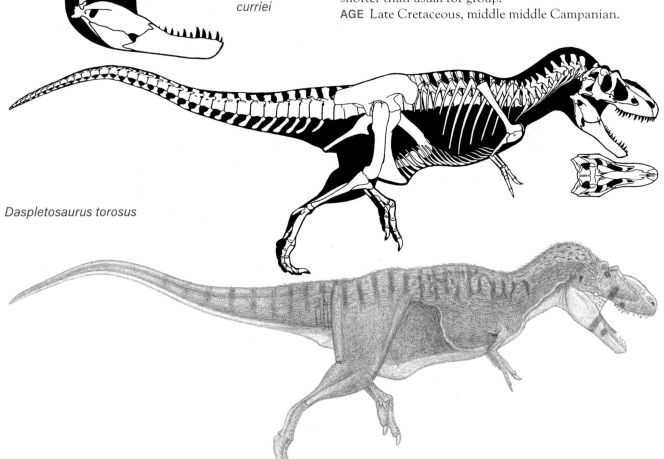

Daspletosaurus torosus

DISTRIBUTION AND FORMATION/S Alberta; upper Oldman.

HABITAT Well-watered, forested floodplain with coastal swamps and marshes, cool winters.

HABITS Stout build indicates this genus was specialized to cope with horned ceratopsids, and armored ankylosaurs when available, expanding the resource base it could prey on, although more vulnerable hadrosaurs were probably still common prey.

NOTES May be more than one species; may be direct ancestor of *D. wilsoni*.

Daspletosaurus wilsoni

—— 9 m (30 ft) TL, 2.5 tonnes

FOSSIL REMAINS Some skulls and skeletons of varying completeness including juveniles.

ANATOMICAL CHARACTERISTICS Similar to *D. torosus*, except head shallower, tip of snout rounded.

AGE Late Cretaceous, late middle Campanian.

DISTRIBUTION AND FORMATION/S Montana, Alberta?; middle Judith River, lower Dinosaur Park?.

HABITAT Well-watered, forested floodplain with coastal swamps and marshes, cool winters.

NOTES May be *T. torosus*, may be direct ancestor of *D. horneri*.

Daspletosaurus wilsoni

Daspletosaurus

Daspletosaurus horneri

Daspletosaurus horneri

—— 9 m (30 ft) TL, 2.5 tonnes

FOSSIL REMAINS Two or three skulls and partial remains, juvenile to adult.

ANATOMICAL CHARACTERISTICS Similar to *D. torosus*.

AGE Late Cretaceous, early late Campanian.

DISTRIBUTION AND FORMATION/S Montana, Alberta?; upper Two Medicine, middle Dinosaur Park?.

HABITAT Seasonally dry upland woodlands.

NOTES Modest time separation suggests two Montana skulls may be different geotime species, a Dinosaur Park skull may be same species. Daspletosaurs may be ancestral to *Tyrannosaurus* and its Asian relations.

Daspletosaurus? unnamed species

—— 11 m (36 ft) TL, 5 tonnes

FOSSIL REMAINS Minority of skull and skeleton.

ANATOMICAL CHARACTERISTICS Robust. Large brow hornlet, two small incisors at tip of dentary.

AGE Late Cretaceous, middle Campanian.

DISTRIBUTION AND FORMATION/S Montana; lower Judith River.

HABITAT Well-watered, forested floodplain with coastal swamps and marshes, cool winters.

NOTES Originally thought to be a *Tyrannosaurus* from much later lower Hell Creek, may be a distinct genus. Along with *T.? mcraeensis* indicates that western North American tyrannosaurids approached *Tyrannosaurus* size much earlier than previously realized.

Nanuqsaurus hoglundi

—— 8 m (27 ft) TL, 2.5 tonnes

FOSSIL REMAINS Small minority of skull, juvenile.

ANATOMICAL CHARACTERISTICS Insufficient information.

AGE Late Cretaceous, middle Maastrichtian.

DISTRIBUTION AND FORMATION/S Northern Alaska; middle Prince Creek.

HABITAT Well-watered coastal woodland, cool summers, severe winters including heavy snows.

NOTES Based on inadequate remains, is included because indicates presence of a tyrannosaurid in a harsh Arctic environment. Thought to be a dwarf tyrannosaur, other remains indicate was large. Possibly heavily feathered, especially in winter.

Tarbosaurus bataar

—— 9.5 m (31 ft) TL, 4 tonnes

FOSSIL REMAINS A number of skulls and skeletons from juvenile to adult, completely known. Small skin patches.

ANATOMICAL CHARACTERISTICS Overall adult build moderately robust. Heads of even largest examples not especially broad aft, no erect brow hornlet, supraorbital boss knob shaped, not highly prominent, dentary lacks groove on side, 12–13 maxillary and 14–15 dentary teeth not exceptionally large and robust, usually two small incisors at tip of dentary.

AGE Late Cretaceous, late Campanian and/or early Maastrichtian.

DISTRIBUTION AND FORMATION/S Mongolia and northern China; lower? Nemegt, Nemegt Svita, Yuanpu, Qiupa, etc.

HABITAT Temperate, well-watered, dense woodlands with seasonal rains and winter snow.

HABITS Lacking large horned ceratopsids in its habitat, not as powerful as *Tyrannosaurus*.

NOTES Examples from stratigraphic levels and/or with anatomy other than original fossil may be a different species; *T. efremovi* has a deeper skull than the more *Tyrannosaurus*-like *T. bataar*. *Raptorex kriegsteini*, which is smallest skeleton in growth series, and *Bagaraatan ostromi* share a short groove on the side of the dentary with juveniles of the same size and are very probably young of this taxon. Very fragmentary *Zhuchengtyrannus magnus* indicates that a tyrannosaurine as large as *T. bataar* lived in the region a little earlier. May be close relative of *Tyrannosaurus*.

Tyrannosaurus? mcraeensis

—— 12 m (40 ft) TL, 7 tonnes

FOSSIL REMAINS Minority of skull.

ANATOMICAL CHARACTERISTICS Display boss above orbit elongated semi-spindle shape, does not project strongly above top of head, aft dentary not deep, two small incisors at tip of dentary.

AGE Late Cretaceous, late Campanian or early Maastrichtian.

DISTRIBUTION AND FORMATION/S New Mexico: lower Hall Lake.

HABITS Possibly not as powerful as late Maastrichtian *Tyrannosaurus*.

NOTES Because is 6 to 7 million years younger than *Tyrannosaurus imperator*, and aft dentary is not as deep, may not be in genus *Tyrannosaurus*. Helps show that titanic tyrannosaurids were on the American scene long before the late Maastrichtian. Type may be ancestral to *T. imperator*, which has similar orbital display boss.

Tyrannosaurus imperator

—— 12 m (40 ft) TL, 7.5 tonnes

FOSSIL REMAINS A number of skulls and skeletons (Sue, Tristan, Trix, G-rex, Cope, etc.), other remains possibly

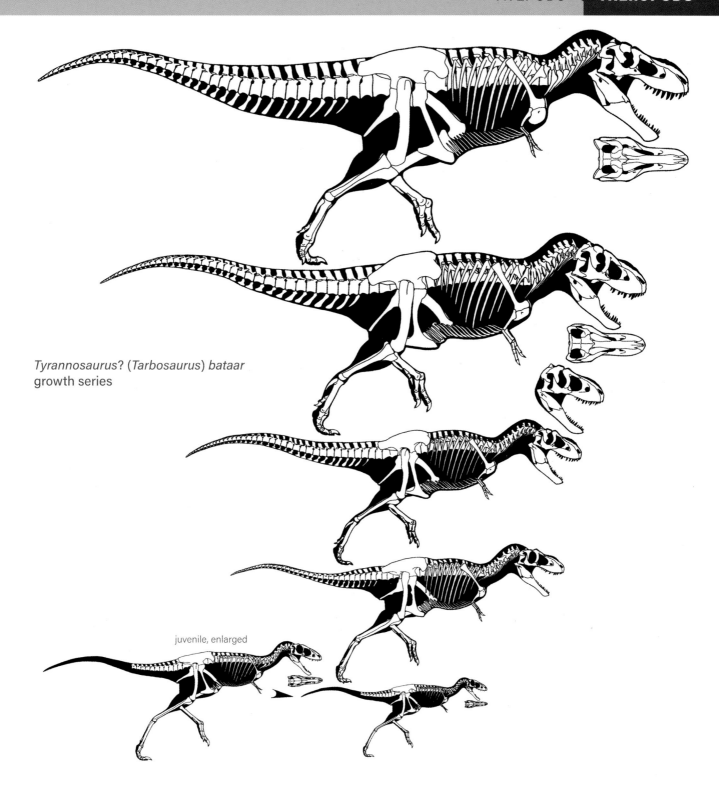

Tyrannosaurus? (Tarbosaurus) bataar
growth series

juvenile, enlarged

including some incomplete juveniles, adults completely known.
ANATOMICAL CHARACTERISTICS Overall adult build robust. Display boss above orbit elongated spindle shape at least in large males, does not project strongly above top of head, usually two small incisors at tip of dentary, otherwise as for *T. rex*.
AGE Late Cretaceous, late Maastrichtian.

DISTRIBUTION AND FORMATION/S Lower Lance and Hell Creek; Montana, Dakotas, Wyoming.
HABITAT Well-watered coastal woodlands, climate cooler than in latest Maastrichtian, possibly chilly in winter.
HABITS Same as for *T. rex*.
NOTES This species retained the ancestral tyrannosaurid condition of robust femora and two incisors and the low, elongated orbital display boss of *T. mcraeensis*; relationships

163

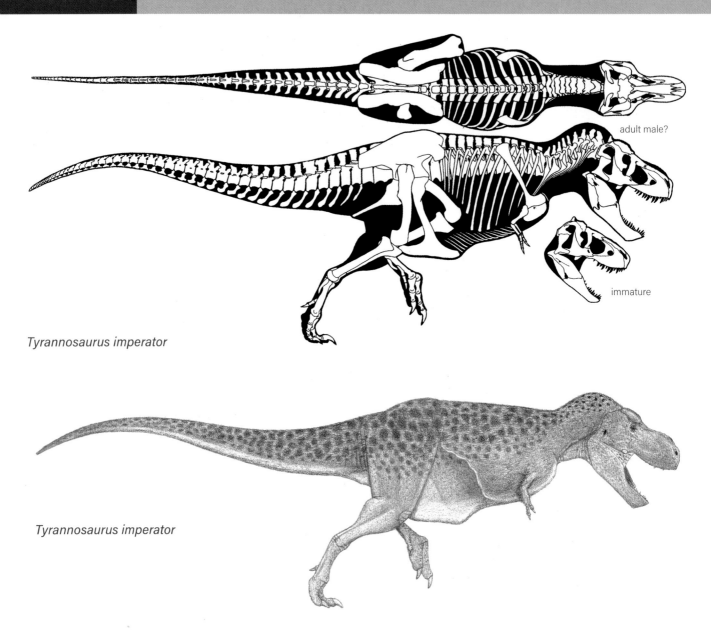

adult male?

immature

Tyrannosaurus imperator

Tyrannosaurus imperator

with Asian tarbosaurs possible. Absence from Canada due to lack of formations north of border at that time. May be ancestral to either or both of the later *Tyrannosaurus* species.

Tyrannosaurus rex
—— 12 m (40 ft) TL, 7.5 tonnes

FOSSIL REMAINS A number of skulls and skeletons (Scotty, Samson, Wy-rex, etc.), other remains possibly including some incomplete juveniles, adults completely known.

ANATOMICAL CHARACTERISTICS Overall adult build robust. Skull much more heavily constructed and stouter than those of other tyrannosaurids, back of skull especially broad to accommodate oversized jaw and neck muscles; no other land predator with as powerful a bite; eyes face more strongly forward, increasing overlap of fields of vision; snout also broad, no erect brow hornlet

at least in adults, display boss above orbit is a knob-shaped disk that projects well above top of head at least in males, lower jaws very deep, lack groove on side of dentary, 11–12 maxillary and 12–14 dentary teeth unusually large and conical even in juveniles to some extent, one small incisor at tip of dentary. Neck very stout. Lower arm and hand very small even in juveniles. Lower leg of juveniles not highly elongated.

AGE Late Cretaceous, latest Maastrichtian.

DISTRIBUTION AND FORMATION/S Montana, Dakotas, Wyoming, Colorado, Alberta, Saskatchewan; upper Lance and Hell Creek, Ferris, Denver, Frenchman, Scollard, Willow Creek.

HABITAT Well-watered coastal woodlands, climate warmer than earlier in Maastrichtian.

HABITS An extreme version of the tyrannosaurid form; healed wounds on adult hadrosaurs and ceratopsids

164

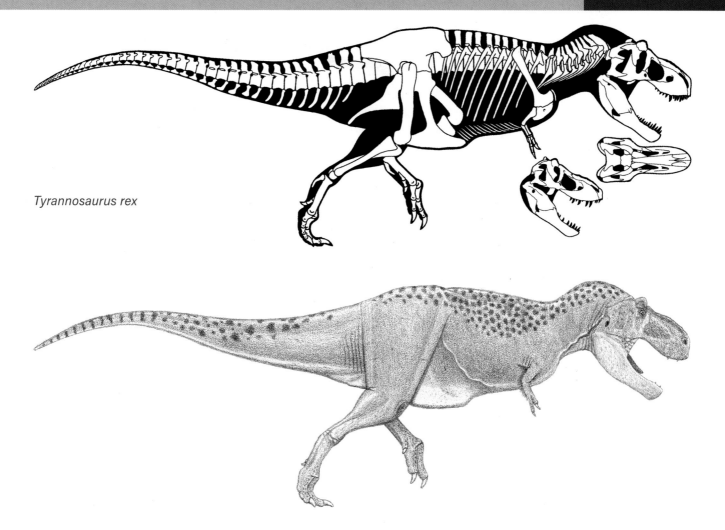

Tyrannosaurus rex

indicate that adults hunted similarly elephant-sized prey on a regular basis, using the tremendously powerful head and teeth to lethally wound victims. Such firepower and size were more than needed to hunt less-dangerous juveniles or just adult hadrosaurs, which made up a minority of the herbivore population dominated by dangerous horned *Triceratops prorsus*. Because the great width of the aft skull probably evolved to maximize power of the jaw muscles for dispatching dangerous prey, the forward-facing eyes with overlapping fields of vision may have been a secondary result.

NOTES Once very rare, high financial value has helped encourage the discovery of dozens of *Tyrannosaurus* fossils, all of which were placed in *T. rex* as continues to be a common practice; so is the perpetual citation of the species name *rex* when such is much less frequently done regarding other dinosaur genera, even in the same presentation. Unusually great variation in the robustness of the fossils, shifts in the number of dentary incisors, strong differences in the shapes of the supraorbital bosses used to distinguish species, and differential pattern of distribution of varying robustness not being uniform throughout the stratigraphic range, as should be true if the robusts and graciles were females and males, indicate

that multiple species evolved over some 1.5 million years. If correct, then the main competitor of robust *T. rex* was the more gracile *T. regina*, with *T. rex* possibly more specialized in preying on massive *Triceratops*. Distinctive supraorbital bosses differentiate the two very close relatives. Because their juvenile anatomy was in many regards that of miniature adults on their way to becoming elephant-sized tyrannosaurids, the smaller armed, less gracile juvenile *Tyrannosaurus* were not highly competitive against similar-sized *Nanotyrannus* and *Stygivenator* that, being specialized, medium-sized hunters, were about twice as abundant in the ecosystem. At the same time, tyrannosaurids being better adapted for being gigantic; adult *Tyrannosaurus* reigned supreme as the dominant big predator.

Tyrannosaurus regina

—— 12 m (40 ft) TL, 7.5 tonnes

FOSSIL REMAINS A number of skulls and skeletons (Wankel, Stan, Black Beauty, Thomas, Rigby, etc.), other remains possibly including some incomplete juveniles, adults completely known.

ANATOMICAL CHARACTERISTICS Overall build of adult skull and skeleton more gracile. Display boss above orbit

Tyrannosaurus

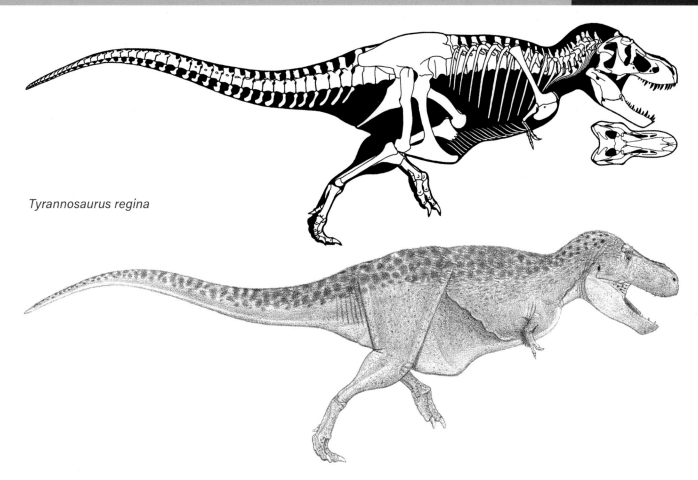

Tyrannosaurus regina

does not project strongly above head. One small incisor at tip of dentary, otherwise similar to *T. rex*.
AGE Late Cretaceous, latest Maastrichtian.
DISTRIBUTION AND FORMATION/S Montana, Dakotas, Wyoming, Colorado, Alberta, Saskatchewan, Utah?; upper Lance and Hell Creek, Ferris, Denver, Frenchman, Scollard, Willow Creek, North Horn?.
HABITAT Well-watered coastal woodlands, climate warmer than earlier in Maastrichtian.
HABITS Same as for *T. rex*, perhaps better adapted for hunting less common edmontosaurs than were more robust *T. rex*.

NOTES The presence of *T. rex* and *T. regina* at the same time and place may be a location where their otherwise separate ranges overlapped, or they may have dwelled together over much of North America—this same principle applies to the numerous other tyrannosaurs found in the region at the end of the Maastrichtian. A skull bone from the North Horn Formation has an orbital display boss of this species' type. Two species of giant *Tyrannosaurus* probably were present in the same region at the final extinction. *T. megagracilis* is probably a juvenile of this species or *T. rex*

— NEOCOELUROSAURS

SMALL TO GIGANTIC PREDATORY AND HERBIVOROUS COELUROSAURS OF THE LATE JURASSIC TO THE MODERN ERA, ALL CONTINENTS.

ANATOMICAL CHARACTERISTICS Highly variable. Heads from large to small, brow hornlet absent, toothed to toothless and beaked, when teeth present serrations tend to be reduced in some manner or absent. Neck moderately to very long. Tails very long to very short. Shoulder girdles usually like that in birds, with horizontal scapula blades and vertical, anterior-facing coracoids, arms very long to short, large, semilunate inner carpal block that further increased wrist flexion usually present, hand usually long, fingers from three to one. Brains enlarged, semiavian in form.
ONTOGENY Growth rates apparently moderate.
HABITS Reproduction generally similar to that of ratites and tinamous; in at least some cases males incubated the eggs and were probably polygamous; egg hatching in a given clutch not synchronous.
NOTES Neocoelurosaurs are more derived than tyrannosauroids; some examples such as *Bicentenaria*, *Sciurumimus*, and *Zuolong* may be basal to this group.

— NEOCOELUROSAUR MISCELLANEA

SMALL NEOCOELUROSAURS OF THE LATE JURASSIC TO THE LATE CRETACEOUS.

ANATOMICAL CHARACTERISTICS Uniform. Heads small to moderately large, elongated, crests absent. Skeletons gracile. Tails long. Arms and hands well developed. Legs and feet long and slender.
HABITS Fast small game hunters, also fishers, similar in function to earlier coelophysids.
NOTES The classic small coelurosaurs were a widespread element in dinosaur faunas, the generalized small canids of their time; relationships often uncertain and group splittable into subdivisions.

Coelurus fragilis
—— 2.5 m (8 ft) TL, 25 kg (50 lb)
FOSSIL REMAINS Majority of skeleton.
ANATOMICAL CHARACTERISTICS Lightly built. Fingers long and slender.
AGE Late Jurassic, late Oxfordian.
DISTRIBUTION AND FORMATION/S Wyoming; lower Morrison.
HABITAT Short wet season, otherwise semiarid with open floodplain prairies and riverine forests.
HABITS Able to pursue faster prey than *Ornitholestes*.
NOTES Remains imply close relatives higher in the Morrison. This and *Tanycolagreus* may form the family Coeluridae, which may be tyrannosauroids.

Tanycolagreus topwilsoni
—— 4 m (13 ft) TL, 120 kg (250 lb)
FOSSIL REMAINS Much of the skull and majority of the skeleton.
ANATOMICAL CHARACTERISTICS Head large, long, subrectangular. Leg long and gracile.
AGE Late Jurassic, late Oxfordian.
DISTRIBUTION AND FORMATION/S Wyoming; lower Morrison.
HABITAT Short wet season, otherwise semiarid with open floodplain prairies and riverine forests.
HABITS Prey included fairly large game.

Bicentenaria argentina
—— 3 m (10 ft) TL, 50 kg (100 lb)
FOSSIL REMAINS Partial skull and skeletal elements.
ANATOMICAL CHARACTERISTICS Insufficient information.

AGE Late Cretaceous, middle Cenomanian.
DISTRIBUTION AND FORMATION/S Western Argentina; upper Candeleros.
HABITAT Short wet season, otherwise semiarid with open floodplains and riverine forests.

Sciurumimus albersdoerferi
—— Adult size uncertain
FOSSIL REMAINS Complete juvenile skull and skeleton, extensive external fibers.
ANATOMICAL CHARACTERISTICS Insufficient information due to juvenile status of fossil.
AGE Late Jurassic, late Kimmeridgian.
DISTRIBUTION AND FORMATION/S Southern Germany; upper Rögling.

Zuolong salleei
—— 3 m (10 ft) TL, 50 kg (100 lb)
FOSSIL REMAINS Majority of skull and minority of skeleton.
ANATOMICAL CHARACTERISTICS Standard for group.
AGE Late Jurassic, early Oxfordian.
DISTRIBUTION AND FORMATION/S Northwestern China; upper Shishugou.
HABITAT Well-watered woodlands with short dry season.

Zuolong salleei

Sciurumimus albersdoerferi

Ornitholestes hermanni

—— 2 m (7 ft) TL, 14 kg (30 lb)

FOSSIL REMAINS Nearly complete skull and majority of skeleton.

ANATOMICAL CHARACTERISTICS Head subrectangular, rather small relative to body, teeth on lower jaw restricted to front end. Leg moderately long.

AGE Late Jurassic, late Oxfordian.

DISTRIBUTION AND FORMATION/S Wyoming; lower Morrison.

HABITAT Short wet season, otherwise semiarid with open floodplain prairies and riverine forests.

HABITS Hunted small game as well as fished.

NOTES A classic coelurosaur.

Aniksosaurus darwini

—— 2.5 m (9 ft) TL, 30 kg (65 lb)

FOSSIL REMAINS Several partial skeletons.

ANATOMICAL CHARACTERISTICS Robustly built.

Posterior pelvis broad.

AGE Late Cretaceous, late Cenomanian or Turonian.

DISTRIBUTION AND FORMATION/S Southern Argentina; lower Bajo Barreal.

HABITAT Seasonally wet, well-forested floodplain.

Scipionyx samniticus

—— Adult size uncertain

FOSSIL REMAINS Complete skull and almost complete skeleton, juvenile, some internal organs preserved.

ANATOMICAL CHARACTERISTICS Proportions characteristic for juvenile.

AGE Early Cretaceous, early Albian.

DISTRIBUTION AND FORMATION/S Central Italy; unnamed formation.

HABITS Juveniles hunted small vertebrates, insects.

NOTES May be the juvenile of much larger abelisaur or carcharodontosaur, or of a compsognathid. Intestinal tract has a preavian condition.

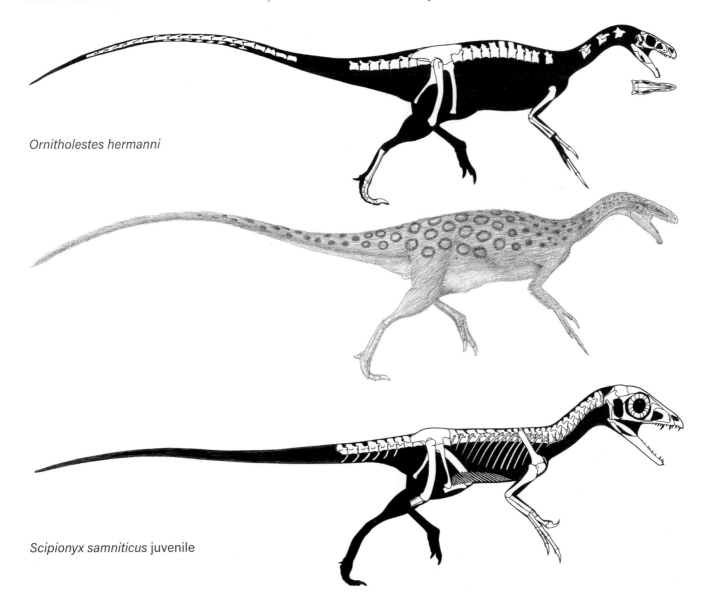

Ornitholestes hermanni

Scipionyx samniticus juvenile

— COMPSOGNATHIDS

SMALL PREDATORY NEOCOELUROSAURS LIMITED TO THE LATE JURASSIC AND EARLY CRETACEOUS OF EURASIA AND SOUTH AMERICA.

ANATOMICAL CHARACTERISTICS Uniform. In most regards standard for small neocoelurosaurs. Necks moderately long. Tails quite long. Hands strongly asymmetrical because thumb and claw unusually stout and outer finger slender. Boots on pubes large; legs moderately long.

HABITS Ambushed and chased small game, also fish in some cases. Thumb an important weapon for hunting and/or combat within species.

NOTES A common and widely distributed group of small neocoelurosaurs, the foxes of their time.

Juravenator starki
— Adult size uncertain

FOSSIL REMAINS Nearly complete juvenile skull and skeleton, small skin patches, external fibers.

ANATOMICAL CHARACTERISTICS Head subrectangular, snout fairly deep, indentation in snout, teeth large. Scapula blade slender. Skin with small scales on leg and tail, some of body covered by fibers, full extent of either uncertain.

AGE Late Jurassic, middle Kimmeridgian.

DISTRIBUTION AND FORMATION/S Southern Germany; lower Painten.

HABITAT Found as drift in lagoonal deposits near probably arid, brush-covered islands.

HABITS Large teeth indicate it hunted fairly large animals; kink in upper jaw indicates it also fished.

NOTES Indicates that some dinosaurs bore both bare body skin and feathers, like some birds.

Compsognathus longipes
— Adult size uncertain

FOSSIL REMAINS Nearly complete skull and skeleton, possibly juvenile.

ANATOMICAL CHARACTERISTICS Head shallow, subtriangular, teeth small.

AGE Late Jurassic, middle or late Kimmeridgian.

DISTRIBUTION AND FORMATION/S Southern Germany; Painten, level uncertain.

HABITAT Found as drift in lagoonal deposits near probably arid, brush-covered islands.

NOTES The second dinosaur known from a largely complete skull and skeleton. Large head relative to larger-bodied *C. corallestris* suggests is a juvenile, but head is not large compared with those of other adult compsognathids. Gut contents are those of a dismembered lizard.

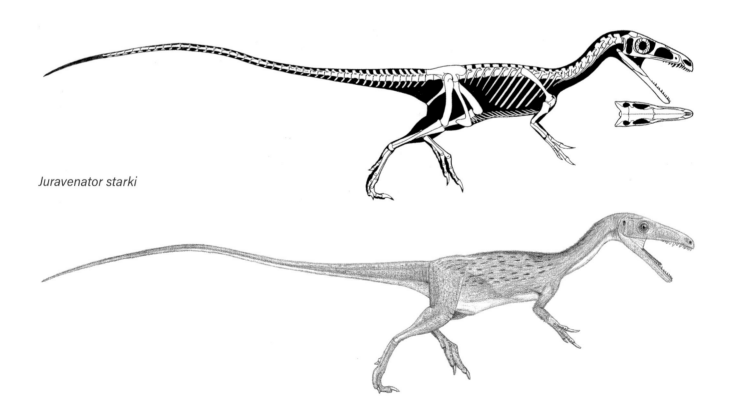

Juravenator starki

Compsognathus corallestris
—— 1.2 m (4 ft) TL, 2.1 kg (4.5 lb)
FOSSIL REMAINS Skull and majority of skeleton.
ANATOMICAL CHARACTERISTICS Head small, shallow,
subtriangular, teeth small.
AGE Late Jurassic, early Tithonian.
DISTRIBUTION AND FORMATION/S Southern France;
unnamed formation.
HABITAT Found as drift in lagoonal deposits near
probably arid, brush-covered islands.
NOTES Usual placement in earlier *C. longipes* from
different island problematic. Gut contents are of small
lizard-like animals.

"*Ubirajara jubatus*"
—— Adult size uncertain
FOSSIL REMAINS Partial skeleton, possibly juvenile,
extensive external fibers.

ANATOMICAL CHARACTERISTICS Mane of long bristle
fibers over neck and back, pair of elongated ribbon
feathers on side of each shoulder.
AGE Early Cretaceous, Aptian.
DISTRIBUTION AND FORMATION/S Northeastern Brazil;
Crato.
HABITS Used long shoulder feathers for display, may
have begun to reproduce before fully grown.
NOTES Dispute over national ownership of fossil has led
to name being potentially officially withdrawn.

Xunmenglong yingliangis
—— Adult size uncertain
FOSSIL REMAINS Minority of skeleton, possibly juvenile.
ANATOMICAL CHARACTERISTICS Insufficient
information.
AGE Early Cretaceous, early Hauterivian.
DISTRIBUTION AND FORMATION/S Northeastern China;
lower Huajiying.

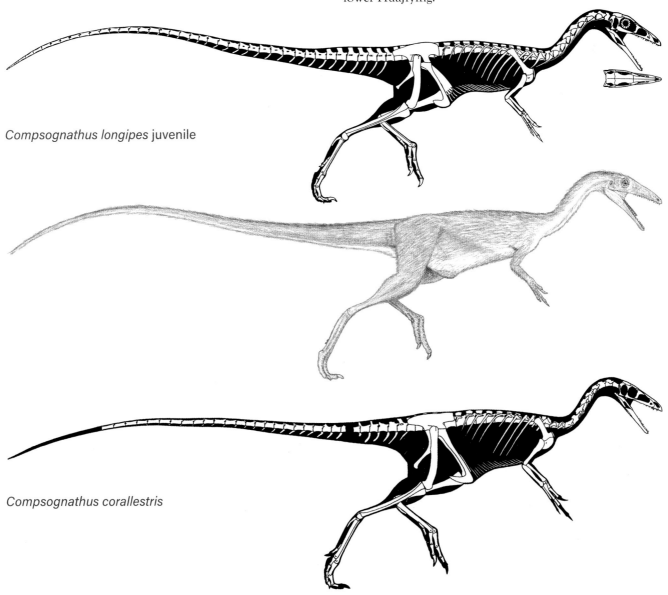

Compsognathus longipes juvenile

Compsognathus corallestris

Sinosauropteryx prima
—— 1 m (3 ft) TL, 1.2 kg (2.5 lb)

FOSSIL REMAINS A few complete skulls and skeletons, extensive external fibers, internal organs.

ANATOMICAL CHARACTERISTICS Snout subtriangular, teeth small. Arm shorter and thumb and claw stouter than in other compsognathids. Simple protofeathers on most of head except front of snout and most of body except hands and feet; protofeathers atop head and body, possible bandit mask, and dark bands on tail dark brown or reddish brown. Elongated eggs 4 cm (1.5 in) long.

AGE Early Cretaceous, latest Barremian and earliest Aptian.

DISTRIBUTION AND FORMATION/S Northeastern China; middle Yixian.

HABITAT Well-watered volcanic highland forest-and-lake district, winters chilly with snow.

NOTES Internal remains once interpreted as eggs instead are intestines. The dinosaur whose life appearance is best understood including coloration, and the one whose appearance may have been restored close to correctly.

Sinosauropteryx? unnamed species?
—— 1 m (3 ft) TL, 1.3 kg (2.8 lb)

FOSSIL REMAINS Nearly complete skull and skeleton, some external fibers.

ANATOMICAL CHARACTERISTICS Head subtriangular, teeth large. Tail rather short. Arm and hand rather small. Leg long. Simple protofeathers over most of body including tuft at end of tail, protofeathers atop head, body, and tail dark brown or reddish brown. Tail not banded.

AGE Early Cretaceous, latest Barremian and / or earliest Aptian.

DISTRIBUTION AND FORMATION/S Northeastern China; middle Yixian.

HABITAT Well-watered volcanic highland forest-and-lake district, winters chilly with snow.

NOTES Usual placement in *S. prima* problematic unless is a smaller-armed gender morph.

Sinocalliopteryx gigas
—— 2.3 m (7.5 ft) TL, 21 kg (45 lb)

FOSSIL REMAINS Complete skull and skeleton, extensive external fibers.

ANATOMICAL CHARACTERISTICS Head subtriangular, small paired crestlets atop snout, teeth fairly large. Tail rather short. Leg long. Simple protofeathers over most of body including upper feet, especially long at hips, tail base, and thigh, forming tuft at end of tail.

AGE Early Cretaceous, latest Barremian and / or earliest Aptian.

Sinosauropteryx prima and *Confuciusornis sanctus*

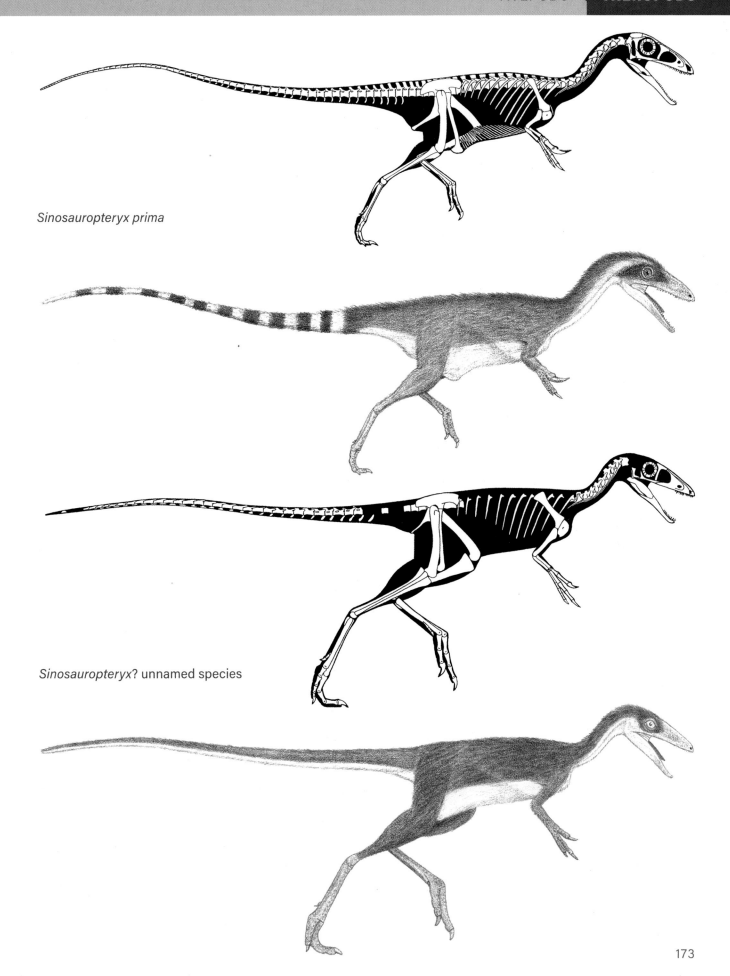

Sinosauropteryx prima

Sinosauropteryx? unnamed species

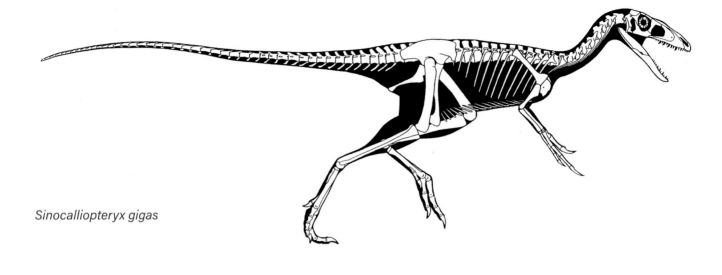

Sinocalliopteryx gigas

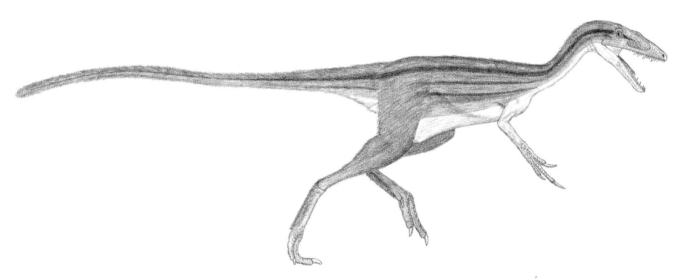

DISTRIBUTION AND FORMATION/S Northeastern China; middle Yixian.
HABITAT Well-watered volcanic highland forest-and-lake district, winters chilly with snow.
HABITS Hunted larger prey than smaller *Huaxiagnathus*. Foot feathers probably for display.
NOTES Gut contents include birds and small dromaeosaurids.

Huaxiagnathus orientalis
—— 1.7 m (5.5 ft) TL, 6 kg (13 lb)
FOSSIL REMAINS Nearly complete skull and skeleton.
ANATOMICAL CHARACTERISTICS Head subrectangular, front fairly deep, teeth not very large.
AGE Early Cretaceous, latest Barremian and/or earliest Aptian.
DISTRIBUTION AND FORMATION/S Northeastern China; middle Yixian.
HABITAT Well-watered volcanic highland forest-and-lake district, winters chilly with snow.
HABITS Hunted larger prey than smaller *Sinosauropteryx*.

Aristosuchus pusillus
—— 2 m (6 ft) TL, 8 kg (18 lb)
FOSSIL REMAINS Minority of skeleton.
ANATOMICAL CHARACTERISTICS Insufficient information.
AGE Early Cretaceous, Barremian.
DISTRIBUTION AND FORMATION/S Southern England; Wessex.

Mirischia asymmetrica
—— 2 m (6 ft) TL, 8 kg (18 lb)
FOSSIL REMAINS Minority of skeleton. Some internal organs preserved.
ANATOMICAL CHARACTERISTICS Standard for group.
AGE Early Cretaceous, probably Albian.
DISTRIBUTION AND FORMATION/S Eastern Brazil; Santana.
NOTES Found as drift in nearshore marine sediments.

Huaxiagnathus orientalis

— HAPLOCHEIRIDS

SMALL NEOCOELUROSAURS LIMITED TO THE LATE JURASSIC OF ASIA.

ANATOMICAL CHARACTERISTICS Teeth small, bladed, serrated, limited in number and to front third or half of jaws. Scapula blades slender, arms and hands moderately long. Feet moderately elongated.
NOTES Have been considered basal alvarezsaurs in part because of stout thumb of *Haplocheirus*, but at least as likely to be more closely related to ornithomimosaurs. Limited distribution may reflect lack of sufficient sampling.

Aorun zhaoi
—— Adult size uncertain
FOSSIL REMAINS Majority of juvenile skull and minority of skeleton.
ANATOMICAL CHARACTERISTICS Thumb not enlarged.
AGE Late Jurassic, late Callovian.
DISTRIBUTION AND FORMATION/S Northwestern China; lower Shishugou.
HABITAT Well-watered woodlands with short dry season.
NOTES May not belong to group.

Haplocheirus sollers
—— 2 m (6 ft) TL, 21 kg (45 lb)
ANATOMICAL CHARACTERISTICS Thumb enlarged, other two fingers quite slender.

Aorun zhaoi

FOSSIL REMAINS Nearly complete skull and skeleton.
AGE Late Jurassic, early Oxfordian.
DISTRIBUTION AND FORMATION/S Northwestern China; upper Shishugou.
NOTES If not an alvarezsaur, enlarged thumb was an independent development.

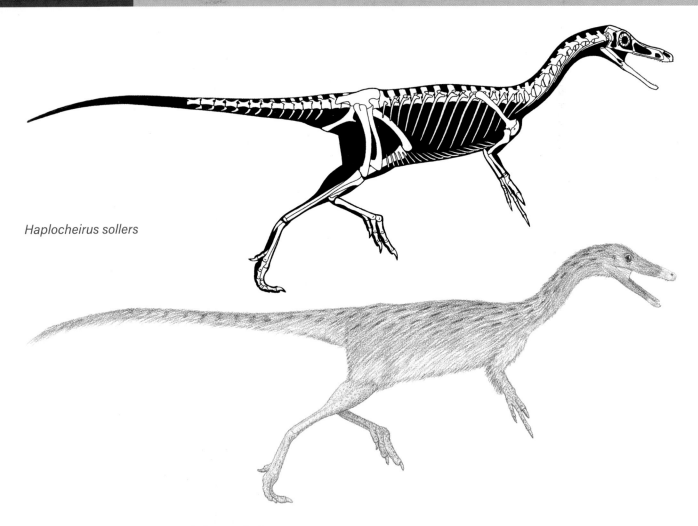

Haplocheirus sollers

— ORNITHOMIMOSAURS

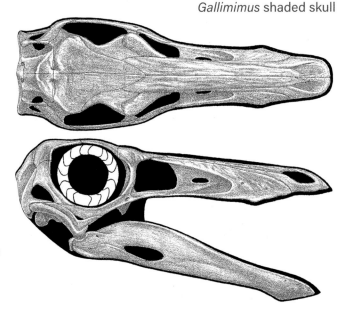

Gallimimus shaded skull

SMALL TO GIGANTIC NONPREDATORY NEOCOELUROSAURS OF THE CRETACEOUS OF NORTHERN HEMISPHERE AND AFRICA.

ANATOMICAL CHARACTERISTICS Usually uniform. Usually not heavily built. Heads small, shallow, and narrow, teeth reduced or absent and shallow, blunt beak present, eyes face partly forward, and some degree of stereo vision possible, extra joint in lower jaw absent. Necks long, at least fairly slender, ribs so short they do not overlap, increasing flexibility of neck. Scapula blades slender, semilunate carpal blocks absent and wrists not highly flexible, arms and hands long and slender. Legs long, toes short, claws not sharp. Brains semiavian in structure and size, olfactory bulbs reduced. Gizzard stones sometimes present.

HABITAT WELL-WATERED AREAS.

HABITS Small, slender skulls, unhooked beaks, and lightly constructed necks bar these from being predators. Possibly omnivorous, combining some small animals and insects with plant material gathered with assistance of long arms and hands. Main defense speed, also kicks from powerful legs and slashing with large hand claws.

NOTES The dinosaurs most similar to ostriches and other big ratites. Fragmentary remains imply possible presence in Australia. A frequent and sometimes common element of Cretaceous faunas.

Gallimimus muscle study

—BASO-ORNITHOMIMOSAURS

SMALL TO LARGE ORNITHOMIMOSAURS OF THE CRETACEOUS OF EURASIA.

ANATOMICAL CHARACTERISTICS Not as gracile as ornithomimids, pelves usually not as large, feet not as compressed, hallux usually still present. May be splittable into a larger number of subdivisions.

NOTES Baso-ornithomimosaurs are basal ornithomimosaurs excluding ornithomimids.

Nqwebasaurus thwazi
— 1 m (3 ft) TL, 1 kg (2.5 lb)

FOSSIL REMAINS Minority of skull and majority of skeleton.

ANATOMICAL CHARACTERISTICS Hand moderately long, thumb enlarged. Boot on pubis small, leg very long and gracile.

AGE Late Jurassic or Early Cretaceous.

DISTRIBUTION AND FORMATION/S Southern South Africa; upper Kirkwood.

NOTES Indicates that at least early ornithomimosaurs were present in Southern Hemisphere, and perhaps in the Late Jurassic.

Pelecanimimus polyodon
— 2.5 m (8 ft) TL, 30 kg (60 lb)

FOSSIL REMAINS Complete skull and front part of skeleton, some soft tissues.

ANATOMICAL CHARACTERISTICS Snout long and tapering, small hornlets above orbits, hundreds of tiny teeth concentrated in front of jaws. Fingers subequal in length, claws nearly straight. Small soft crest at back of head, throat pouch, no feathers preserved on limited areas of smooth, unscaly skin.

AGE Early Cretaceous, late Barremian.

DISTRIBUTION AND FORMATION/S Central Spain; Calizas de la Huérguina.

HABITS Teeth may have been for cutting plants and/or filtering small organisms, throat pouch may have been for containing fish. Hornlets and crest for display within the species.

NOTES Found as drift in nearshore marine sediments.

Hexing qingyi
— Adult size uncertain

FOSSIL REMAINS Complete skull and minority of skeleton, possibly immature.

ANATOMICAL CHARACTERISTICS A few small teeth at front end of lower jaw. Thumb much shorter than other fingers, claws fairly short and curved.

AGE Early Cretaceous, latest Barremian.

DISTRIBUTION AND FORMATION/S Northeastern China; lowermost Yixian.

HABITAT Well-watered volcanic highland forest-and-lake district, winters chilly with snow.

Pelecanimimus polyodon

Hexing qingyi

Shenzhousaurus orientalis

—— 1.6 m (5 ft) TL, 10 kg (20 lb)
FOSSIL REMAINS Complete skull and majority of skeleton.
ANATOMICAL CHARACTERISTICS A few small, conical teeth at front end of lower jaw. Thumb not as long as other fingers, claws nearly straight.
AGE Early Cretaceous, latest Barremian.
DISTRIBUTION AND FORMATION/S NORTHEASTERN CHINA; LOWER YIXIAN.
HABITAT Well-watered volcanic highland forest-and-lake district, winters chilly with snow.

Nedcolbertia justinhofmanni

—— Adult size uncertain
FOSSIL REMAINS Minority of several skeletons, immature.
ANATOMICAL CHARACTERISTICS Leg long and gracile.
AGE Early Cretaceous, late Valanginian.
DISTRIBUTION AND FORMATION/S Utah; lower Cedar Mountain.
HABITAT Short wet season, otherwise semiarid with floodplain prairies, open woodlands, and riverine forests.

Arkansaurus fridayi

—— 4.5 m (15 ft), 250 kg (500 lb)
FOSSIL REMAINS Fragmentary skeleton.
ANATOMICAL CHARACTERISTICS Insufficient information.
AGE Early Cretaceous, Albian or Aptian.
DISTRIBUTION AND FORMATION/S Arkansas; Trinity Group.
HABITAT Coastal plain.

Harpymimus okladnikovi

—— 3 m (10 ft) TL, 50 kg (110 lb)
FOSSIL REMAINS Nearly complete skull and majority of skeleton.
ANATOMICAL CHARACTERISTICS A few small teeth at tip of lower jaw. Thumb not as long as other fingers, claws gently curved.
AGE Early Cretaceous, late Albian.
DISTRIBUTION AND FORMATION/S Mongolia; Shinekhudag.

Harpymimus okladnikovi

Beishanlong grandis

—— 7 m (23 ft) TL, 550 kg (1,200 lb)
FOSSIL REMAINS Minority of skeletons.

ANATOMICAL CHARACTERISTICS Fairly robustly built.
AGE Early Cretaceous, probably Aptian or Albian.
DISTRIBUTION AND FORMATION/S Central China; lower Xinminpu.

Garudimimus brevipes

—— 2.5 m (8 ft) TL, 30 kg (60 lb)
FOSSIL REMAINS Complete skull and majority of skeleton.
ANATOMICAL CHARACTERISTICS Toothless and beaked.
AGE Early Late Cretaceous.
DISTRIBUTION AND FORMATION/S Mongolia; Bayanshiree.

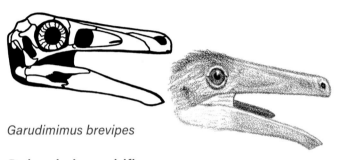

Garudimimus brevipes

Deinocheirus mirificus

—— 11.5 m (38 ft) TL, 5.5 tonnes
FOSSIL REMAINS Complete skull, majority of a skeleton and minority of two others.
ANATOMICAL CHARACTERISTICS Head not as small relative to skeleton as in other ornithomimosaurs, very slender and narrow with very long beak flaring out to a small duck bill, eye sockets not large, lower jaw very deep, jaws weakly muscled, teeth absent. Neck fairly stout. Trunk vertebrae articulated in a very strong arc up from hip and down to shoulders, vertebral spines form tall sail just in front of hips. Tail tipped with a small fused pygostyle. Hips broader than in other ornithomimosaurs. Arm 2.5 m (9 ft) long, rather slender, fingers subequal in length, claws blunt-tipped hooks. Hip very large and deep, leg robust but not massive, feet moderately long, toes short and ending with blunt, hooflike claws.
AGE Late Cretaceous, late Campanian and early Maastrichtian?
DISTRIBUTION AND FORMATION/S Mongolia; at least lower Nemegt.
HABITAT Temperate, well-watered, dense woodlands with seasonal rains and winter snow.
HABITS That original fossil is from lower Nemegt and some others from later suggests latter remains are different species. Omnivore that fed on softer vegetation and aquatic creatures as indicated by fish remains in apparent stomach contents. Deep jaws indicate strong tongue and possible suction action. Arms could be used for gathering vegetation, possibly digging. Better able to defend itself against predators with large, clawed arms than smaller, faster ornithomimosaurs.

Deinocheirus mirificus

NOTES Examples from levels other than original fossil may be different species. Full form only recently realized from more complete remains long after discovery of isolated gigantic arms in the 1960s. This, *Beishanlong*, and *Garudimimus* may form family Deinocheiridae.

Deinocheirus mirificus

—ORNITHOMIMIDS

MEDIUM-SIZED ORNITHOMIMOSAURS OF THE CRETACEOUS, LIMITED TO THE NORTHERN HEMISPHERE.

ANATOMICAL CHARACTERISTICS Highly uniform. Gracile build. Toothless and beaked. Fingers subequal in length, claws at least fairly long and not strongly curved. Trunk compact. Tails shorter and lighter than standard for theropods. Pelves very large and leg very long, so leg muscles exceptionally well developed, feet very long and strongly compressed from side to side, hallux completely lost, so speed potential very high. At least some examples had large feathers on arms.
HABITS Main defense very high speed and maneuverability.
ONTOGENY Growth rates moderately rapid, adult size reached in a few years in medium-sized and large examples.
NOTES Partial remains show that some western North American examples first approached half a tonne by late Campanian.

Aepyornithomimus tugrikinensis
—— 3 m (10 ft) TL, 90 kg (200 lb)
FOSSIL REMAINS Fragmentary skeleton.
ANATOMICAL CHARACTERISTICS Somewhat robustly built.

AGE Late Cretaceous, middle Campanian.
DISTRIBUTION AND FORMATION/S Mongolia; lower Djadokhta.
HABITAT Desert with dunes and oases.

Kinnareemimus khonkaenensis
—— Size not available
FOSSIL REMAINS Minority of skeleton.
ANATOMICAL CHARACTERISTICS Insufficient information.
AGE Early Cretaceous, Valanginian or Hauterivian.
DISTRIBUTION AND FORMATION/S Northeastern Thailand; Sao Khua.

Archaeornithomimus asiaticus
—— Adult size uncertain
FOSSIL REMAINS Minority of skeleton.
ANATOMICAL CHARACTERISTICS Insufficient information.
AGE Late Cretaceous.
DISTRIBUTION AND FORMATION/S China; Iren Dabasu.
HABITAT Seasonally wet-dry woodlands.

Sinornithomimus dongi
—— 2.5 m (8 ft) TL, 50 kg (110 lb)
FOSSIL REMAINS Over a dozen skulls and skeletons, many complete, juvenile to adult, completely known.

ANATOMICAL CHARACTERISTICS Head somewhat shorter and skeleton not quite as gracile, as in most later ornithomimids. Hand elongated, thumb shorter than other fingers. Leg long.
AGE Late Cretaceous, Turonian.
DISTRIBUTION AND FORMATION/S Northern China; Ulansuhai.

Anserimimus planinychus
—— 3 m (10 ft) TL, 90 kg (200 lb)
FOSSIL REMAINS Partial skeleton.
ANATOMICAL CHARACTERISTICS Hand moderately elongated, fairly heavily built, thumb shorter than other fingers.
AGE Late Cretaceous, early Maastrichtian.
DISTRIBUTION AND FORMATION/S Mongolia; Nemegt Svita.
HABITAT Temperate, well-watered, dense woodlands with seasonal rains and winter snow.

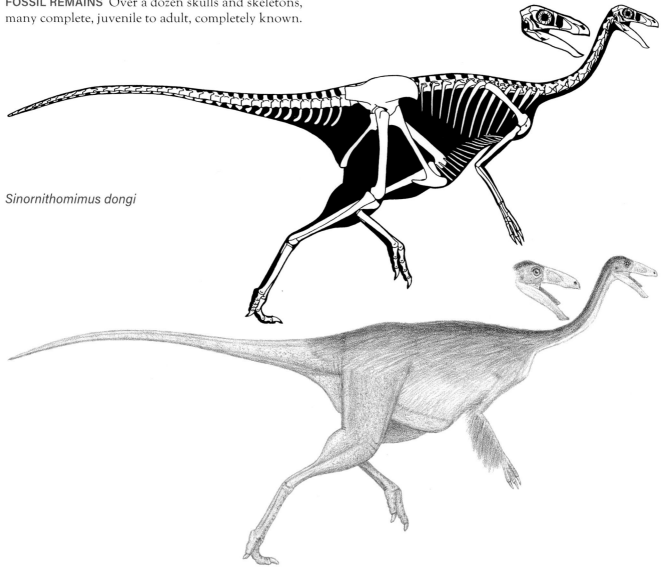

Sinornithomimus dongi

Gallimimus bullatus

—— 6 m (20 ft) TL, 510 kg (1,100 lb)

FOSSIL REMAINS Several complete skulls and skeletons, juveniles to adult, completely known.

ANATOMICAL CHARACTERISTICS Beak elongated. Hand not strongly elongated, thumb shorter than other fingers, shorter armed and legged than other advanced ornithomimids.

AGE Late Cretaceous, late Campanian? and/or early Maastrichtian.

DISTRIBUTION AND FORMATION/S Mongolia; at least upper Nemegt.

HABITAT Temperate, well-watered, dense woodlands with seasonal rains and winter snow.

HABITS Presumably not quite as swift as longer-legged ornithomimids. Suggestion that was a filter feeder because of fluted inner beak surface highly problematic.

NOTES That original fossil is from upper Nemegt suggests earlier remains are different species.

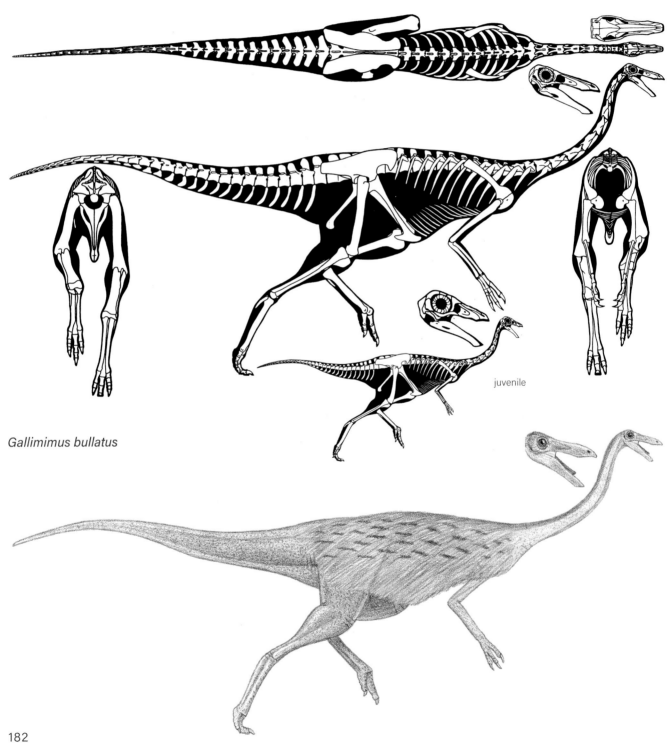

Gallimimus bullatus

juvenile

Qiupalong henanensis

—— 2.5 m (8 ft) TL, 50 kg (100 lb)

FOSSIL REMAINS Minority of skeleton.

ANATOMICAL CHARACTERISTICS Insufficient information.

AGE Late Late Cretaceous.

DISTRIBUTION AND FORMATION/S Eastern China; Quipa.

NOTES Placement of North American remains in same genus very dubious.

Rativates evadens

—— 3 m (10 ft) TL, 90 kg (100 lb)

FOSSIL REMAINS Partial skull and skeleton.

ANATOMICAL CHARACTERISTICS Leg long.

AGE Late Cretaceous, late middle Campanian.

DISTRIBUTION AND FORMATION/S Alberta; lower Dinosaur Park.

HABITAT Well-watered, forested floodplain with coastal swamps and marshes, cool winters.

Struthiomimus altus

—— 4 m (13 ft) TL, 180 kg (400 lb)

FOSSIL REMAINS Several skulls and skeletons of varying completeness.

ANATOMICAL CHARACTERISTICS Hand elongated, thumb shorter than other fingers, claws fairly robust and curved. Leg long.

AGE Late Cretaceous, late middle and/or early to middle late Campanian.

DISTRIBUTION AND FORMATION/S Alberta; Dinosaur Park, level uncertain.

HABITAT Well-watered, forested floodplain with coastal swamps and marshes, cool winters.

NOTES The adequacy of the remains on which S. altus is based is problematic; genus Struthiomimus is based on a largely complete fossil, stratigraphic levels of both are uncertain. May be an ancestor of following Struthiomimus unnamed species.

Struthiomimus? unnamed species

—— 4.4 m (14 ft) TL, 250 kg (500 lb)

FOSSIL REMAINS Several skulls and skeletons of varying completeness.

ANATOMICAL CHARACTERISTICS Hand elongated, thumb shorter than other fingers, claws fairly robust and curved. Leg long.

AGE Late Cretaceous, early Maastrichtian.

DISTRIBUTION AND FORMATION/S Alberta; upper Horseshoe Canyon.

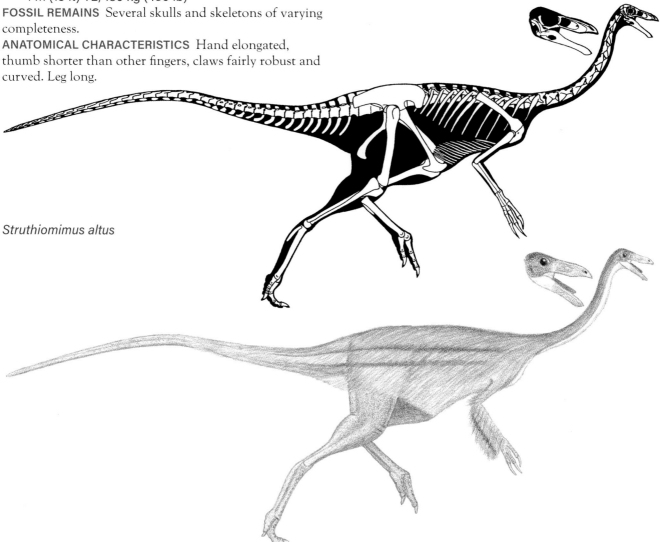

Struthiomimus altus

HABITAT Well-watered, forested floodplain with coastal swamps and marshes, cool winters.
NOTES That this belonged to the same genus as much earlier *S. altus* is problematic.

Ornithomimus? unnamed species or *samueli*
—— 3.4 m (11 ft) TL, 200 kg (210 lb)
FOSSIL REMAINS Complete skull and majority of skeleton, skull and partial skeleton, other remains.
ANATOMICAL CHARACTERISTICS Hand elongated, fingers probably nearly equal in length because thumb metacarpal longer than others, claws long, nearly straight, and delicate. Leg long. Short- and medium-length feathers covering most of body and tail, large feathers on arm, no feathers on leg from thigh on down.

AGE Late Cretaceous, late middle and/or early late Campanian.
DISTRIBUTION AND FORMATION/S Alberta; lower and/or middle Dinosaur Park.
HABITAT Well-watered, forested floodplain with coastal swamps and marshes, cool winters.
NOTES Placement in much later, more robust, fragmentary genus *Ornithomimus* problematic, as is placement in later *O.? edmontonicus* even if is in same genus. Original incomplete and damaged middle Dinosaur Park Formation *samueli* fossil may or may not be same taxon as nearly complete fossil from lower in formation. May be ancestor(s) of "*O.*" *edmontonicus*.

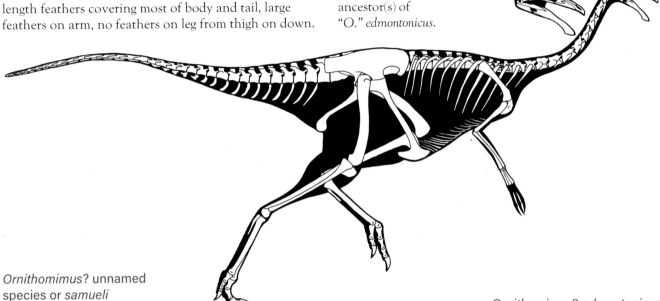

Ornithomimus? unnamed species or *samueli*

Ornithomimus? *edmontonicus*

Ornithomimus? edmontonicus
—— 3.8 m (16 ft) TL, 210 kg (450 lb)

FOSSIL REMAINS Skulls and skeletons of varying completeness, feathers as per prior entry.
ANATOMICAL CHARACTERISTICS Hand elongated, fingers nearly equal in length because thumb metacarpal longer than others, claws long, nearly straight, and delicate. Leg long. Short and medium-length feathers covering most of body and tail, large feathers on arm, no feathers on leg from thigh on down.
AGE Late Cretaceous, latest Campanian and/or early Maastrichtian.
DISTRIBUTION AND FORMATION/S Alberta; lower Horseshoe Canyon.
HABITAT Well-watered, forested floodplain with coastal swamps and marshes, cool winters.
NOTES Usual placement in much later, more robust, fragmentary genus *Ornithomimus* problematic. Ratite-like body plumage previously predicted by this illustrator.

Ornithomimus velox
—— 2.5 m (8 ft) TL, 70 kg (130 lb)

FOSSIL REMAINS Minority of skeleton.
ANATOMICAL CHARACTERISTICS Hand moderately elongated, fairly heavily built, fingers probably nearly equal in length because thumb base metacarpal longer than others. Leg fairly robust.
AGE Late Cretaceous, latest Maastrichtian.
DISTRIBUTION AND FORMATION/S Colorado; upper Denver.
HABITAT Well-watered, forested floodplain with coastal swamps and marshes, cool winters.
HABITS Presumably not quite as swift as slenderer-limbed ornithomimids.
NOTES Fossil appears to be an adult. Because is so fragmentary cannot be shown at this time that earlier ornithomimids up to 10 million years older with similar hand anatomy are in this genus. Because ornithomimids with the two types of hands with other details and otherwise very similar are found at same levels in the late

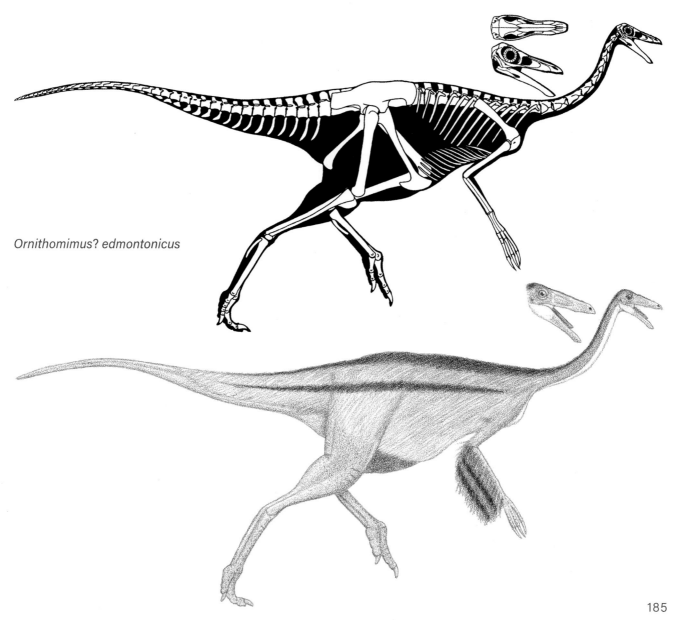

Ornithomimus? edmontonicus

Campanian and Maastrichtian of western North America, it is possible that they are the genders of the same species at a given time. If so, then *Ornithomimus* has priority over *Struthiomimus*, but the one genus lasting so long remains problematic.

Dromiceiomimus brevitertius

—— 3.5 m (11.5) TL, 160 kg (350 lb)

FOSSIL REMAINS Partial skeletons.

ANATOMICAL CHARACTERISTICS Leg very long because of elongated shank.

AGE Late Cretaceous, latest Campanian.

DISTRIBUTION AND FORMATION/S Alberta; lower Horseshoe Canyon.

HABITAT Well-watered, forested floodplain with coastal swamps and marshes, cool winters.

NOTES Often placed in *Ornithomimus edmontonicus* even though this species has taxonomic priority, or in genus *Dromiceiomimus* even though all western North American Campanian–Maastrichtian ornithomimids show limited variation at species level. If is a gender dimorph of contemporary *O.? edmontonicus*, then this taxon's name has priority at least at species level, and perhaps at genus level for former, and all "*Ornithomimus*" prior to latest Maastrichtian. May be direct ancestor of *D. ingens*.

Dromiceiomimus ingens

—— 3.6 m (11.5) TL, 165 kg (360 lb)

FOSSIL REMAINS Partial skeletons.

ANATOMICAL CHARACTERISTICS Leg very long because of elongated shank.

AGE Late Cretaceous, early Maastrichtian.

DISTRIBUTION AND FORMATION/S Alberta; upper Horseshoe Canyon.

HABITAT Well-watered, forested floodplain with coastal swamps and marshes, cool winters.

NOTES Usual placement in similar but earlier *D. brevitertius* problematic. May not be smaller than other western North American ornithomimids.

Dromiceiomimus? or Struthiomimus? unnamed species

—— 5 m (16 ft) TL, 400 kg (800 lb)

FOSSIL REMAINS Partial skeleton.

ANATOMICAL CHARACTERISTICS Hand elongated, thumb shorter than other fingers, claws fairly robust and curved. Leg long because of elongated shank.

AGE Late Cretaceous, late Maastrichtian.

DISTRIBUTION AND FORMATION/S Colorado, Wyoming; lower Lance.

HABITAT Well-watered coastal woodlands.

NOTES Usual placement of fossil in very fragmentary *Struthiomimus sedens* of unknown level of Lance is problematic, as is placement in much earlier genus, placement in somewhat earlier *Dromiceiomimus* less so, length of shank relative to femur is near boundary of two genera for a large animal.

—AIRFOILANS

SMALL TO GIGANTIC PREDATORY AND HERBIVOROUS NEOCOELUROSAURS OF THE MIDDLE JURASSIC TO THE MODERN ERA, ALL CONTINENTS.

ANATOMICAL CHARACTERISTICS Highly variable. Brains expanding and becoming more avian in form. Wrist flexion increased.

NOTES Airfoilans are neocoelurosaurs with forelimb-borne airfoils of some form, or ancestors with same that are in the clade that includes extant birds. Broadly similar but not identical to inconsistently defined maniraptors and paravians.

—SCANSORIOPTERYGIDS

SMALL AIRFOILANS OF THE MIDDLE / LATE JURASSIC OF ASIA.

ANATOMICAL CHARACTERISTICS Heads short and broad, lower jaws shallow, a few procumbent, pointed teeth at front of jaws. Necks medium length. Trunks unusually shallow because pubis quite short. Tails not very long, very abbreviated in at least some adults. Small ossified sternal plate present in at least some examples. Arms and hands elongated, hands strongly asymmetrical because outer finger is a hyperelongated, extra-lateral, very elongated strut in at least some examples, arms and hands appear to support wing membranes, finger claws large. Pelves shallow, at least in juveniles, legs not elongated, hallux partly reversed.

HABITS Very probably strongly arboreal, trunk hugging facilitated by flattened trunk. Some form of flight, at least gliding or possibly marginally powered, apparently present. Probably insectivorous.

NOTES Relationships with other neocoelurosaurs uncertain. Close relationship with oviraptorosaurs may be contradicted by presence of long tail in some examples. Appears to have been a brief evolutionary experiment in dinosaur flight that lost out to the birdlike aveairfoilans, the earliest known examples of which—*Anchiornis*, *Eosinopteryx*, *Serikornis*, *Xiaotingia*—lived in the same habitat.

Scansoriopteryx heilmanni
— 0.5 m (1 ft) TL, 0.17 kg (0.4 lb)

FOSSIL REMAINS Probably minority of an adult, and majority of two juvenile skeletons, all with complete or partial skulls, some feather fibers.

ANATOMICAL CHARACTERISTICS Shallow midline crest on snout, teeth small. Tail not abbreviated at least in juveniles. Arm and hand highly elongated, extra strut present in hand.

AGE Late Jurassic, middle Oxfordian.

DISTRIBUTION AND FORMATION/S Northeastern China; middle Tiaojishan.

HABITAT Well-watered forest and lake district.

NOTES The name *Scansoriopteryx* appears to have edged out *Epidendrosaurus ningchengensis* in the race for priority. *Yi qi*—the shortest dinosaur name—is probably the adult form of this species. Lack of preserved extra hand strut in juveniles may be due to lack of sufficient growth and/or ossification. Competes with *Ambopteryx*, *Anchiornis*, *Caihong*, *Epidexipteryx*, and *Serikornis* for title of smallest known nonavian dinosaur; four of the contenders lived in the same formation.

Epidexipteryx hui
— 0.3 m (1 ft) TL, 0.25 kg (0.5 lb)

FOSSIL REMAINS Complete skull and majority of skeleton, some feather fibers.

ANATOMICAL CHARACTERISTICS Teeth of lower jaw large and procumbent. Tail abbreviated. Small ossified sternal plates present. Pubis not retroverted. Arm feathers apparently short, four very long banded feathers trail from tail, simpler feathers cover much of body.

AGE Middle or Late Jurassic, Callovian or Oxfordian.

DISTRIBUTION AND FORMATION/S Northeastern China; Tiaojishan.

HABITAT Well-watered forest and lake district.

HABITS Long tail feathers for display within the species.

NOTES Smaller wing suggests lesser flight performance than *Scansoriopteryx* and *Ambopteryx*. Had probably the shortest known tail among nonavian dinosaurs.

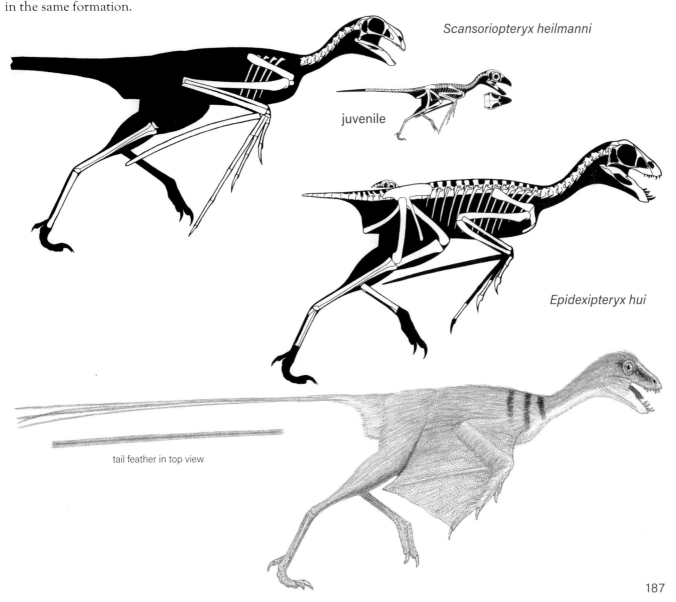

Scansoriopteryx heilmanni

juvenile

Epidexipteryx hui

tail feather in top view

Ambopteryx longibrachium

—— 0.3 m (1 ft) TL, 0.25 kg (0.5 lb)

FOSSIL REMAINS Complete skull and skeleton, some feather fibers and other soft tissues.

ANATOMICAL CHARACTERISTICS Tail abbreviated, tipped with a slender fused pygostyle. Arm and hand highly elongated, extra strut present in hand. Pubis probably retroverted.

AGE Late Jurassic, Oxfordian.

DISTRIBUTION AND FORMATION/S Northeastern China; Yanliao.

HABITAT Well-watered forest and lake district.

NOTES Skull too damaged to restore.

— AVEAIRFOILANS

SMALL TO GIGANTIC PREDATORY AND HERBIVOROUS AIRFOILANS OF THE LATE JURASSIC TO THE MODERN ERA, ALL CONTINENTS.

ANATOMICAL CHARACTERISTICS Highly variable. Heads toothed to toothless and beaked; when teeth are present, serrations tend to be reduced in some manner or absent. Tails very long to very short. Shoulder girdles usually birdlike, with slender horizontal scapula blades and vertical coracoids with outer surface facing forward, furculas often large, arms very long to short, wrists usually with a large, half-moon-shaped semilunate carpal block that allowed arm to be more folded, hands usually long. Brains enlarged, semiavian in form. Pennaceous feathers often present. Overall appearance usually very birdlike.

ONTOGENY Growth rates apparently moderate.

HABITS Reproduction generally similar to that of ratites and tinamous; in at least some cases males incubated the eggs and were probably polygamous; egg hatching in a given clutch not synchronous. Some examples preserved in birdlike curled-up sleeping posture.

NOTES Aveairfoilans are airfoilans with avian shoulder girdles, or ancestors with same that are in the clade that includes extant birds. Prone to evolving and especially losing flight, perhaps repeating cycle in some cases.

— AVEAIRFOILAN MISCELLANEA

Yixianosaurus longimanus

—— 1 m (3 ft) TL, 1 kg (2.2 lb)

FOSSIL REMAINS Arms.

ANATOMICAL CHARACTERISTICS Hands elongated, finger claws large and strongly hooked.

AGE Early Cretaceous, latest Barremian and/or earliest Aptian.

DISTRIBUTION AND FORMATION/S Northeastern China; Yixian.

HABITAT Well-watered volcanic highland forest-and-lake district, winters chilly with snow.

HABITS Well-developed arms suitable for handling prey and climbing.

NOTES Relationships very uncertain, may not be aveairfoilan.

Bradycneme draculae

—— 1 m (3.3 ft) TL, 5 kg (10 lb)

FOSSIL REMAINS Fragmentary skeleton(s).

ANATOMICAL CHARACTERISTICS Insufficient information

AGE Late Cretaceous, early Maastrichtian.

DISTRIBUTION AND FORMATION/S Romania; Sanpetru.

HABITAT Forested island.

NOTES May include *Heptasteornis andrewsi*. Once thought by some to be giant Mesozoic owls, may be alvarezsaurs.

Balaur bondoc

—— 2.5 m (9 ft) TL, 15 kg (30 lb)

FOSSIL REMAINS Partial skeleton.

ANATOMICAL CHARACTERISTICS Robustly built. Upper hand elements fused, outer finger very reduced. Pubis very retroverted, foot short, broad, inner toe large so foot effectively tetradactyl, inner two toes hyperextendable with large claws.

AGE Late Cretaceous, middle Maastrichtian.

DISTRIBUTION AND FORMATION/S Romania; Sebeş.

HABITAT Forested island.

HABITS Probably secondarily flightless. Possibly omnivorous or herbivorous, in which case double sickle claws used for defense and climbing rather than primarily for predation.

NOTES Originally thought to be a predatory dromaeosaur, more probably a near-bird.

— DEINONYCHOSAURS

SMALL TO MEDIUM-SIZED PREDATORY AND OMNIVOROUS AVEAIRFOILANS OF THE LATE JURASSIC TO THE END OF THE DINOSAUR ERA, ON MOST CONTINENTS.

ANATOMICAL CHARACTERISTICS Fairly variable. Eyes face partly forward and some degree of stereo vision possible, tooth serrations reduced or absent. Tails slender, bases very flexible, especially upward. Arms and hands well developed, sometimes very long, finger claws large hooks. Second toe hyperextendable and/or claw at least somewhat enlarged.

HABITS Very agile, sophisticated predators and omnivores, prey varying from insects and small game to big game. Climbing ability generally good, especially in smaller species, longer-armed species, and juveniles; hyperextendable toe could be used as hook and spike during climbing in species living in areas with trees. Second toe claw used as weapon to dispatch prey—especially when claw and species it was borne by were large, to pin down prey, and in defense. Two-toed trackways confirm that hyperextendable claw was normally carried clear of ground; relative scarcity of such trackways indicates most deinonychosaurs did not spend much time patrolling shorelines.

NOTES Is not known whether deinonychosaurs were a distinct group or a stage within aveairfoilans, membership also uncertain. Presence of large sternal plates, ossified sternal ribs, and ossified uncinate processes in most flightless deinonychosaurs indicates they were secondarily flightless.

— DEINONYCHOSAUR MISCELLANEA

NOTES Neither the placement of these aveairfoilans in the deinonychosaurs nor their placement within the group is certain.

Caihong juji

— 0.4 m (1.3 ft) TL, 0.3 kg (0.7 lb)

FOSSIL REMAINS Complete skull and skeleton, significantly damaged, extensive feathers.

ANATOMICAL CHARACTERISTICS Head long, shallow, subrectangular, apparent long, low paired crests atop snout, teeth fairly large, bladed. Tail medium length. Arm much shorter than leg. Pubis long and vertical. Toe claws not strongly curved. Well-developed head feather crest, arm primary feathers symmetrical, moderately long on arm, same on leg, tail fan very broad and long, short feathers on toes, feathers black, also iridescent at least on head, chest, tail base.

AGE Late Jurassic, early Oxfordian.

DISTRIBUTION AND FORMATION/S Northeastern China; lower Tiaojishan.

HABITAT Well-watered forest and lake district.

HABITS Bladed teeth indicates hunted fairly substantial game. Lack of strongly curved toe claws indicates was not highly arboreal. Arm wing too small and primary feathers too symmetrical for flight, at most a parachuting or very modest gliding ability was present, potentially secondarily flightless. Large tail fan primarily for display.

NOTES Size, shape and orientation of possible snout crests and other head proportions uncertain because of distortion of preserved skull, same for skeletal details. Competes with *Ambopteryx*, *Anchiornis*, *Epidexipteryx*, *Scansoriopteryx*, and *Serikornis* for title of smallest known nonavian dinosaur; four of the contenders lived in the same formation.

Hesperornithoides miessleri

— 1.75 m (6 ft) TL, 12 kg (25 lb)

FOSSIL REMAINS Majority of damaged skull and minority of skeleton.

ANATOMICAL CHARACTERISTICS Lightly built, head probably subtriangular, some teeth are large blades. Tail lacks ossified rods. Arm and hand moderately large. Leg very long, sickle claw small

AGE Late Jurassic, early and/or early middle Kimmeridgian.

DISTRIBUTION AND FORMATION/S Wyoming; middle Morrison.

HABITAT Short wet season, otherwise semiarid with open floodplain prairies and riverine forests.

HABITS Nonflier. Could dispatch comparably large prey with large teeth.

NOTES May be a basal troodontid.

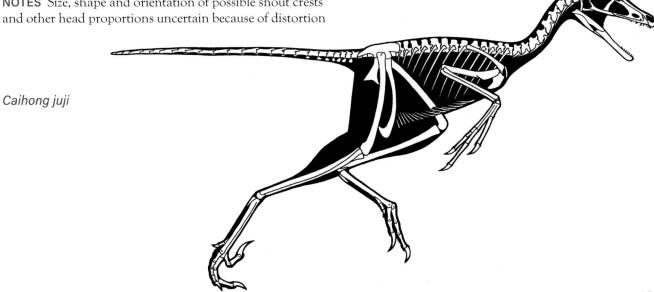

Caihong juji

Fujianvenator prodigiosus
— 0.5 m (1.5 ft) TL, 0.3 kg (0.6 lb)
FOSSIL REMAINS Majority of skeleton.
ANATOMICAL CHARACTERISTICS Large ossified sternum, possible ossified sternal ribs, arm and hand very long, hand and fingers slender, thumb claw large. Pubis orientation uncertain, lower leg long and slender, status of second toe claw unknown.
AGE Late Jurassic, latest Kimmeridgian.
DISTRIBUTION AND FORMATION/S Southeastern China; middle Nanyuan.
HABITS Probable powered flier. Fast runner. Probably predaceous but not certain.
NOTES Large sternum and long arms and hands indicate powered flight, but slender hand indicates such was not powerful, may have been incipiently secondarily flight reduced.

Richardoestesia (or *Ricardoestesia*) *gilmorei*
— 2 m (3.5 ft) TL, 10 kg (20 lb)
FOSSIL REMAINS Minority of skull.
ANATOMICAL CHARACTERISTICS Lower jaw very slender.

AGE Late Cretaceous, late middle and/or early to middle late Campanian.
DISTRIBUTION AND FORMATION/S Alberta; Dinosaur Park, level uncertain.
HABITAT Well-watered, forested floodplain with coastal swamps and marshes, cool winters.
HABITS Hunted small game, fished.
NOTES May be a dromaeosaurid or troodontid. Numerous remains imply type was common in other late Late Cretaceous habitats.

Imperobatar antarcticus
— 3.5 m (12 ft) TL, 70 kg (150 lb)
FOSSIL REMAINS Minority of skull and skeleton.
ANATOMICAL CHARACTERISTICS Sickle claw absent.
AGE Late Cretaceous, early Maastrichtian.
DISTRIBUTION AND FORMATION/S James Ross Island; middle Snow Hill Island.
HABITAT Coastal polar forests with cold, dark winters.
NOTES Some size estimates have been excessive. One of two Mesozoic avepods known from Antarctic region. Relationships obscure.

— ARCHAEOPTERYGIANS

SMALL, LARGE-ARMED PREDATORY DEINONYCHOSAURS LIMITED TO THE LATER JURASSIC OF EURASIA.

ANATOMICAL CHARACTERISTICS Lightly built. Heads not large, subtriangular, teeth small, not highly numerous, unserrated. Trunks deep because pubes long and not strongly retroverted; ossified sternal plates, sternal ribs, and

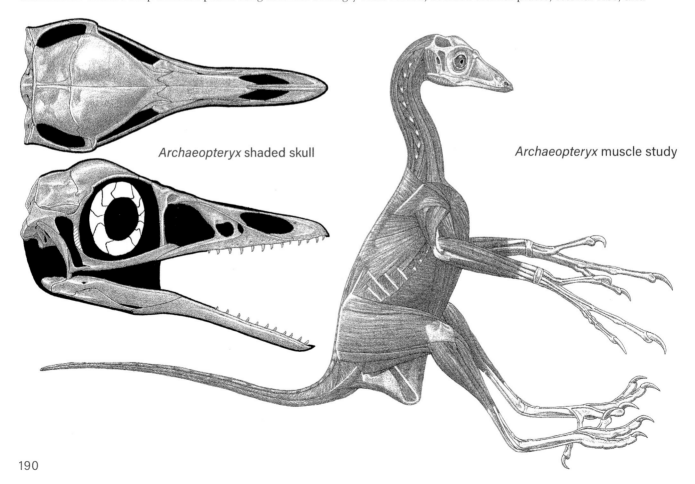

Archaeopteryx shaded skull

Archaeopteryx muscle study

uncinate processes on ribs absent. Tails medium length. Arms and hands elongated. Legs elongated, second toes slender and claw somewhat enlarged, maybe hyperextendable.

HABITS Digestive tract contents confirm hunted small game and fished, first known dinosaurs to form gastric pellets. Flight performance minimal to significant.

NOTES Extent of group and relationships of these early aveairfoilans are uncertain.

Anchiornis huxleyi
—— 0.5 m (1.5 ft) TL, 0.3 kg (0.7 lb)

FOSSIL REMAINS Complete skulls and skeletons, extensive feathers, gastric pellets.

ANATOMICAL CHARACTERISTICS Head short, subtriangular. Arm not as long as leg. Finger and toe claws moderately curved. Well-developed head feather crest, arm primary feathers symmetrical, moderately long on arm, same on leg, feather edges somewhat irregular, short feathers on toes, most feathers dark gray or black, head feathers speckled reddish brown, head crest partly brown or reddish brown, broad whitish bands on arm and leg feathers interrupted by narrow dark irregular bands, primary feather tips black.

AGE Late Jurassic, middle Oxfordian.

DISTRIBUTION AND FORMATION/S Northeastern China; middle Tiaojishan.

HABITAT Well-watered forest and lake district.

HABITS Somewhat arboreal. Arm wing too small and primary feathers too symmetrical for strong flight, at most a parachuting or very modest gliding ability was present, potentially secondarily flightless.

NOTES Probably includes *Auromis xui*. May be a troodont.

Anchiornis huxleyi

Anchiornis huxleyi

Eosinopteryx brevipenna?

—— 1 m (3 ft) TL, 1 kg (2.2 lb)

FOSSIL REMAINS Complete skull and skeleton possibly juvenile, minority of skeleton, extensive feathers.

ANATOMICAL CHARACTERISTICS Head short, subrectangular. Arm not as long as leg. Finger and toe claws not strongly curved, claw on second toe markedly larger than others. Primary feathers on arm symmetrical, may be symmetrical pennaceous feathers on upper foot, feather edges somewhat irregular.

AGE Late Jurassic, middle Oxfordian.

DISTRIBUTION AND FORMATION/S Northeastern China; middle Tiaojishan.

HABITAT Well-watered forest and lake district.

HABITS Not strongly arboreal. Arm wing too small and primary feathers too symmetrical for flight, absence of tail feathers indicates same, at most a parachuting ability was present, potentially secondarily flightless.

NOTES May be the juvenile of *Pedopenna daohugouensis* if from same middle level of formation. May be an anchiornid or troodont. If not, a juvenile is among smallest known dinosaurs.

Serikornis sungei

—— 0.5 m (1.5 ft) TL, 0.2 kg (0.5 lb)

FOSSIL REMAINS Complete badly damaged skull and skeleton, extensive feathers.

ANATOMICAL CHARACTERISTICS Head short, probably subtriangular. Neck fairly long, tail long. Arm shorter than leg. Primary feathers on arm symmetrical, pennaceous feathers apparently absent on leg and tail, short feathers on toes.

AGE Late Jurassic, Oxfordian.

DISTRIBUTION AND FORMATION/S Northeastern China; Tiaojishan.

HABITAT Well-watered forest and lake district.

HABITS Arm wing too small and primary feathers too symmetrical for flight, absence of tail feathers indicates same, at most a parachuting ability was present, potentially secondarily flightless.

NOTES Damage to skull precludes skeletal restoration. May be an anchiornid or troodont. Competes with *Ambopteryx*, *Anchiornis*, *Caihong*, *Epidexipteryx*, and *Scansoriopteryx* for title of smallest known nonavian dinosaur; four of the contenders lived in the same formation.

Xiaotingia zhengi

—— 0.6 m (2 ft) TL, 0.6 kg (1.4 lb)

FOSSIL REMAINS Majority of skull and skeleton, poorly preserved feathers.

ANATOMICAL CHARACTERISTICS Head short, subtriangular, teeth blunt. Arm not as long as leg. Finger and toe claws strongly curved. Long pennaceous feathers on leg, otherwise preservation of feathers insufficient for assessment.

AGE Late Jurassic, Oxfordian.

DISTRIBUTION AND FORMATION/S Northeastern China; Tiaojishan.

HABITAT Well-watered forest and lake district.

HABITS Strongly arboreal. Arm wing too small for significant flight, at most a parachuting ability was present, potentially secondarily flightless.

NOTES May be a dromaeosaur or troodont, or the closest known relative of *Archaeopteryx*.

Ostromia crassipes

—— 0.4 m (1.3 ft) TL, 0.55 m (1.8 ft) wingspan, 0.5 kg (1.1 lb)

FOSSIL REMAINS Minority of skeleton (Haarlem), with feathers.

ANATOMICAL CHARACTERISTICS Insufficient information.

AGE Late Jurassic, earliest? Tithonian.

DISTRIBUTION AND FORMATION/S Southern Germany; middle? Painten.

HABITAT Found as drift in lagoonal deposits near

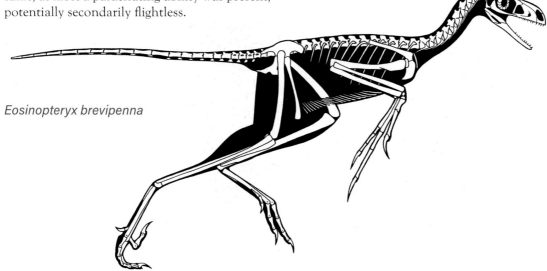

Eosinopteryx brevipenna

probably arid, brush- and mangrove-covered islands then immediately off the northeastern coast of North America.
NOTES Long placed in *Archaeopteryx*, which it shared its habitat with, but now considered more basal. Indicates Solnhofen protobird fauna more diverse than previously realized, as does the more advanced flier *Alcmonavis*.

Archaeopteryx siemensii

—— 0.4 m (1.3 ft) TL, 0.55 m (1.8 ft) wingspan, 0.23 kg (0.5 lb)
FOSSIL REMAINS Two nearly complete skulls and skeletons (Berlin, Schamhaupten), at least some immature, extensive feathers, nearly completely known.
ANATOMICAL CHARACTERISTICS Head subtriangular, snout pointed, teeth subconical, unserrated. Trunk shallower because pubis more retroverted. Tail modest in length. Arm longer and more strongly built than leg. Hallux fairly large and semireversed, toe claw curvature modest at least in juveniles. Most of body covered by short feathers, arm and hand supporting well-developed, broad chord wings made of asymmetrical, crisp-edged feathers. Lower leg supporting a modest-sized feather airfoil, tail supporting a long set of feather vanes forming an airfoil.
AGE Late Jurassic, middle Kimmeridgian.
DISTRIBUTION AND FORMATION/S Southern Germany; lowermost Altmühltal, lowermost Painten.

HABITAT Found as drift in lagoonal deposits near probably arid, brush- and mangrove-covered islands then immediately off the northeastern coast of North America.
HABITS Probably semiarboreal. Capable of low-grade powered flight and gliding probably somewhat inferior to that of *Sapeornis*, immature condition of fossils may be leading to underestimation of flight abilities because adult wing elements may have been more fused, etc. Legs could not splay out nearly flat so feathers probably used as auxiliary rudders and air brakes. May have been able to swim with wings. Defense included climbing and flight, and claws.
NOTES Long known as the first bird, "Urvogel," *Archaeopteryx* is now known to be one among an array of later Jurassic dinobirds. Fossils placed in genus span over 1.5 million years and are unlikely to be one or two species, of which *A. lithographica* has priority, that last most or all of that time; species may have appeared in stratigraphic sequence but multiple species at a given time are possible, and/or some species may have been present for somewhat longer time than presented here. *A. siemensii* is best-preserved species. Some contend that all fossils are juveniles and that maximum mass was over 25% heavier. This and *Xiaotingia* may form family Archaeopterygidae. May be direct ancestor of *A. grandis*.

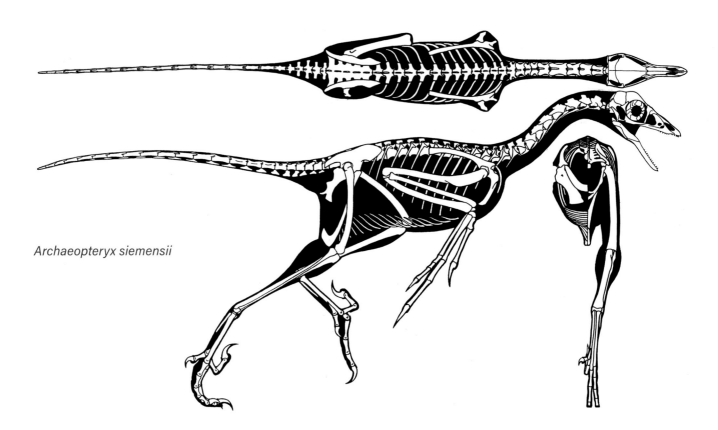

Archaeopteryx siemensii

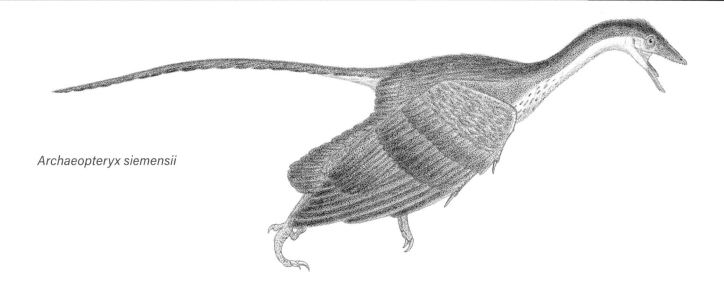

Archaeopteryx siemensii

Archaeopteryx siemensii and *Pterodactylus antiquus*

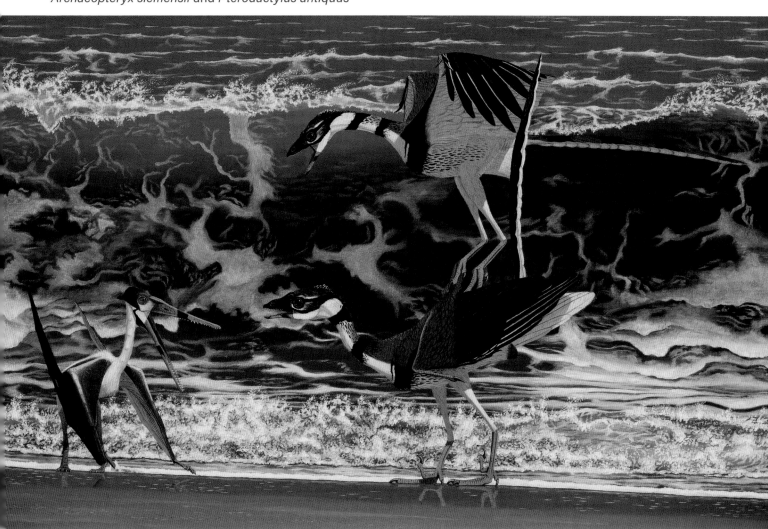

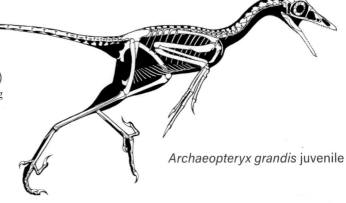

Archaeopteryx grandis juvenile

Archaeopteryx grandis

—— 0.5 m (1.7 ft) TL, 0.7 m (2.3 ft) wingspan, 0.45 kg (1 lb)
FOSSIL REMAINS Several skulls and skeletons of varying completeness (Solnhofen, Thermopolis, Altmühl, Eichstätt), at least some immature, extensive feathers.
ANATOMICAL CHARACTERISTICS Largely same as for *A. siemensii*. Toe claw curvature modest at least in juveniles.
AGE Late Jurassic, latest middle Kimmeridgian.
DISTRIBUTION AND FORMATION/S Southern Germany; middle Altmühltal.
HABITAT Found as drift in lagoonal deposits near probably arid, brush- and mangrove-covered islands then immediately off the northeastern coast of North America.
HABITS Largely same as for *A. siemensii*.
NOTES Placement in genus *Wellnhoferia* problematic, *A. recurva* is probably the juvenile of this species. Unusual fourth toe of Solnhofen fossil probably individual anomaly. May be direct ancestor of *A. lithographica*.

Archaeopteryx lithographica

—— 0.45 m (1.5 ft) TL, 0.6 m (2 ft) wingspan, 0.34 kg (0.75 lb)
FOSSIL REMAINS Several partial skulls and skeletons (London, Munich, Maxberg), at least some immature, extensive feathers.
ANATOMICAL CHARACTERISTICS Largely same as for *A. siemensii*. Toe claw curvature strong at least in some adults.
AGE Late Jurassic, earliest late Kimmeridgian.
DISTRIBUTION AND FORMATION/S Southern Germany; upper Altmühltal.
HABITAT Found as drift in lagoonal deposits near probably arid, brush- and mangrove-covered islands then immediately off the northeastern coast of North America.
HABITS Largely same as for *A. siemensii*, perhaps more arboreal.

NOTES Original fossil is a feather that may or may not be the same taxon as the skeletons from this same level; feather color patterns restored for this species based on feather correspondingly problematic, more so for other *Archaeopteryx* species. Probably includes *A. bavarica*. May be direct ancestor of *A. albersdoerferi*.

Archaeopteryx or unnamed genus albersdoerferi

—— 0.35 m (1.7 ft) TL, 0.7 m (2.3 ft) wingspan, 0.15 kg (0.35 lb)
FOSSIL REMAINS Partial skull and skeleton (Daiting), probably adult.
ANATOMICAL CHARACTERISTICS Extensive fusion of various elements.
AGE Late Jurassic, middle late Kimmeridgian.
DISTRIBUTION AND FORMATION/S Southern Germany; Mörnsheim, level uncertain.
HABITAT Found as drift in lagoonal deposits near probably arid, brush- and mangrove-covered islands then immediately off the northeastern coast of North America.
HABITS Extensive fusion of wrist and upper hand suggests superior flight abilities or is an adult feature.
NOTES May not belong to earlier *Archaeopteryx*.

—ALCMONAVIANS

SMALL POSSIBLE DEINONYCHOSAURS OF THE LATE JURASSIC OF EUROPE.

ANATOMICAL CHARACTERISTICS Central finger bases flattened to better support fully developed, broad chord wings.
NOTES Relationships uncertain, may not be deinonychosaurs, may be basal birds.

Alcmonavis poeschli

—— 0.6 m (2 ft) TL, 0.8 m (2.8 ft) wingspan, 0.8 kg (1.7 lb)
FOSSIL REMAINS Minority of skeleton (Mühlheim).
ANATOMICAL CHARACTERISTICS As for group.
AGE Late Jurassic, middle late Kimmeridgian.
DISTRIBUTION AND FORMATION/S Southern Germany; lower Mörnsheim.

HABITAT Found as drift in lagoonal deposits near probably arid, brush- and mangrove-covered islands then immediately off the northeastern coast of North America.
HABITS Markedly better-powered flier than contemporary *Archaeopteryx*.
NOTES Earliest known dinosaur with a flattened finger base for an improved outer wing.

— DROMAEOSAURIDS

SMALL- TO MEDIUM-SIZED FLYING AND FLIGHTLESS PREDATORY DEINONYCHOSAURS OF THE CRETACEOUS, ON MOST CONTINENTS.

ANATOMICAL CHARACTERISTICS Fairly variable. Teeth bladed, serrations limited to back edge, quadratojugal usually inverted T shape. Neck ribs so short they do not overlap, increasing flexibility of neck. Arms large to very large. Tails usually long to very long and ensheathed in very long and slender ossified rods that are extensions of forward processes of the vertebrae. Large ossified sternal plates, sternal ribs, and uncinate processes present. Large sickle claw on robust hyperextendable toes. Olfactory bulbs enlarged.

HABITS Archpredators equipped to ambush, pursue, and dispatch relatively large prey by use of sickle claw as a primary weapon, especially when large and borne by large species, as well as to pin down prey. Sickle claw also facilitated climbing taller prey. Leaping performance when arboreal or attacking prey excellent. Flight performance from well developed to none; the retention of the pterosaur-like ossified tail rods in flightless dromaeosaurids is further evidence that they were secondarily flightless. Juveniles of large species with longer arms may have possessed some flight ability.

NOTES Teeth imply that small members of the group may have evolved by the Late Jurassic. Fragmentary remains indicate presence in Australia. Flightless examples descended from flying dromaeosaurids, perhaps multiple times leading to different subgroups, and flight may have reevolved in some cases.

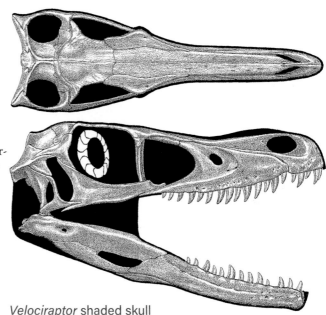

Velociraptor shaded skull

Velociraptor muscle study

— DROMAEOSAURID MISCELLANEA

Yurgovuchia doellingi
—— 2.5 m (6 ft) TL, 20 kg (40 lb)
FOSSIL REMAINS Minority of skeleton.
ANATOMICAL CHARACTERISTICS Robustly built.
AGE Early Cretaceous, late Berriasian.

DISTRIBUTION AND FORMATION/S Utah; lowermost Cedar Mountain.
HABITAT Short wet season, otherwise semiarid with floodplain prairies and open woodlands, and riverine forests.

Vectiraptor greeni
—— 2.5 m (6 ft) TL, 20 kg (40 lb)
FOSSIL REMAINS Fragmentary skeleton.
ANATOMICAL CHARACTERISTICS Insufficient information.
AGE Early Cretaceous, middle Barremian.
DISTRIBUTION AND FORMATION/S Southern England; upper Wessex.

Tianyuraptor ostromi
—— 2 m (6.5 ft) TL, 11 kg (25 lb)
FOSSIL REMAINS Majority of skull and nearly complete skeleton, complete skull and majority of skeleton, extensive feathers.
ANATOMICAL CHARACTERISTICS Head moderate size. Trunk shallow because pubis short. Arm not elongated. Leg fairly elongated, toe claws strongly curved. Most of body covered by short feathers, arm and hand supported well-developed, modest-sized winglets with irregular edge feathers, no pennaceous feathers on leg, tail supported a long set of feather vanes.
AGE Early Cretaceous, middle Barremian.
DISTRIBUTION AND FORMATION/S Northeastern China; upper Yixian.
HABITAT Well-watered volcanic highland forest-and-lake district, winters chilly with snow.

HABITS Substantial arboreal abilities. Modest-sized sickle claw indicates did not hunt particularly large prey. Winglets too small for significant flight, primary function probably display.
NOTES Probably includes *Zhenyuanlong suni*; if so, squared-off snout may be a male attribute, more pointed tip female. May be a close relative of microraptorines, possibly secondarily flightless.

Daurlong wangi
—— 1.15 m (3.8 ft) TL, 5.5 kg (12 lb)
FOSSIL REMAINS Nearly complete skull and skeleton, some feathers, internal tissues.
ANATOMICAL CHARACTERISTICS Head robustly built, deep; snout square tipped, middle upper jaw teeth large. Tail short. Arm not elongated. Leg long partly because main toes long.
AGE Early Cretaceous, latest Barremian and/or earliest Aptian.
DISTRIBUTION AND FORMATION/S Northern China; Longjiang.
HABITS Largely terrestrial predator, large teeth indicate hunted fairly large game.
NOTES Details of back portion of skull obscure. Only known short-tailed dromaeosaur.

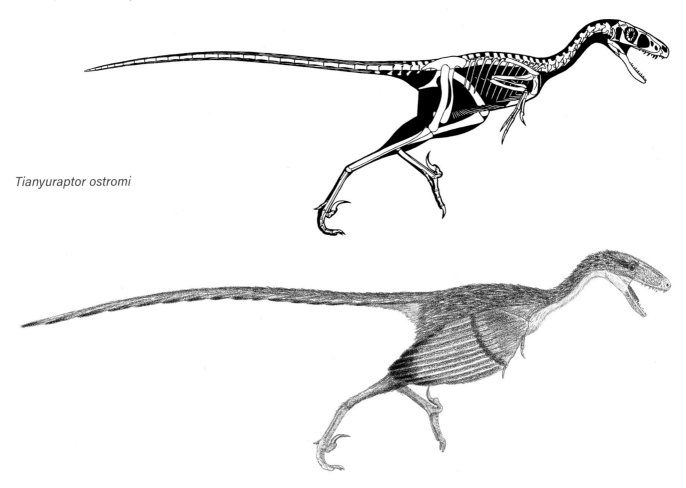

Tianyuraptor ostromi

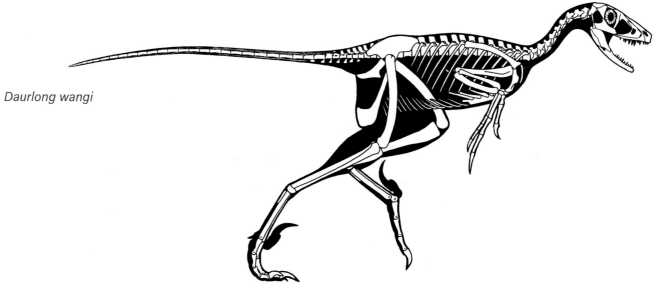

Daurlong wangi

Luanchuanraptor henanensis
—— 1.1 m (3.5 ft) TL, 2 kg (4 lb)
FOSSIL REMAINS Minority of skull and skeleton.
ANATOMICAL CHARACTERISTICS Insufficient
information.

AGE Late Cretaceous.
DISTRIBUTION AND FORMATION/S Central China;
Qiupa.
NOTES May be an Asian unenlagiine.

—— MICRORAPTORINES

SMALL FLYING DROMAEOSAURIDS OF THE CRETACEOUS OF THE NORTHERN HEMISPHERE.

ANATOMICAL CHARACTERISTICS Lightly built. Tooth serrations reduced in various ways. Arms and hands very large, outer, upper hand bone curved, central fingers stiffened and bases flattened to better support fully developed, broad chord wings made of asymmetrical crisp edged feathers. Hip sockets more upwardly directed than usual, legs very long, supported well-developed second wing made of asymmetrical feathers that extended onto upper foot, head of femur more spherical than in other theropods, sickle claws well developed, other toe claws strongly curved. Part of heads and most of bodies covered by short, simple feathers, tails supported a long set of feather vanes, forming an airfoil, inner wing leading edge propatagium present.
HABITAT Well-watered forests and lakes.
HABITS Very large, stiff foot feathers not well suited for running, and strongly curved toe claws indicate strong arboreality. Better development of sternum, sternal ribs, and uncinates, more streamlined trunk, modified upper hand and central finger, larger outer arm wing, extra leg wing, and pterosaur-like tail indicate microraptorines were better-powered fliers than *Archaeopteryx* and *Sapeornis*. Leg appears to have been more splayable sideways than normal in theropods because hip socket faces more upward, femoral head is more spherical, and legs are often splayed sideways in articulated fossils (unlike most articulated theropods, including *Archaeopteryx*, which are usually preserved on their sides), but hind wings were not flappable and possibly provided extra wing area during glides or soaring, and air brakes when landing or ambushing prey from the air.
NOTES Similar limb design indicates all microraptorines had forewings and hind wings, preserved only in *Sinornithosaurus zhaoianus* (wing feathers are missing from a number of other Yixian bird species). *Zhongjianosaurus yangi* and *Shanag ashile* may be basal birds. Flightless dromaeosaurids descended from microraptorines or similar aerial dromaeosaurids. Limitation of these basal dromaeosaurids to the Northern Hemisphere may reflect lack of sufficient sampling.

Sinornithosaurus (or *Graciliraptor*) *lujiatunensis*
—— 1 m (3 ft) TL, 1.5 kg (3.5 lb)
FOSSIL REMAINS Minority of skull and skeleton.
ANATOMICAL CHARACTERISTICS Hand slender.
AGE Early Cretaceous, middle Barremian.

DISTRIBUTION AND FORMATION/S Northeastern China;
lowermost Yixian.
HABITAT Well-watered volcanic highland forest-and-lake
district, winters chilly with snow.
NOTES Placement in distinct genus problematic.

Sinornithosaurus millenii

—— 1.2 m (4 ft) TL, 3 kg (6.5 lb)

FOSSIL REMAINS Nearly complete skull and majority of skeleton, poorly preserved feathers.

ANATOMICAL CHARACTERISTICS Head large, long, and shallow, jugal shallow and postorbital bar probably incomplete, quadratojugal inverted T shape with short vertical process. Trunk shallow because pubis double retroverted. Furcula and sternal plates large. Tail not very long. Feathers irregularly black and rufous.

AGE Early Cretaceous, middle Barremian.

DISTRIBUTION AND FORMATION/S Northeastern China; middle Yixian.

HABITAT Well-watered volcanic highland forest-and-lake district, winters chilly with snow.

HABITS Could hunt similarly large prey.

NOTES Poor preservation of feathers has caused many to problematically presume it was not winged despite flattening of central finger. The genus appears to have included multiple fairly similar species over millions of years from the base of the Yixian to the top of the Jiufotang formations.

Sinornithosaurus (= Changyuraptor) yangi

—— 1 m (3.4 ft) TL, 2.5 kg (5 lb)

FOSSIL REMAINS Complete but poorly preserved skull and skeleton, some feathers.

ANATOMICAL CHARACTERISTICS Tail not very long, tail feathers exceptionally long. Trunk shallow because pubis double retroverted. Furcula and sternal plates large.

AGE Early Cretaceous, middle Barremian.

DISTRIBUTION AND FORMATION/S Northeastern China; Yixian, level uncertain.

HABITAT Well-watered volcanic highland forest-and-lake district, winters chilly with snow.

HABITS Similar to S. millenii.

NOTES Placement in distinct genus problematic. Establishes that large microraptorines were winged fliers.

Sinornithosaurus (= Microraptor) zhaoianus

—— 0.85 m (2.8 ft) TL, 0.85 m (2.8 ft) wingspan, 1 kg (2 lb)

FOSSIL REMAINS Large number of complete and partial, usually damaged skulls and skeletons, extensive well-preserved feathers, adult to juvenile.

ANATOMICAL CHARACTERISTICS Head not proportionally large, subtriangular, jugal shallow and postorbital bar probably incomplete, teeth less bladelike and less serrated than in S. millenii. Neck rather short, tail very long. Trunk shallow because pubis double retroverted. Furcula and sternal plates large. Leg quite long. Finger and toe claws strongly curved. Leg feathers preserved, overall feathers black with iridescence.

AGE Early Cretaceous, late Barremian.

DISTRIBUTION AND FORMATION/S Northeastern China; lower Jiufotang.

HABITAT Well-watered volcanic highland forest-and-lake district, winters chilly with snow.

HABITS Strongly curved claws indicate strong arboreality. Both lesser size and less bladed and serrated teeth indicate that this attacked smaller prey than S. millenii; gut contents include small fish, birds, mammals.

NOTES Placement in distinct genus problematic, may include S. gui and S. hanqingi. Second only to Sinosauropteryx in being accurately restorable among dinosaurs, the orientation of the leg feathers posing the greatest uncertainty.

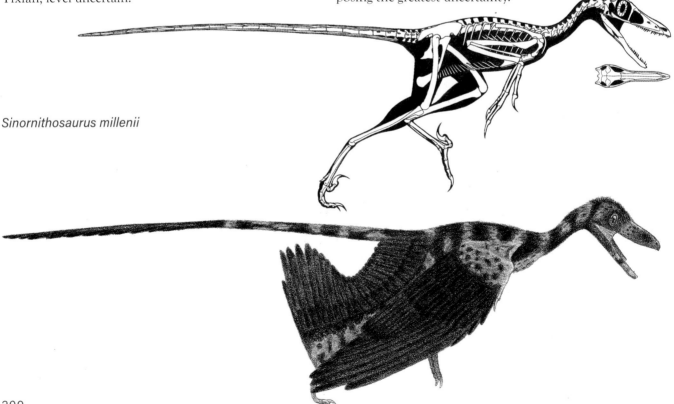

Sinornithosaurus millenii

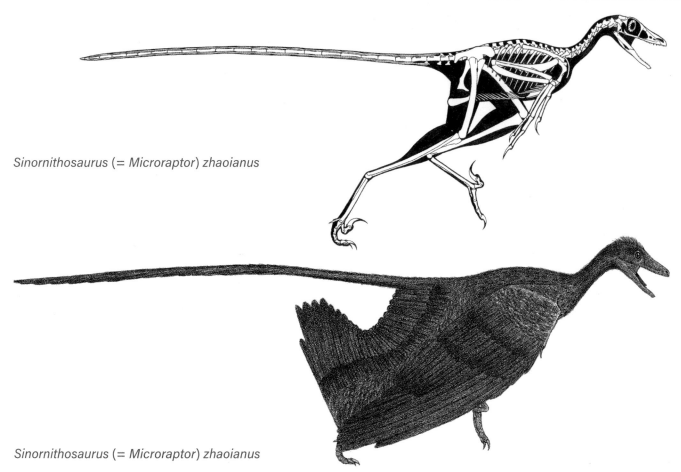

Sinornithosaurus (= *Microraptor*) *zhaoianus*

Sinornithosaurus (= *Microraptor*) *zhaoianus*

Sinornithosaurus (= Microraptor or Cryptovolans) pauli

—— 0.9 m (3 ft) TL, 0.95 m (3.1 ft) wingspan, 1.2 kg (2.5 lb)
FOSSIL REMAINS A few skulls and skeletons, extensive feathers.
ANATOMICAL CHARACTERISTICS Tail not very long. Trunk shallow because pubis double retroverted. Furcula and sternal plates large. Leg not very long.
AGE Early Cretaceous, early Aptian?
DISTRIBUTION AND FORMATION/S Northeastern China; upper? Jiufotang.
HABITAT Well-watered volcanic highland forest-and-lake district, winters chilly with snow.
NOTES Placement in distinct genus problematic.

Jeholraptor haoiana

—— 1.2 m (4 ft) TL, 3 kg (7 lb)
FOSSIL REMAINS Majority of skull and skeleton, immature.
ANATOMICAL CHARACTERISTICS Head fairly robust and deep, jugal robust, and postorbital bar complete and broad, quadratojugal L shaped. Furcula and sternal plates not very large, hand not very slender. Two main toes fairly short, sickle claw not large, toe claws strongly curved.
AGE Early Cretaceous, middle Barremian.
DISTRIBUTION AND FORMATION/S Northeastern China; upper Yixian.
HABITAT Well-watered volcanic highland forest-and-lake district, winters chilly with snow.
HABITS Strongly curved toe claws indicate strong arboreality. Smaller furcula, sternum, and less gracile

hand suggest was not as good a flier as *Sinornithosaurus*, hunted fairly large game.
NOTES Usual placement in earlier *S. millenii* incorrect because of very different quadratojugals and other reasons. L-shaped quadratojugals in this and *Wulong* suggest they are more closely related to one another than to other microraptorines and probably form their own subgroup.

Wulong bohaiensis

—— Adult size uncertain
FOSSIL REMAINS Nearly complete skull and skeleton, extensive feathers, juvenile.
ANATOMICAL CHARACTERISTICS Jugal robust and postorbital bar complete, quadratojugal more L shaped. Hand not very slender.
AGE Early Cretaceous, early Aptian.
DISTRIBUTION AND FORMATION/S Northeastern China; upper Jiufotang.
HABITAT Well-watered volcanic highland forest-and-lake district, winters chilly with snow.
NOTES Details of back of skull obscure. Immature status hinders full analysis, large head is a juvenile feature.

Hesperonychus elizabethae

—— 1 m (3 ft) TL, 1.5 kg (3.5 lb)
FOSSIL REMAINS Minority of a few skeletons.
ANATOMICAL CHARACTERISTICS Trunk shallow because pubis is double retroverted. Well-preserved hip socket directed more upward than in nonmicroraptor dinosaurs, indicating leg could be splayed more strongly sideways.

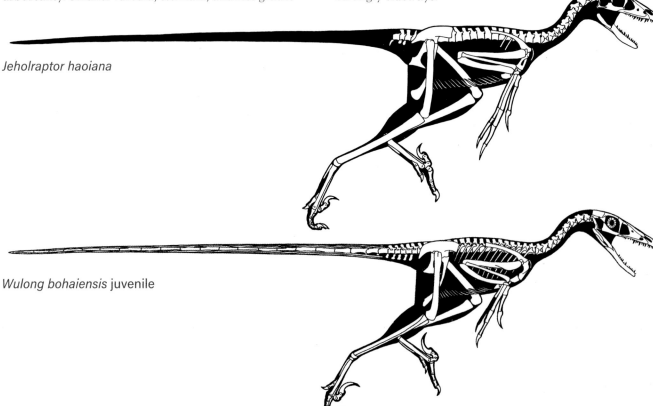

Jeholraptor haoiana

Wulong bohaiensis juvenile

AGE Late Cretaceous, early late and middle late Campanian.
DISTRIBUTION AND FORMATION/S Alberta; at least middle and upper Dinosaur Park.

HABITAT Well-watered, forested floodplain with coastal swamps and marshes, cool winters.
NOTES *Hesperonychus* implies that microraptorines or close relatives survived into the late Late Cretaceous.

— HALSZKARAPTORINES

SMALL DROMAEOSAURIDS LIMITED TO THE LATE CRETACEOUS OF ASIA.

ANATOMICAL CHARACTERISTICS Lightly built. Heads shallow, snouts somewhat duck-like, quadratojugal apparently not inverted T-shaped, teeth fairly numerous, small, and unserrated. Necks fairly long and slender. Trunks shallow partly because ribs swept strongly back and out. Tails lack ossified rods. Arms short, outer fingers longest. Sickle claws modest sized.
HABITS Possibly partly aquatic surface floaters, fishers, small game, insects.

Mahakala omnogovae
—— 0.5 m (1.7 ft) TL, 0.1 (2 lb)
FOSSIL REMAINS Minority of skull and partial skeleton.
ANATOMICAL CHARACTERISTICS Upper foot at least elongated.
AGE Late Cretaceous, middle Campanian.
DISTRIBUTION AND FORMATION/S Mongolia; lower Djadokhta.
HABITAT Desert with dunes and oases.
HABITS Less aquatic than other group members.

Halszkaraptor escuilliei
—— 0.9 m (3 ft) TL, 0.6 kg (1.4 lb)
FOSSIL REMAINS Majority of skull and skeleton.
ANATOMICAL CHARACTERISTICS Leg rather short.
AGE Late Cretaceous, middle Campanian.

DISTRIBUTION AND FORMATION/S Mongolia; lower Djadokhta.
HABITAT Desert with dunes and oases.

Natovenator? polydontus
—— 0.7 m (2.3 ft) TL, 0.3 kg (1.2 lb)
FOSSIL REMAINS Majority of skull and partial skeleton.
ANATOMICAL CHARACTERISTICS Standard for group.
AGE Late Cretaceous, latest Campanian and/or earliest Maastrichtian.
DISTRIBUTION AND FORMATION/S Mongolia; Barun Goyot.
HABITAT Semidesert with some dunes and oases.
NOTES May be same species and/or genus as fragmentary *Hulsanpes perlei*, may be same genus as *Halszkaraptor*.

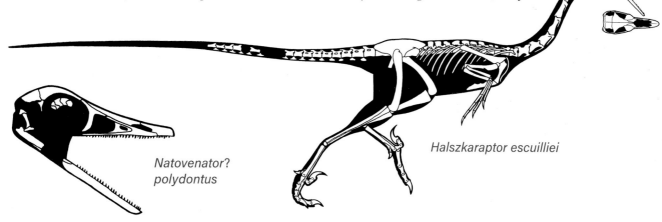

Natovenator? polydontus

Halszkaraptor escuilliei

— UNENLAGIINES

SMALL TO MEDIUM-SIZED FLYING AND FLIGHTLESS DROMAEOSAURIDS LIMITED TO THE LATE CRETACEOUS OF THE SOUTHERN HEMISPHERE.

ANATOMICAL CHARACTERISTICS Variable. Tails may lack ossified rods. Pubes vertical.
HABITS Appear to have been more prone to fishing.
NOTES Unenlagiines indicate that dromaeosaurids experienced a radiation of distinctive forms in the Southern Hemisphere that may have included fliers; losses and/or independent evolution of flight probably occurred.

Buitreraptor gozalezorum

Buitreraptor gonzalezorum

— 1.5 m (5 ft) TL, 3.6 kg (8 lb)

FOSSIL REMAINS Majority of skull and skeleton.
ANATOMICAL CHARACTERISTICS Head very long, shallow, and narrow, especially snout and lower jaw, teeth small, numerous, nonserrated. Arm and hand very elongated and very slender. Leg long and gracile, sickle claw fairly small.
AGE Late Cretaceous, early Cenomanian.
DISTRIBUTION AND FORMATION/S Western Argentina; lower Candeleros.
HABITAT Short wet season, otherwise semiarid with open floodplains and riverine forests.
HABITS Hunted small game, probably fished using long snout and long arms and hands. Main defenses high speed and sickle claw.

Unenlagia comahuensis

— 3.5 m (12 ft) TL, 50 kg (100 lb)

FOSSIL REMAINS Minority of skeleton.
ANATOMICAL CHARACTERISTICS Sickle claw medium sized.
AGE Late Cretaceous, late Turonian.
DISTRIBUTION AND FORMATION/S Western Argentina; Portezuelo.
HABITAT Well-watered woodlands with short dry season.
HABITS Probably able to dispatch fairly large prey. May have fished like other unenlagiines.
NOTES May include *U. paynemili* and *Neuquenraptor argentinus*.

Austroraptor cabazai

— 6 m (20 ft) TL, 250 kg (500 lb)

FOSSIL REMAINS Majority of skull and minority of skeleton.

Austroraptor cabazai

ANATOMICAL CHARACTERISTICS Head very long, shallow, especially snout and lower jaw, teeth fairly small, numerous, conical. Upper arm fairly short (rest unknown).
AGE Late Cretaceous, late Campanian.
DISTRIBUTION AND FORMATION/S Central Argentina; Allen.
HABITAT Semiarid coastline.
HABITS Probably fished, terrestrial prey also plausible.
NOTES Head form parallels that of spinosaurs.

Ypupiara lopai

— 3 m (10 ft) TL, 30 kg (60 lb)

FOSSIL REMAINS Fragmentary skull.
ANATOMICAL CHARACTERISTICS Head very long, shallow, teeth modest sized.
AGE Late Cretaceous, Maastrichtian.
DISTRIBUTION AND FORMATION/S Southeastern Brazil; Marilia.

Overoraptor chimentoi

— 3 m (10 ft) TL, 30 kg (60 lb)

FOSSIL REMAINS Minority of skeleton.
ANATOMICAL CHARACTERISTICS Arm not highly elongated.
AGE Late Cretaceous, Cenomanian.
DISTRIBUTION AND FORMATION/S Western Argentina; Huincul.
HABITAT Short wet season, otherwise semiarid with open floodplains and riverine forests.
HABITS Nonflier.
NOTES Placement in unenlagiines not certain; not yet known whether the head was similar to the elongated form seen in larger unenlagiines.

Rahonavis ostromi

— 0.7 m (2.2 ft) TL, 1 kg (2 lb)

FOSSIL REMAINS Partial skeleton.
ANATOMICAL CHARACTERISTICS Arm very large. Pubis vertical. Large sickle claw on hyperextendable toe. Quill nodes on trailing edge of upper arm indicate large flight feathers.
AGE Late Cretaceous, Campanian.

DISTRIBUTION AND FORMATION/S Northwestern Madagascar; Maevarano.
HABITAT Seasonally dry floodplain with coastal swamps and marshes.
HABITS Diet may have included aquatic and/or terrestrial small prey, sickle claw possibly used to help dispatch larger prey. Capable of powered flight superior to that of *Archaeopteryx* and *Sapeornis*. Good climber and leaper. Defense included climbing and flight as well as sickle claws.
NOTES Placement in unenlagiines not certain; not yet known whether the head was similar to the elongated form seen in larger unenlagiines.

— DROMAEOSAURINES

SMALL TO LARGE DROMAEOSAURIDS OF THE CRETACEOUS OF THE NORTHERN HEMISPHERE.

ANATOMICAL CHARACTERISTICS Fairly variable. Strongly built. Front of snouts short. Teeth large, frontmost D-shaped in cross section.
HABITS Strong skulls and large, strong teeth indicate that dromaeosaurines used their heads to wound prey more than other dromaeosaurids.
NOTES Relationships within group uncertain, probably splittable into a number of subdivisions.

Utahraptor ostrommaysorum
— 6 m (20 ft) TL, 300 kg (600 lb)
FOSSIL REMAINS Numerous skeletal parts, juvenile to adult.
ANATOMICAL CHARACTERISTICS Very robustly built. Tail not highly elongated, flexible. Sickle toe claw large.
AGE Early Cretaceous, late Valanginian.
DISTRIBUTION AND FORMATION/S Utah; lower Cedar Mountain.
HABITAT Short wet season, otherwise semiarid with floodplain prairies and open woodlands, and riverine forests.
HABITS Not especially fast, an ambush predator that preyed on large dinosaurs.

Bambiraptor feinbergi
— 1.3 m (4 ft) TL, 7 kg (15 lb)
FOSSIL REMAINS Almost complete skull and skeleton, less complete skeleton.
ANATOMICAL CHARACTERISTICS Lightly built. Subrectangular head fairly deep. Arm and hand quite long. Pubis moderately retroverted, leg long, sickle claw large.
AGE Late Cretaceous, late middle Campanian.
DISTRIBUTION AND FORMATION/S Montana; middle Two Medicine.
HABITAT Seasonally dry upland woodlands.
HABITS Probably a generalist able to use head, arms, and sickle claw to handle and wound prey of various sizes, including small ornithopods and protoceratopsids. Long arms indicate good climbing ability and may be compatible with limited flight ability, especially in juveniles.

Achillobator giganticus
— 5 m (16 ft) TL, 250 kg (500 lb)
FOSSIL REMAINS Minority of skull and skeleton.
ANATOMICAL CHARACTERISTICS Head fairly deep. Pubis vertical, sickle claw large.

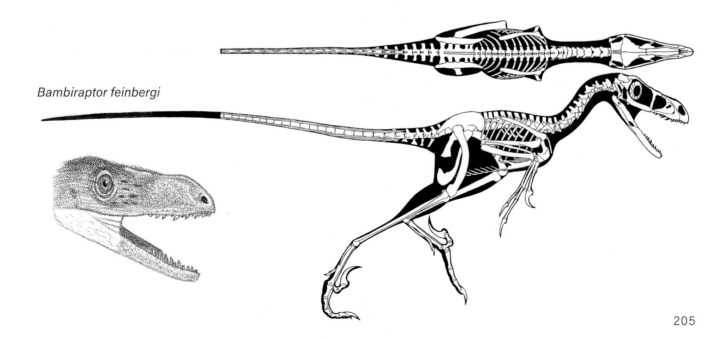

Bambiraptor feinbergi

AGE Early Late Cretaceous.
DISTRIBUTION AND FORMATION/S Mongolia;
Bayanshiree.
HABITS Preyed on large dinosaurs.

Atrociraptor marshalli
—— 2 m (6.5 ft) TL, 15 kg (30 lb)
FOSSIL REMAINS Partial skull and small portion of
skeleton.
ANATOMICAL CHARACTERISTICS Head deep, teeth stout.
AGE Late Cretaceous, latest Campanian and/or early
Maastrichtian.
DISTRIBUTION AND FORMATION/S Alberta, Montana;
lower Horseshoe Canyon.
HABITAT Well-watered, forested floodplain with coastal
swamps and marshes, cool winters.
HABITS Able to attack relatively large prey; used strong
head and teeth to wound prey more than usual for
dromaeosaurs.

Saurornitholestes langstoni
—— 2 m (6.5 ft) TL, 17 kg (35 lb)
FOSSIL REMAINS Complete skull and majority of
skeleton, other material.
ANATOMICAL CHARACTERISTICS Overall build fairly
robust. Head subrectangular, lower jaw fairly deep. Arm
somewhat short. Sickle claw large, slender.
AGE Late Cretaceous, late Campanian.
DISTRIBUTION AND FORMATION/S Alberta, Montana?:
lower and middle Dinosaur Park, upper Two Medicine?
HABITAT Well-watered, forested floodplain with coastal
swamps and marches and drier uplands.
NOTES Placement of Alberta and Montana remains in
same taxon problematic.

Dromaeosaurus albertensis
—— 2 m (7 ft) TL, 15 kg (30 lb)
FOSSIL REMAINS Majority of skull, skeletal fragments.
ANATOMICAL CHARACTERISTICS Head broad and
robust. Teeth large, stout, front teeth semi D-shaped in
cross section.
AGE Late Cretaceous, late middle and/or early to middle
late Campanian.

Dromaeosaurus albertensis

DISTRIBUTION AND FORMATION/S Alberta; Dinosaur
Park, level uncertain.
HABITAT Well-watered, forested floodplain with coastal
swamps and marshes, cool winters.
HABITS Able to attack relatively large prey.
NOTES Is not certain a large sickle claw belongs to this
species. Not common in its habitat.

Adasaurus mongoliensis
—— 2 m (7 ft) TL, 15 kg (30 lb)
FOSSIL REMAINS Partial skull and skeleton.
ANATOMICAL CHARACTERISTICS Head and body
robustly built. Pubis moderately retroverted, sickle claw
not large.
AGE Late Cretaceous, latest Campanian and/or early
Maastrichtian.
DISTRIBUTION AND FORMATION/S Mongolia; middle or
upper Nemegt.
HABITAT Temperate, well-watered, dense woodlands with
seasonal rains and winter snow.
HABITS Able to attack relatively large prey. Did not use
sickle claw as attack weapon as much as other similar-
sized dromaeosaurids.

Acheroraptor temertyorum and/or Dakotaraptor steini
—— 6 m (20 ft) TL, 350 kg (700 lb)
FOSSIL REMAINS Minority of juvenile skull and
minority of adult skeleton.

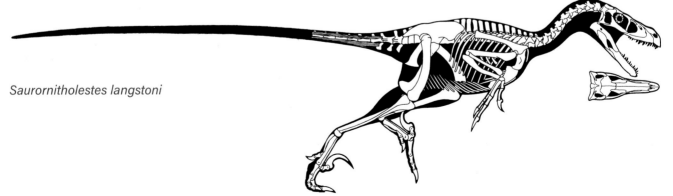

Saurornitholestes langstoni

ANATOMICAL CHARACTERISTICS Lightly built. Leg long, sickle claw very large. Quill nodes on trailing edge of upper arm indicate large feather array.
AGE Late Cretaceous, latest Maastrichtian.
DISTRIBUTION AND FORMATION/S South Dakota; upper Hell Creek.

HABITAT Well-watered coastal woodlands, climate warmer than earlier in Maastrichtian.
HABITS Sickle a very powerful weapon.
NOTES May be a velociraptorine. *Acheroraptor temertyorum* may be the juvenile of later named *Dakotaraptor steini*.

— VELOCIRAPTORINES

MEDIUM-SIZED DROMAEOSAURIDS LIMITED TO THE LATE CRETACEOUS OF THE NORTHERN HEMISPHERE.

ANATOMICAL CHARACTERISTICS Fairly strongly but not heavily built. Front of snouts markedly elongated. Teeth large, frontmost D-shaped in cross section. Sickle claws large.
HABITS Large sickle claws too big to be only for pinning prey, were primary weapons.

Unnamed genus and/or species?

—— 3.5 m (12 ft) TL, 70 kg (150 lb)
FOSSIL REMAINS Partial skeletons.
ANATOMICAL CHARACTERISTICS Pubis moderately retroverted, sickle claw large and not strongly curved.
AGE Early Cretaceous, early Albian?.
DISTRIBUTION AND FORMATION/S Montana; middle Cloverly.
HABITAT Short wet season, otherwise semiarid with floodplain prairies, open woodlands, and riverine forests.
HABITS Probably a generalist that ambushed and pursued small to big game. Juveniles relatively longer armed than adults and may have had limited flight ability.
NOTES Usually placed in later *Deinonychus antirrhopus*, may be direct ancestor of and in same genus as latter.

Deinonychus antirrhopus

—— 3.3 m (11 ft) TL, 65 kg (145 lb)
FOSSIL REMAINS Majority of several skulls and partial skeletons.
ANATOMICAL CHARACTERISTICS Arm fairly long. Head lightly built, very large and long, subtriangular, snout arched in one fossil, may be depressed in another, robustness of lower jaw variable. Leg moderately long, sickle claw large and strongly curved.

AGE Early Cretaceous, middle? Albian.
DISTRIBUTION AND FORMATION/S Montana; upper Cloverly.
HABITAT Short wet season, otherwise semiarid with floodplain prairies, open woodlands, and riverine forests.
HABITS Probably a generalist that ambushed and pursued small to big game. Juveniles relatively longer armed than adults and may have had limited flight ability.
NOTES One of the classic dromaeosaurids, the primary basis of the *Jurassic Park* "raptors." Maxilla and snout not short and deep as often incorrectly restored. If snout tops were arched in some individuals and depressed in others, may represent genders or species.

Kansaignathus sogdianus

—— 2 m (7 ft) TL, 21 kg (45 lb)
FOSSIL REMAINS Fragmentary skull and possible skeleton.
ANATOMICAL CHARACTERISTICS Lower jaw strongly upcurved.
AGE Late Cretaceous, Santonian.
DISTRIBUTION AND FORMATION/S Tajikistan; Ialovachsk.

Deinonychus antirrhopus (see also next page)

Deinonychus antirrhopus

Kuru kulla
—— 2 m (7 ft) TL, 21 kg (45 lb)
FOSSIL REMAINS Minority of skull and skeleton.
ANATOMICAL CHARACTERISTICS Fairly robustly built, head large, sickle claw modest sized.
AGE Late Cretaceous, latest Campanian and / or earliest Maastrichtian.
DISTRIBUTION AND FORMATION/S Mongolia; Barun Goyot.
HABITAT Semidesert with some dunes and oases.
HABITS Probably used head as weapon and sickle claw less so than other velociraptorines.

Shri devi
—— 2 m (7 ft) TL, 21 kg (45 lb)
FOSSIL REMAINS Minority of skull and skeleton.
ANATOMICAL CHARACTERISTICS Lightly built. Ilium elongated, pubis strongly retroverted, sickle claw large.
AGE Late Cretaceous, latest Campanian and / or earliest Maastrichtian.
DISTRIBUTION AND FORMATION/S Mongolia; Barun Goyot.
HABITAT Semidesert with some dunes and oases.

Velociraptor mongoliensis
—— 2.5 m (8 ft) TL, 28 kg (60 lb)

FOSSIL REMAINS A number of complete and partial skulls and skeletons, juvenile to adult, completely known.
ANATOMICAL CHARACTERISTICS Lightly built especially head, latter long, snout extra long and shallow, strongly depressed in juveniles and less so in adults. Arm fairly long. Pubis strongly retroverted. Sickle claw large. Quill nodes on trailing edge of upper arm indicate large feather array.
AGE Late Cretaceous, middle Campanian.
DISTRIBUTION AND FORMATION/S Mongolia, northern China; lower Djadokhta.
HABITAT Desert with dunes and oases.
HABITS Probably a generalist that ambushed and pursued small to big game.
NOTES The other classic dromaeosaurid. Probably includes the contemporary *V. osmolskae*. Famous fighting pair preserves a *Velociraptor* and *Protoceratops* locked in comortal combat. May be direct ancestor of *V. mangas*.

Velociraptor (= Tsaagan) mangas
—— 2 m (7 ft) TL, 21 kg (45 lb)

FOSSIL REMAINS Two? nearly complete skulls, one with majority of skeleton.
ANATOMICAL CHARACTERISTICS Head lightly built, snout not as shallow and depressed as that of *V. mongoliensis*.
AGE Late Cretaceous, late Campanian.
DISTRIBUTION AND FORMATION/S Mongolia, northern China?; upper Djadokhta, upper Wulansuhai?.
HABITAT Desert with dunes and oases.
HABITS Similar to *Velociraptor*.

Velocriraptor (= Tsaagan) mangas

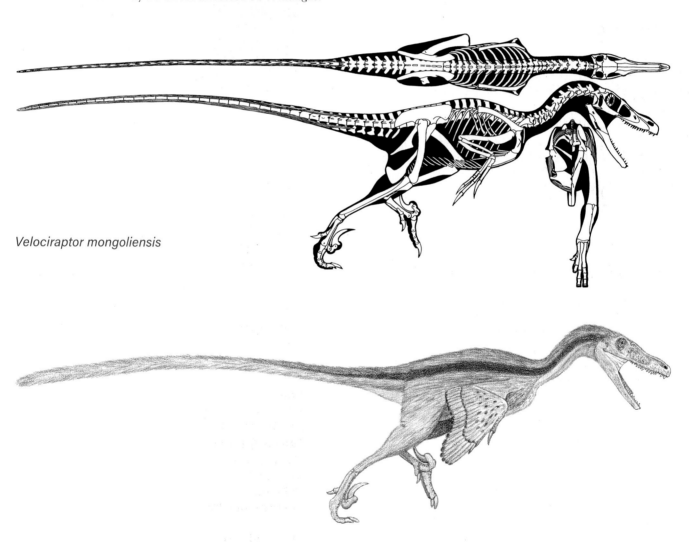

Velociraptor mongoliensis

Velociraptor (= Tsaagan) mangas

immature

NOTES Not different enough from *Velociraptor* to be different genus. May include *Linheraptor exquisitus*; if so, differing degrees of depression of snout may represent genders.

Boreonykus certekorum
—— 1.5 m (5 ft) TL, 5 kg (10 lb)
FOSSIL REMAINS Fragmentary skull and skeleton(s).
ANATOMICAL CHARACTERISTICS Insufficient information.
AGE Late Cretaceous, late Campanian.
DISTRIBUTION AND FORMATION/S Alberta: middle Wapati.
HABITAT Chilly winters.

Dineobellator notohesperus
—— 3 (10 ft) TL, 60 kg (130 lb)
FOSSIL REMAINS Fragmentary skull and of skeleton.
ANATOMICAL CHARACTERISTICS Greater skeletal flexibility. Quill nodes on trailing edge of lower arm indicate large feather array.
AGE Late Cretaceous, early Maastrichtian.
DISTRIBUTION AND FORMATION/S New Mexico; lower Ojo Alamo.
NOTES May be most anatomically sophisticated known terrestrial dromaeosaurid in terms of skeletal action.

— TROODONTIDS

SMALL- TO MEDIUM-SIZED OMNIVOROUS DEINONYCHOSAURS FROM THE LATE JURASSIC UNTIL THE END OF THE DINOSAUR ERA OF THE NORTHERN HEMISPHERE.

ANATOMICAL CHARACTERISTICS Fairly variable. Lightly built. Eyes face strongly forward and often very large; postorbital bars from robust to incomplete, teeth numerous, small, especially at front of upper jaw. Neck ribs usually so short they do not overlap, increasing flexibility of neck. Ossified sternal ribs and uncinates not present. Tails not as long as in dromaeosaurids. Ossified sternums absent. Arms not elongated. Pubes vertical or somewhat retroverted, leg long and gracile, sickle claws not greatly enlarged, but usually larger than other toe claws. Eggs moderately elongated, tapering.
HABITAT HIGHLY VARIABLE, FROM DESERTS TO POLAR FORESTS.
HABITS Running performance very high, leaping and climbing ability poor compared with that of other deinonychosaurs. Pursuit predators that focused on smaller game and could fish but could use sickle claws to dispatch larger prey, especially larger species with large claws, and also use claws to pin down prey. Omnivorous in that also consumed significant plant material. Examples with very large eyes possibly more nocturnal than other dinosaurs. Eggs laid in pairs subvertically in rings, probably by more than one female in each nest, partly exposed so they could be brooded and incubated by adults sitting in center. Juveniles not highly developed so may have received care in or near nest.
NOTES Group name problematic because is based on teeth that are similar to those of pachycephalosaurs. Some researchers consider late Middle/early Late Jurassic *Anchiornis*, *Eosinopteryx*, and *Xiaotingia* of that age to be the earliest known members of this group. May have evolved from anchiornians or relations of limited flight performance, which was then lost. Incomplete postorbital bars and/or retroverted pubes in some examples may indicate close relationships to birds, or at least partly independent developments.

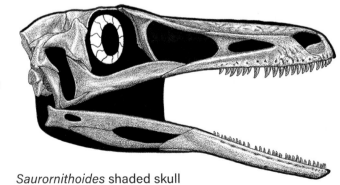

Saurornithoides shaded skull

— BASO-TROODONTS

SMALL TROODONTIDS FROM THE LATE JURASSIC AND CRETACEOUS OF THE NORTHERN HEMISPHERE.

ANATOMICAL CHARACTERISTICS Highly uniform. Heads short, subrectangular.
NOTES May include above late Middle/early Late Jurassic potential troodontids.

Hesperornithoides miessleri
—— 1 m (3 ft) TL, 3 kg (6 lb)
FOSSIL REMAINS Partial skull and skeleton.
ANATOMICAL CHARACTERISTICS Teeth and serrations fairly large.
AGE Late Jurassic, early and/or early middle Kimmeridgian.
DISTRIBUTION AND FORMATION/S Wyoming; middle Morrison.
HABITAT Short wet season, otherwise semiarid with open floodplain prairies and riverine forests.

Jinfengopteryx elegans
—— 0.5 m (1.7 ft) TL, 0.5 kg (1 lb)
FOSSIL REMAINS Complete skull and skeleton, feathers.

ANATOMICAL CHARACTERISTICS Head lightly built, postorbital bar incomplete. Arms fairly large. Pubis vertical. Well-developed pennaceous feathers line entire tail.
AGE Late Late Jurassic or early Early Cretaceous.
DISTRIBUTION AND FORMATION/S Northeastern China; Qiaotou.
HABITAT Well-watered forests and lakes, winters chilly with snow.
HABITS Diet mainly small game and insects. Roundish objects in belly region may be large seeds or nuts.
NOTES May be a juvenile. Originally thought to be a bird close to *Archaeopteryx*; is the earliest certain troodontid known from skeletal remains.

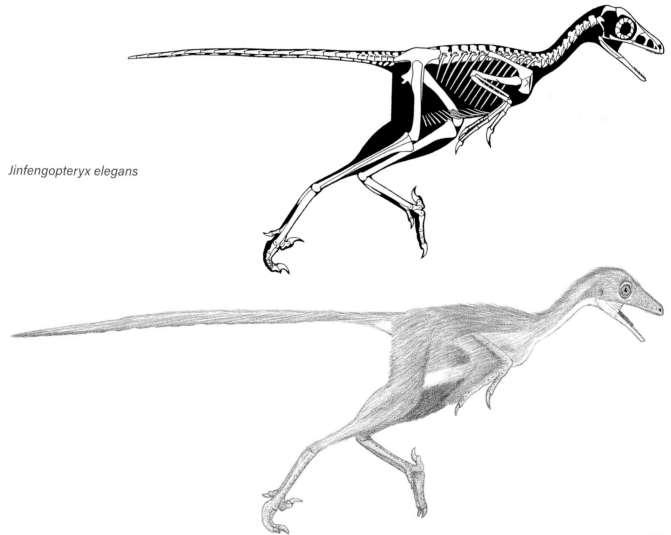

Jinfengopteryx elegans

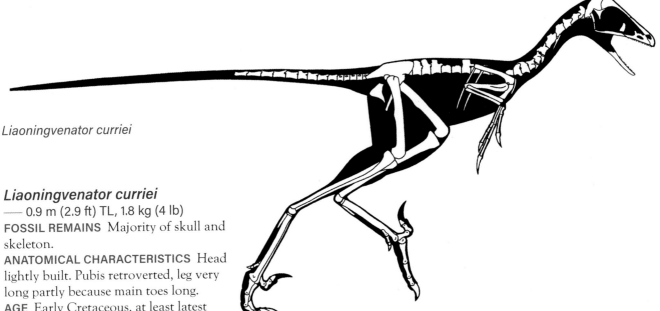

Liaoningvenator curriei

Liaoningvenator curriei

—— 0.9 m (2.9 ft) TL, 1.8 kg (4 lb)

FOSSIL REMAINS Majority of skull and skeleton.

ANATOMICAL CHARACTERISTICS Head lightly built. Pubis retroverted, leg very long partly because main toes long.

AGE Early Cretaceous, at least latest Barremian.

DISTRIBUTION AND FORMATION/S Northeastern China; lower Yixian.

HABITAT Well-watered volcanic highland forest-and-lake district, winters chilly with snow.

HABITS Diet mainly small game and insects.

Mei long

—— 0.45 m (1.4 ft) TL, 0.5 kg (1 lb)

FOSSIL REMAINS Several nearly complete skulls and skeletons.

ANATOMICAL CHARACTERISTICS Head lightly built, postorbital bar incomplete, teeth small, unserrated. Arm fairly large. Pubis slightly retroverted.

AGE Early Cretaceous, at least earliest Barremian.

DISTRIBUTION AND FORMATION/S Northeastern China; lower Yixian.

HABITAT Well-watered volcanic highland forest-and-lake district, winters chilly with snow.

HABITS Diet mainly small game and insects.

NOTES Some specimens preserved in birdlike curled up sleeping posture.

Sinovenator changii

—— 2 m (6.5 ft) TL, 27 kg (60 lb)

FOSSIL REMAINS Numerous skulls and skeletons, complete to partial skull, adult and juvenile.

ANATOMICAL CHARACTERISTICS Lightly built. Postorbital bar complete, teeth fairly small, serrations absent in front teeth, small and limited to back edge on rest. Trunk short. Pelvis quite deep, pubis somewhat retroverted, leg very long.

AGE Early Cretaceous, at least latest Barremian.

DISTRIBUTION AND FORMATION/S Northeastern China; lower Yixian.

HABITAT Well-watered volcanic highland forest-and-lake district, winters chilly with snow.

Mei long

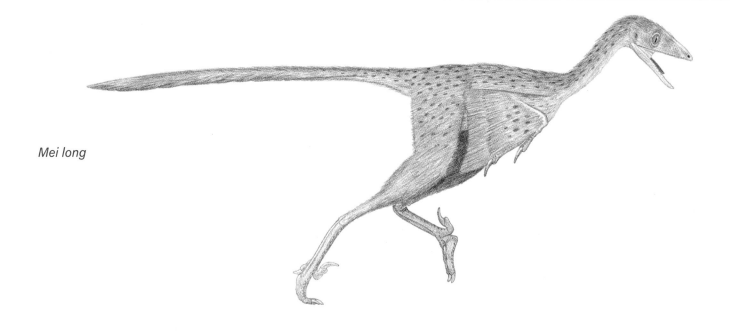

Mei long

Sinovenator changii

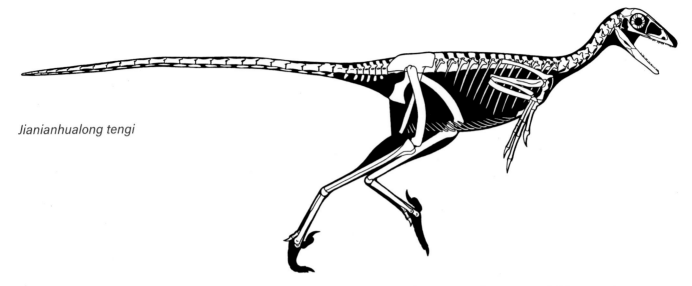

Jianianhualong tengi

Jianianhualong tengi

—— Adult size not certain

FOSSIL REMAINS Majority of skull and skeleton, possibly juvenile.

ANATOMICAL CHARACTERISTICS Postorbital bar complete, teeth fairly small, serrations absent on front teeth, small and limited to back edge on rest. Tail fairly long. Pelvis not deep, pubis somewhat retroverted, leg rather short.

AGE Early Cretaceous, middle Barremian.

DISTRIBUTION AND FORMATION/S Northeastern China; upper Yixian.

HABITAT Well-watered volcanic highland forest-and-lake district, winters chilly with snow.

Daliansaurus liaoningensis

—— 0.7 m (2.5 ft) TL, 1.5 kg (3 lb)

FOSSIL REMAINS Partial skull and majority of skeleton.

ANATOMICAL CHARACTERISTICS Teeth fairly large, serrations absent in front teeth, small and limited to back edge on rest. Outer toe claw as large as second.

AGE Early Cretaceous, at least latest Barremian.

DISTRIBUTION AND FORMATION/S Northeastern China; lower Yixian.

HABITAT Well-watered volcanic highland forest-and-lake district, winters chilly with snow.

—— TROODONTID MISCELLANEA

NOTES TROODONTIDS THAT ARE NEITHER BASO-TROODONTINES NOR TROODONTINES, RELATIONSHIPS UNCERTAIN.

Sinornithoides youngi

—— 1.1 m (3.5 ft) TL, 2.9 kg (5.5 lb)

FOSSIL REMAINS Partial skull and complete skeleton, not fully mature.

ANATOMICAL CHARACTERISTICS Head long and shallow, teeth and denticles small. Arm fairly small. Pubis vertical.

HABITS That not just hyperextendable second toe was large indicates different hunting and other habits from small basal troodonts.

Almas ukhaa

—— Adult size uncertain

FOSSIL REMAINS Majority of two skulls and partial skeletons.

ANATOMICAL CHARACTERISTICS Teeth small. Pubis retroverted.

AGE Late Cretaceous, late Campanian.

DISTRIBUTION AND FORMATION/S Mongolia; upper Djadokhta, Ulansuhai?

NOTES From the same age Ulansuhai Formation, *Papiliovenator neimengguensis* may be more mature member of this species, or at least genus.

Almas ukhaa

AGE Early Cretaceous.

DISTRIBUTION AND FORMATION/S Northern China; Ejinhoro.

NOTES Specimen preserved in birdlike curled up sleeping posture.

Sinornithoides youngi

Geminiraptor suarezarum
—— 1.5 (5 ft) TL, 8 kg (18 lb)
FOSSIL REMAINS Minority of skull.
ANATOMICAL CHARACTERISTICS Insufficient information.
AGE Early Cretaceous, late Berriasian.
DISTRIBUTION AND FORMATION/S Utah; lowermost Cedar Mountain.
HABITAT Short wet season, otherwise semiarid with floodplain prairies and open woodlands, and riverine forests.

Sinusonasus magnodens
—— 1 m (3.3 ft) TL, 2.5 kg (5.5 lb)
FOSSIL REMAINS Partial skull and majority of skeleton.
ANATOMICAL CHARACTERISTICS Head long and shallow, serrations absent in front teeth, small and limited to back edge on rest, teeth relatively large.
AGE Early Cretaceous, at least latest Barremian.
DISTRIBUTION AND FORMATION/S Northeastern China; lower Yixian.
HABITAT Well-watered volcanic highland forest-and-lake district, winters chilly with snow.
HABITS Attacked bigger game than *Sinovenator*, smaller game than *Sinornithosaurus*.

Xixiasaurus henanensis
—— 1.5 (5 ft) TL, 8 kg (18 lb)
FOSSIL REMAINS Partial skull and minority of skeleton.
ANATOMICAL CHARACTERISTICS Head fairly shallow, teeth fairly large, sharp, unserrated.
AGE Late Cretaceous, probably Campanian.
DISTRIBUTION AND FORMATION/S EASTERN CHINA; MAJIACUN.

Byronosaurus jaffei
—— 1.5 m (5 ft) TL, 8 kg (18 lb)
FOSSIL REMAINS Partial skull and minority of skeleton.
ANATOMICAL CHARACTERISTICS Snout very long and very shallow, a little depressed, eyes not especially large, teeth sharp, unserrated.
AGE Late Cretaceous, late Campanian.
DISTRIBUTION AND FORMATION/S Mongolia; upper Djadokhta.
HABITAT Desert with dunes and oases.
HABITS Hunted small game, possibly fished, probably less nocturnal than most troodonts.
NOTES *Almas ukhaa* may be juvenile of this species.

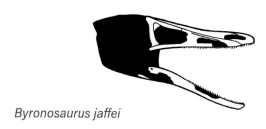

Byronosaurus jaffei

— TROODONTINES

MEDIUM-SIZED TROODONTIDS LIMITED TO THE LATE CRETACEOUS OF THE NORTHERN HEMISPHERE.

ANATOMICAL CHARACTERISTICS Uniform. Heads long and shallow, postorbital bars robust. Legs long.
NOTES May include above late Middle / early Late Jurassic potential troodonts.

Talos sampsoni
—— 2 m (7.5 ft) TL, 15 kg (30 lb)
FOSSIL REMAINS Minority of skeleton.
ANATOMICAL CHARACTERISTICS Leg very slender.
AGE Late Cretaceous, late Campanian.
DISTRIBUTION AND FORMATION/S Utah; middle Kaiparowits.

Philovenator curriei
—— Adult size uncertain
FOSSIL REMAINS Minority of skeleton, juvenile.
ANATOMICAL CHARACTERISTICS Insufficient information.
AGE Late Cretaceous, late Campanian.
DISTRIBUTION AND FORMATION/S Northern China; Wulansuhai.
HABITS Similar to *Velociraptor*.
NOTES May be same genus as *Linhevenator*.

Linhevenator tani
—— 2.1 m (7 ft) TL, 20 kg (40 lb)
FOSSIL REMAINS Poorly preserved partial skull and skeleton.

ANATOMICAL CHARACTERISTICS Arm short and robust. Sickle claw large.
AGE Late Cretaceous, Campanian.
DISTRIBUTION AND FORMATION/S Mongolia; Ulansuhai.
HABITS Strong arm and large sickle-claw foot weapon indicate hunted bigger game than other troodonts.

Gobivenator mongoliensis
—— 1.7 m (5.5 ft) TL, 10 kg (22 lb)
FOSSIL REMAINS Majority of skull and skeleton.
ANATOMICAL CHARACTERISTICS Snout semitubular, teeth small and unserrated. Arm fairly small. Pubis vertical.
AGE Late Cretaceous, late Campanian.
DISTRIBUTION AND FORMATION/S Mongolia; upper Djadokhta.
HABITAT Desert with dunes and oases.
HABITS Hunted both small and large game, possibly fished. Possibly more nocturnal than most theropods.
NOTES The first nearly completely known troodont.

Troodont hatchlings

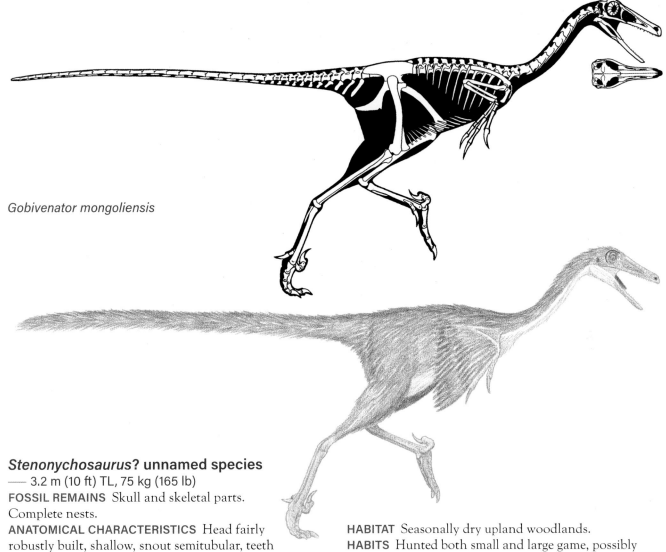

Gobivenator mongoliensis

Stenonychosaurus? unnamed species

—— 3.2 m (10 ft) TL, 75 kg (165 lb)

FOSSIL REMAINS Skull and skeletal parts. Complete nests.

ANATOMICAL CHARACTERISTICS Head fairly robustly built, shallow, snout semitubular, teeth with large denticles. Pubis vertical. Elongated eggs 18 cm (7 in) long.

AGE Late Cretaceous, middle and/or late Campanian.

DISTRIBUTION AND FORMATION/S Montana; upper Two Medicine.

HABITAT Seasonally dry upland woodlands.

HABITS Hunted both small and large game, possibly fished.

NOTES Usually placed in *Troodon formosus*, which is based on inadequate remains; is uncertain whether this is same genus or species as *S. inequalis*.

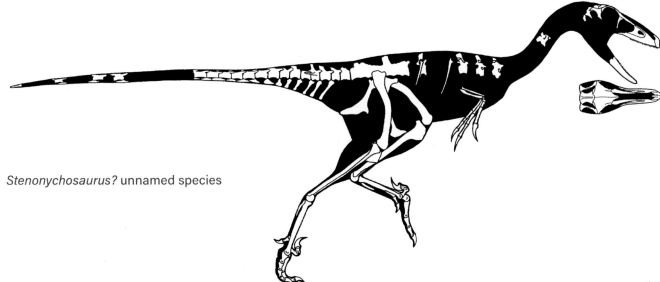

Stenonychosaurus? unnamed species

Stenonychosaurus? unnamed species

Stenonychosaurus? inequalis?
—— 2.5 m (8 ft) TL, 35 kg (70 lb)
FOSSIL REMAINS Minorities of a few skulls and skeletons.
ANATOMICAL CHARACTERISTICS Head fairly robustly built, shallow, snout semitubular, teeth with large denticles.
AGE Late Cretaceous, late middle Campanian.
DISTRIBUTION AND FORMATION/S Alberta; lower Dinosaur Park.
HABITAT Well-watered, forested floodplain with coastal swamps and marshes, cool winters.
HABITS Hunted both small and large game, possibly fished.
NOTES Based on questionably adequate remains, lack of information on its pubis hinders determining whether is close to either *Latenivenatrix* or to *Stenonychosaurus* unnamed species; may be same genus as *Saurornithoides*.

Latenivenatrix mcmasterae
—— 5 m (16 ft) TL, 250 kg (500 lb)
FOSSIL REMAINS Minorities of a few skulls and skeletons.
ANATOMICAL CHARACTERISTICS Pubis somewhat retroverted.
AGE Late Cretaceous, middle late Campanian.
DISTRIBUTION AND FORMATION/S Alberta; upper Dinosaur Park.
HABITAT Well-watered, forested floodplain with coastal swamps and marshes, cool winters.

HABITS Hunted larger game than smaller troodonts.
NOTES Probably not same genus as earlier, different *Stenonychosaurus inequalis*. Largest known troodont.

Saurornithoides mongoliensis
—— Adult size uncertain
FOSSIL REMAINS Majority of badly damaged skull and minority of skeleton.
ANATOMICAL CHARACTERISTICS Head fairly robustly built, shallow, snout semitubular, teeth with large denticles. Pubis vertical.
AGE Late Cretaceous, middle Campanian.
DISTRIBUTION AND FORMATION/S Mongolia; lower Djadokhta.
HABITAT Desert with dunes and oases.
HABITS Similar to *Stenonychosaurus*.
NOTES May have been smaller than *S. junior*. *Archaeornithoides deinosauriscus* may be juvenile of this taxon. *Saurornithoides* may be same genus as *Stenonychosaurus*. May be an ancestor of *S. junior*.

Saurornithoides (or *Zanabazar*) *junior*
—— 2.3 m (7.5 ft) TL, 20 kg (45 lb)
FOSSIL REMAINS Majority of a skull and minority of the skeleton.

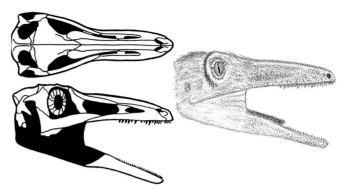

Saurornithoides (or *Zanabazar*) *junior*

ANATOMICAL CHARACTERISTICS Head fairly robustly built, shallow, snout semitubular, teeth with large denticles.
AGE Late Cretaceous, latest Campanian and/or early Maastrichtian.
DISTRIBUTION AND FORMATION/S Mongolia; middle or upper Nemegt.
HABITAT Temperate, well-watered, dense woodlands with seasonal rains and winter snow.
HABITS Similar to *S. mongoliensis*.
NOTES Available evidence insufficient to make separate genus from *Saurornithoides*. May include fragmentary *Borogovia* and *Tochisaurus* at least at genus level.

— THERIZINOSAURIFORMS
SMALL TO GIGANTIC FLYING AND FLIGHTLESS HERBIVOROUS AVEAIRFOILANS OF THE CRETACEOUS OF THE NORTHERN HEMISPHERE.

ANATOMICAL CHARACTERISTICS Variable. Heads somewhat elongated, blunt upper beak, extra joint in lower jaw absent, teeth small, blunt, leaf shaped, not serrated. Neck ribs usually so short they do not overlap, increasing flexibility of neck. Tails from very long to very short. Arms long, semilunate carpal from well to poorly developed. Feet not narrow, three to four load-bearing toes. Gastroliths often present.
HABITS Herbivores.
NOTES Therizinosauriforms are jeholornids, therizinosaurians, and avians and their common ancestor, operative only if three groups form a clade that excludes all other dinosaurs except oviraptorosaurs.

— JEHOLORNIDS
SMALL FLYING AVEAIRFOILANS OF THE EARLY CRETACEOUS OF ASIA.

ANATOMICAL CHARACTERISTICS Beaks well developed, two stout upper teeth, lower jaw teeth restricted to front. Tails long and slender. Furculas and sternal plates large, arms longer and much more strongly built than legs, outer upper hand bones curved, outer and central fingers stiffened and base flattened to better support fully developed, long wings made of asymmetrical feathers. Pubes moderately retroverted. Most of body covered by short feathers, base of tails supported feather fan, tails ended with palm-like feather frond, at least in some cases.
HABITS Powered flight fairly well developed. Feather fan at tail base aerodynamic, feather frond at tail tip predominantly for display. Seeds and flowering plant leaves found in guts of some fossils.
NOTES A very therizinosaur-like skull strongly suggests a close relationship. Although may not be not the direct ancestors of therizinosauroids, jeholornids may represent the form of long-headed, long-tailed, flying basal birds from which therizinosaurs evolved.

Jeholornis prima

Jeholornis curvipes
—— 0.7 m (2.5 ft) TL, 0.8 kg (1.8 lb)
FOSSIL REMAINS Complete but poorly preserved skulls with skeletons.
ANATOMICAL CHARACTERISTICS As for group.
AGE Early Cretaceous, middle Barremian.
DISTRIBUTION AND FORMATION/S Northeastern China; upper Yixian.
HABITAT Well-watered volcanic highland forest-and-lake district, winters chilly with snow.
NOTES May be direct ancestor of *J. prima*.

Jeholornis prima
—— 0.7 m (2.5 ft) TL, 0.8 kg (1.8 lb)
FOSSIL REMAINS Skulls and skeletons, extensive feathers.
ANATOMICAL CHARACTERISTICS As for group.
AGE Early Cretaceous, late Barremian.
DISTRIBUTION AND FORMATION/S Northeastern China; lower Jiufotang.
HABITAT Well-watered volcanic highland forest-and-lake district, winters chilly with snow.
NOTES Probably includes *J. palmapenis*, possibly as a gender morph.

—— THERIZINOSAURIANS

SMALL TO GIGANTIC HERBIVOROUS FLIGHTLESS AVEAIRFOILANS OF THE CRETACEOUS OF THE NORTHERN HEMISPHERE.

ANATOMICAL CHARACTERISTICS Variable. Heads small, cheeks very probably present, jaw gap limited. Necks long and slender. Trunks tilted upward from retroverted and therefore horizontal pelves and tails, bellies large. Tails from very long to very short. Furculas and sternals not large, arms long but shorter and weaker than leg, half-moon carpal blocks from well to poorly developed, finger claws large. Feet not narrow, three to four load-bearing toes, toe claws enlarged. Feathers simple.

HABITS Predominantly browsing herbivores, picked up small animals on occasion. Too slow to readily escape predatory theropods, main defense long arms and hand claws as well as kicks from clawed feet.

ENERGETICS Energy levels and food consumption probably low for dinosaurs.

NOTES The most herbivorous of the theropods, therizinosaurs are so unusual in their form that before sufficient remains were found it was uncertain they were avepods, relationships to prosauropods once being an alternative. The redevelopment of a complete inner toe within the group is an especially unusual evolutionary reversal, partly paralleled in *Chilesaurus* and *Balaur*. Relationships within avepods remain uncertain. May not be aveairfoilans, may not have descended from fliers, may have descended from very early fliers near base of aveairfoilans. If are secondarily flightless aveairfoilans, may have descended from long-skulled herbivorous early birds more than once, from fliers long tailed and then short tailed. Or tail reduction occurred independently in therizinosaurs if they never had flying ancestors as per conventional theory. The modest extent of initial reversals from therizinosaur-like fliers like jeholornids to the first flightless therizinosaurs, the presence of a number of flight-related features in early therizinosaurs, and their progressive reduction as the group evolved toward massive ground herbivores are significantly compatible with an origin among fliers. Loss of flight may have occurred more than once, leading to appearance of different subgroups. Lack of pennaceous feathers is paralleled in birds with long history of flightlessness, such as ratites.

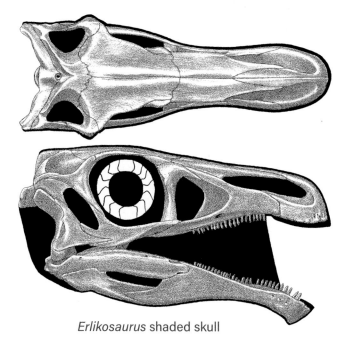

Erlikosaurus shaded skull

— BASO-THERIZINOSAURIANS

MEDIUM-SIZED THERIZINOSAURIANS LIMITED TO THE EARLY CRETACEOUS.

ANATOMICAL CHARACTERISTICS Trunks only modestly tilted up, gastralia flexible. Lunate carpals well developed, finger claws hooked. Front of pelves not strongly flared out sideways, legs fairly long, hind foot still tridactyl with inner toe still a short hallux, claws not greatly enlarged.

HABITS Better runners than more advanced therizinosaurs.

Fukuivenator paradoxus
— 2 m (6.5 ft) TL, 15 kg (30 lb)
FOSSIL REMAINS Majority of skull and skeleton.
ANATOMICAL CHARACTERISTICS Teeth sharp, some large, recurved. Tail long. Shoulder girdle not birdlike. Orientation of pubis uncertain.
AGE Early Cretaceous, Barremian or Aptian.
DISTRIBUTION AND FORMATION/S Central Japan; Kitadani.

Falcarius utahensis
— 4 m (13 ft) TL, 120 kg (250 lb)
FOSSIL REMAINS Minority of skull and almost complete skeletal remains known from dozens of partial fossils, juvenile to adult.
ANATOMICAL CHARACTERISTICS Tail very long. Shoulder girdle birdlike. Pubis orientation possibly vertical.
AGE Early Cretaceous, late Berriasian.
DISTRIBUTION AND FORMATION/S Utah; lowermost Cedar Mountain.

HABITAT Short wet season, otherwise semiarid with floodplain prairies, open woodlands, and riverine forests.
NOTES Lack of pelvis from a single individual leaves orientation of pubis problematic.

Jianchangosaurus yixianensis
— Total length uncertain, 26 kg (60 lb)
FOSSIL REMAINS Complete skull and majority of skeleton, some feathers.
ANATOMICAL CHARACTERISTICS Tip of lower jaw downturned. Tail not abbreviated but length uncertain. Shoulder girdle not birdlike. Pubis moderately retroverted. Leg long.
AGE Early Cretaceous, latest Barremian and/or earliest Aptian.
DISTRIBUTION AND FORMATION/S Northeastern China; Yixian, level uncertain.
HABITAT Well-watered volcanic highland forest-and-lake district, winters chilly with snow.
HABITS Best runner among therizinosaurs.

Falcarius utahensis

Jianchangosaurus yixianensis

—THERIZINOSAUROIDS
SMALL TO LARGE THERIZINOSAURIANS OF THE CRETACEOUS.

ANATOMICAL CHARACTERISTICS Fairly variable. Tip of lower jaws downturned. Skeletons robustly built. Tails short. Shoulder girdles not birdlike, arms moderately long, lunate carpal blocks less well developed, fingers not very long, finger claws large hooks. Front of pelves enlarged and flared sideways, and pubes retroverted to support bigger belly, feet short and broad with four complete toes, toe claws not very enlarged. Known eggs subspherical.
HABITS Buried nests and lack of evidence of brooding indicate little or no parental care. Well-developed hatchlings probably able to leave nest immediately on hatching.

Beipiaosaurus inexpectus
—— 1.8 m (6 ft) TL, 50 kg (100 lb)

FOSSIL REMAINS Skull, two partial skeletons, feathers.
ANATOMICAL CHARACTERISTICS Head shallow, sharply tapering toward front. Tail tipped with a small fused pygostyle. Arrays of long, tapering, band-like feathers atop and beneath back of head; simple feathers on much of body.
AGE Early Cretaceous, latest Barremian and/or earliest Aptian.
DISTRIBUTION AND FORMATION/S Northeastern China; middle Yixian.
HABITAT Well-watered volcanic highland forest-and-lake district, winters chilly with snow.
HABITS Feather arrays for display.
NOTES Is not certain that skull and front half of skeleton belong to this particular species.

Lingyuanosaurus sihedangensis
—— Adult size uncertain

FOSSIL REMAINS Minority of skeleton, probably immature.
ANATOMICAL CHARACTERISTICS Insufficient information.

AGE Early Cretaceous, from middle / late Barremian or early Aptian.
DISTRIBUTION AND FORMATION/S Northeastern China; Yixian or Jiufotang.
HABITAT Well-watered volcanic highland forest-and-lake district, winters chilly with snow.

Martharaptor greenriverensis
—— Adult size uncertain

FOSSIL REMAINS Minority of skeleton, possible juvenile.
ANATOMICAL CHARACTERISTICS Insufficient information.
AGE Early Cretaceous, late Valanginian.
DISTRIBUTION AND FORMATION/S Utah; lower Cedar Mountain.
HABITAT Short wet season, otherwise semiarid with floodplain prairies, open woodlands, and riverine forests.

Beipiaosaurus inexpectus

Alxasaurus elesitaiensis

Alxasaurus elesitaiensis
—— 4 m (13 ft) TL, 500 kg (1,000 lb)
FOSSIL REMAINS Majority of several skeletons.
ANATOMICAL CHARACTERISTICS Standard
for group.
AGE Early Cretaceous, Barremian or Albian.
DISTRIBUTION AND FORMATION/S Northern China;
Bayingobi.

Unnamed genus *bohlini*
—— 6 m (20 ft) TL, 1.5 tonnes
FOSSIL REMAINS Partial skeleton.
ANATOMICAL CHARACTERISTICS Insufficient
information.
AGE Early Cretaceous, Albian.

DISTRIBUTION AND FORMATION/S Northern China;
lower Zhonggou.
NOTES Whether this is a therizinosauroid or
therizinosaurid is uncertain. Originally placed in the
much later and different *Nanshiungosaurus*.

Suzhousaurus megatherioides
—— 6 m (20 ft) TL, 1.5 tonnes
FOSSIL REMAINS Partial skeleton.
ANATOMICAL CHARACTERISTICS Standard for group.
AGE Early Cretaceous, Aptian or Albian.
DISTRIBUTION AND FORMATION/S Northern China;
lower Xinminpu.
NOTES Whether this is a therizinosauroid or
therizinosaurid is uncertain.

—— THERIZINOSAURIDS
MEDIUM-SIZED TO GIGANTIC THERIZINOSAUROIDS OF THE LATE CRETACEOUS.

ANATOMICAL CHARACTERISTICS Uniform. Tip of lower jaws downturned. Skeletons more robustly built. Trunks
more strongly tilted up. Tails short. Shoulder girdles not birdlike, semilunate carpal blocks more poorly developed so
wrist flexion reduced, fingers not very long but bearing very large claws. Front of pelves further enlarged and flared
sideways to support bigger belly, feet broad with four toes, toe claws large.
HABITS Strong upward tilt of trunk indicates these were high-level browsers.
NOTES The dinosaur group most similar to the recent giant ground sloths.

Erliansaurus bellamanus
—— 4 m (13 ft) TL, 1 tonne
FOSSIL REMAINS Partial skeleton.
ANATOMICAL CHARACTERISTICS Standard for group.
AGE Late Cretaceous, Santonian.
DISTRIBUTION AND FORMATION/S Northern China;
Iren Dabasu.
HABITAT Seasonally wet-dry woodlands.
NOTES Whether this is a therizinosauroid or
therizinosaurid is uncertain.

Neimongosaurus yangi
—— 3 m (10 ft) TL, 350 kg (800 lb)
FOSSIL REMAINS Minority of skull and skeleton.
ANATOMICAL CHARACTERISTICS Insufficient
information.
AGE Late Cretaceous, probably Campanian.
DISTRIBUTION AND FORMATION/S Northern China;
Iren Dabasu.
HABITAT Seasonally wet-dry woodlands.
NOTES Whether this is a therizinosauroid or
therizinosaurid is uncertain.

Nanshiungosaurus brevispinus

—— 5 m (16 ft) TL, 1.5 tonnes
FOSSIL REMAINS Partial skeleton.
ANATOMICAL CHARACTERISTICS Standard for group.
AGE Late Cretaceous, Campanian.
DISTRIBUTION AND FORMATION/S Northern China; Yuanpu.

Nothronychus graffami

—— 4.2 m (14 ft) TL, 1 tonne
FOSSIL REMAINS Nearly complete skeleton.
ANATOMICAL CHARACTERISTICS Gastralia inflexible. Finger claws strongly hooked, toe claws not enlarged.
AGE Late Cretaceous, early Turonian.
DISTRIBUTION AND FORMATION/S Utah; Tropic Shale.
HABITAT Coastal swamps and marshes.
NOTES Found as drift in nearshore marine sediments. Inflexibility of gastralia caused by lack of major changes in volume of belly in a nongorging, constantly feeding herbivore. May be direct ancestor of *N. mckinleyi.*

Nothronychus mckinleyi

—— 5 m (17 ft) TL, 1.7 tonnes
FOSSIL REMAINS Minority of skeleton.
ANATOMICAL CHARACTERISTICS Same as for *N. graffami.*
AGE Late Cretaceous, middle Turonian.
DISTRIBUTION AND FORMATION/S New Mexico; lower Moreno Hill.
HABITAT Coastal swamps and marshes.

Segnosaurus galbinensis

—— 6 m (20 ft) TL, 3 tonnes
FOSSIL REMAINS Minority of skull and skeleton.
ANATOMICAL CHARACTERISTICS Cheeks not as extensive as those of *Erlikosaurus.* Front of pelvis greatly enlarged, toe claws enlarged.

AGE Early Late Cretaceous.
DISTRIBUTION AND FORMATION/S Mongolia; Bayenshiree Svita.
HABITS Used large clawed feet for defense as well as hands.
NOTES *Enigmosaurus mongoliensis* may be the same as this species or *Erlikosaurus.*

Erlikosaurus andrewsi

—— 4.5 m (15 ft) TL, 1.2 tonnes
FOSSIL REMAINS Complete skull.
ANATOMICAL CHARACTERISTICS Teeth smaller and more numerous than in *Segnosaurus,* cheeks well developed. Toe claws enlarged.
AGE Early Late Cretaceous.

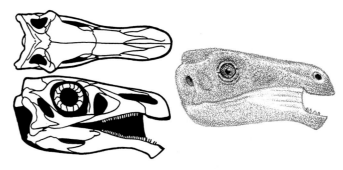

Erlikosaurus andrewsi

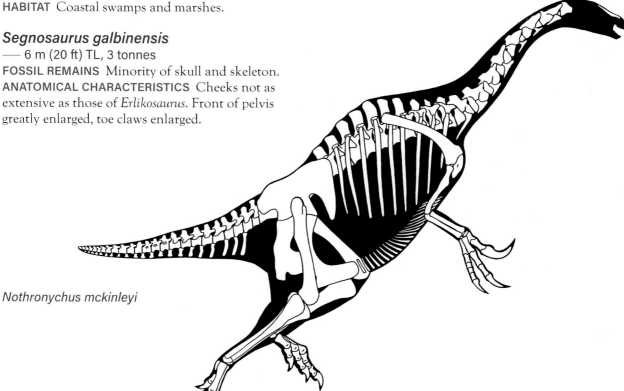

Nothronychus mckinleyi

Erlikosaurus andrewsi

DISTRIBUTION AND FORMATION/S Mongolia; Bayenshiree Svita.
HABITS Used large clawed feet for defense as well as hands.
NOTES May include *Enigmosaurus mongoliensis*.

Therizinosaurus cheloniformis
—— 10 m (33 ft) TL, 5–10 tonnes
FOSSIL REMAINS Arms and some claws, parts of the hind limb.
ANATOMICAL CHARACTERISTICS Arm up to 3.5 m (11 ft) long, bearing very long, saber-shaped claws that were 0.7m (over 2 ft) in length without their original horn sheaths.
AGE Late Cretaceous, late Campanian and/or earliest Maastrichtian.
DISTRIBUTION AND FORMATION/S Mongolia; lower Nemegt.
HABITAT Temperate, well-watered, dense woodlands with seasonal rains and winter snow.
NOTES Lack of remains of main body hinders size estimation, is largest known airfoilan, another example of a very large, high-browsing theropod like *Gigantoraptor* and the *Deinocheirus* that shared its habitat.

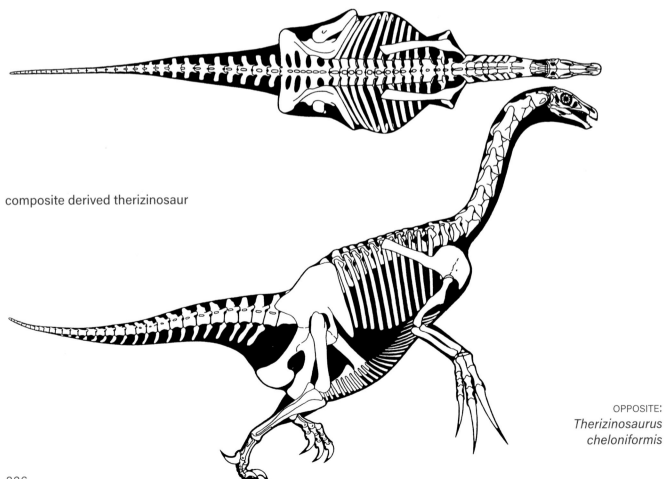

composite derived therizinosaur

OPPOSITE:
*Therizinosaurus
cheloniformis*

— ALVAREZSAURS

SMALL AVEAIRFOILANS FROM THE EARLY CRETACEOUS TO THE END OF THE DINOSAUR ERA, OF ASIA AND WESTERN HEMISPHERE.

ANATOMICAL CHARACTERISTICS Teeth reduced. Scapula blades slender, humeri short, stouter, and distally expanded, thumb longer and much more robust than other fingers. Legs and feet long and slender.
HABITS Main defense high speed.
NOTES Originally thought to be aveairfoilans very close to birds, but opinion that haplocheirids are basal alvarezsaurs suggested otherwise, probability that latter are not related revives probable bird relations and possible loss of flight of alvarezsaurs because of large sterna, fused wrists and hands, and retroverted pubes. If secondarily flightless, probably descended from long-tailed fliers. Limited distribution may reflect insufficient sampling.

— BANNYKIDS

SMALL ALVAREZSAURS OF THE EARLY CRETACEOUS FROM ASIA.

ANATOMICAL CHARACTERISTICS Pectoral crest of humeri not greatly enlarged, outer fingers not greatly reduced.
NOTES Lack of skulls and pubes hinders analysis of habits and relationships.

Tugulusaurus faciles
— 2 m (7 ft) TL, 15 kg (30 lb)
FOSSIL REMAINS Minority of skeleton.
ANATOMICAL CHARACTERISTICS Thumb not enlarged.
AGE Early Cretaceous, late Valanginian.
DISTRIBUTION AND FORMATION/S Northwestern China; lower Lianmuqin.

Bannykus wulatensis
— 2 m (7 ft) TL, 15 kg (30 lb)
FOSSIL REMAINS Partial skeleton.
ANATOMICAL CHARACTERISTICS Thumb enlarged.
AGE Early Cretaceous, Barremian or Aptian.
DISTRIBUTION AND FORMATION/S North-central China; Bayingobi.

— ALVAREZSAURIDS

SMALL ALVAREZSAURS FROM THE EARLY CRETACEOUS TO THE END OF THE DINOSAUR ERA, OF ASIA AND WESTERN HEMISPHERE.

ANATOMICAL CHARACTERISTICS Snouts semitubular, teeth very numerous and small, postorbital bar incomplete as in birds. Necks slender. Arms very short, stout, with greatly enlarged pectoral crest, powerfully muscled, wrists and upper hands fused, hands reduced to a massive thumb and its and robust claw, other fingers very reduced or lost. Pubes strongly retroverted, unbooted, legs and feet very long and slender.
HABITS Extremely stout, powerful arms for breaking into colonial nests, perhaps eggs. Main defense very high speed.
NOTES Reduced outer fingers of little utility and gradually lost. Number of genera may be excessive. Limited distribution may reflect insufficient sampling.

Alvarezsaurus calvoi
— 1 m (3.3 ft) TL, 4 kg (9 lb)
FOSSIL REMAINS Minority of skeleton.
ANATOMICAL CHARACTERISTICS Foot not strongly compressed from side to side.
AGE Late Cretaceous, Santonian.
DISTRIBUTION AND FORMATION/S Western Argentina; Bajo de la Carpa.
HABITAT Semiarid.
NOTES *Achillesaurus manazzonei* probably the adult of this species.

Bonapartenykus ultimus
— 4.5 m (10 ft) TL, 400 kg (1,000 lb)
FOSSIL REMAINS Minority of skeleton, possible eggs.
ANATOMICAL CHARACTERISTICS Insufficient information.

AGE Late Cretaceous, early Maastrichtian.
DISTRIBUTION AND FORMATION/S Central Argentina; upper Allen.
HABITAT Semiarid coastline.

Dzharaonyx eski
— Adult size uncertain
FOSSIL REMAINS Minority of skeleton, possibly juvenile.
ANATOMICAL CHARACTERISTICS Insufficient information.
AGE Late Cretaceous, middle or late Turonian.
DISTRIBUTION AND FORMATION/S Uzbekistan; Bissekty.

Patagonykus puertai
— 1.9 m (6 ft) TL, 28 kg (60 lb)
FOSSIL REMAINS Minority of skeleton.

ANATOMICAL CHARACTERISTICS Pubis not strongly retroverted.
AGE Late Cretaceous, Turonian or Coniacian.
HABITAT Well-watered woodlands with short dry season.
DISTRIBUTION AND FORMATION/S Western Argentina; Rio Neuquén.

Albertonykus borealis

—— 1.1 m (3.5 ft) TL, 5 kg (12 lb)
FOSSIL REMAINS Minority of skeleton.
ANATOMICAL CHARACTERISTICS Insufficient information.
AGE Late Cretaceous, middle Maastrichtian.
DISTRIBUTION AND FORMATION/S Alberta; upper Horseshoe Canyon.
HABITAT Well-watered, forested floodplain with coastal swamps and marshes, cool winters.

Kol ghuva

—— 2 m (6 ft) TL, 25 kg (50 lb)
FOSSIL REMAINS Fragmentary skeleton.
ANATOMICAL CHARACTERISTICS Insufficient information.
AGE Late Cretaceous, late Campanian.
DISTRIBUTION AND FORMATION/S Mongolia; upper Djadokhta.
HABITAT Desert with dunes and oases.

Parvicursor remotus

—— Adult size not certain
FOSSIL REMAINS Minority of skeleton, juvenile, possibly minorities of other skeletons.
ANATOMICAL CHARACTERISTICS Pubis strongly retroverted.
AGE Late Cretaceous, latest Campanian and/or earliest Maastrichtian.
DISTRIBUTION AND FORMATION/S Mongolia; lower or middle Barun Goyot.
HABITAT Semidesert with some dunes and oases.
NOTES *Ceratonykus oculatus*, *Khulsanurus magnificus*, and *Ondogurvel alifanovi* may belong to this species.

Qiupanykus zhangi

—— 0.5 m (1.6 ft) TL, 0.5 kg (1.1 lb)
FOSSIL REMAINS Minority of skeleton, possible eggs.
ANATOMICAL CHARACTERISTICS Insufficient information.
AGE Late Cretaceous.
DISTRIBUTION AND FORMATION/S Eastern China; Qiupa.
NOTES Eggs found near remains may have been food source.

Xixianykus zhangi

—— Adult size uncertain
FOSSIL REMAINS Minority of possibly juvenile skeleton.
ANATOMICAL CHARACTERISTICS Insufficient information.
AGE Late Cretaceous, late Coniacian or Santonian.
DISTRIBUTION AND FORMATION/S Eastern China; Majiacun.

Albinykus baatar

—— 0.5 m (1.6 ft) TL, 0.5 kg (1.1 lb)
FOSSIL REMAINS Fragmentary skeleton.
ANATOMICAL CHARACTERISTICS Insufficient information.
AGE Late Cretaceous, Santonian.
DISTRIBUTION AND FORMATION/S Mongolia; Javkhlant.

Shuvuuia deserti

—— 0.6 m (2 ft) TL, 0.8 kg (3 lb)
FOSSIL REMAINS Two nearly complete skulls and several partial skeletons, external fibers.
ANATOMICAL CHARACTERISTICS Pubis strongly retroverted, foot strongly compressed from side to side. Short, hollow fibers on head and body.
AGE Late Cretaceous, late Campanian.
DISTRIBUTION AND FORMATION/S Mongolia; upper Djadokhta.
HABITAT Desert with dunes and oases.

Shuvuuia deserti
(see also next page)

Shuvuuia deserti

composite alvarezsaurid

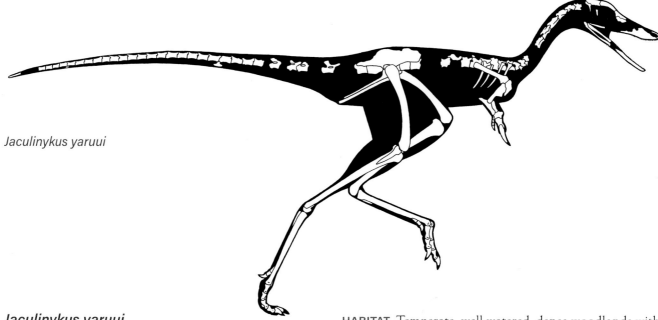

Jaculinykus yaruui

Jaculinykus yaruui

—— 0.9 m (3 ft) TL, 1.6 kg (3.5 lb)
FOSSIL REMAINS Majority of skull and skeleton.
ANATOMICAL CHARACTERISTICS Two fingers. Tail tip ends with a fused rod. Pubis very strongly retroverted, trunk shallow.
AGE Late Cretaceous, latest Campanian and/or earliest Maastrichtian.
DISTRIBUTION AND FORMATION/S Mongolia; uppermost Barun Goyot.
HABITAT Semidesert with some dunes and oases.
NOTES Specimen preserved in birdlike curled-up sleeping posture.

Nemegtonykus citus

—— 1 m (3.3 ft) TL, 4.1 kg (9 lb)
FOSSIL REMAINS Two partial skeletons.
ANATOMICAL CHARACTERISTICS Insufficient information.
AGE Late Cretaceous, late Campanian and/or earliest Maastrichtian.
DISTRIBUTION AND FORMATION/S Mongolia; lower Nemegt.

HABITAT Temperate, well-watered, dense woodlands with seasonal rains and winter snow.

Mononykus olecranus

—— 1 m (3.3 ft) TL, 4 kg (9 lb)
FOSSIL REMAINS Partial skeletons.
ANATOMICAL CHARACTERISTICS Standard for group.
AGE Late Cretaceous, latest Campanian and/or early Maastrichtian.
DISTRIBUTION AND FORMATION/S Mongolia; middle or upper Nemegt.
HABITAT Temperate, well-watered, dense woodlands with seasonal rains and winter snow.

Linhenykus monodactylus

—— 0.5 m (1.6 ft) TL, 0.5 kg (1.1 lb)
FOSSIL REMAINS Minority of skeleton.
ANATOMICAL CHARACTERISTICS Outer finger virtually lost so only thumb is functional.
AGE Late Cretaceous, Campanian.
DISTRIBUTION AND FORMATION/S Mongolia; Ulansuhai.
HABITAT Semidesert with some dunes and oases.

—— OVIRAPTOROSAURIFORMS

SMALL TO LARGE FLYING AND FLIGHTLESS HERBIVOROUS OR OMNIVOROUS AVEAIRFOILANS OF THE CRETACEOUS OF THE NORTHERN HEMISPHERE.

ANATOMICAL CHARACTERISTICS Fairly variable. Heads not large, short and deep, sides of back of head made of slender struts, postorbital bar complete, many bones including lower jaws fused together and extra joint absent, jaw joints highly mobile to allow complex chewing motion, teeth reduced or absent. Necks fairly long, ribs usually so short they do not overlap, increasing flexibility of neck. Trunks short. Tails short. Arms short to very long, fingers three to two. Legs short to very long.
HABITS Omnivorous or herbivorous, picked up small animals at least on occasion.
NOTES Omnivoropterygids, oviraptorosaurs, and avians and their common ancestor, operative only if three groups form a clade that excludes all other dinosaurs.

— OMNIVOROPTERYGIDS

SMALL FLYING OVIRAPTOROSAURS LIMITED TO THE EARLY CRETACEOUS OF ASIA.

ANATOMICAL CHARACTERISTICS Lower jaws shallow, a few procumbent, small, pointed teeth at front of upper jaw. No uncinate processes on ribs. Very short tail tipped with a fused pygostyle. Sternal plates and ossified sternal ribs may be absent, very long arm and hand indicate very large wings, outer finger severely reduced so there are only two fully functional fingers. Pubes a little retroverted, pelves broad, legs short and not as strong as arm, toes long, hallux reversed. Wings very large, tail feather fans present, ankle feathers in some fossils.

HABITS Capable of low-grade powered, gliding, and possibly soaring flight superior to that of *Archaeopteryx*. Good climbers. Defense included climbing and flight. Probably diurnal.

HABITAT Well-watered forests and lakes, winters chilly with snow.

NOTES A very oviraptorosaur-like skull strongly suggests a close relationship. Although probably not the direct ancestors of oviraptorosaurs, omnivoropterygids may represent the parrot-headed, short-tailed flying basal birds that oviraptorosaurs evolved from. Irregular presence of ankle feathers may indicate display rather than aerodynamic function. Specific diets difficult to determine because of the unusual configuration of head and jaws.

Sapeornis unnamed species?

— Adult size uncertain

FOSSIL REMAINS Complete but distorted skull with skeleton, extensive feathers.

AGE Early Cretaceous, latest Barremian and/or earliest Aptian.

DISTRIBUTION AND FORMATION/S Northeastern China; middle Yixian.

HABITAT Well-watered volcanic highland forest-and-lake district, winters chilly with snow.

NOTES Probably a species different from, and may be ancestral to, later *S. chaoyangensis*.

Sapeornis chaoyangensis

— 0.4 m (1.3 ft) TL, 1 kg (2 lb)

FOSSIL REMAINS Several complete skulls and majority of skeletons, feathers, gizzard stones.

AGE Early Cretaceous, early or middle Aptian.

DISTRIBUTION AND FORMATION/S Northeastern China; lower Jiufotang.

HABITAT Well-watered volcanic highland forest-and-lake district, winters chilly with snow.

NOTES A much more common component of the fauna than the preceding species, probably includes *Omnivoropteryx sinousaorum*, *Didactylornis jii*, and a number of other species.

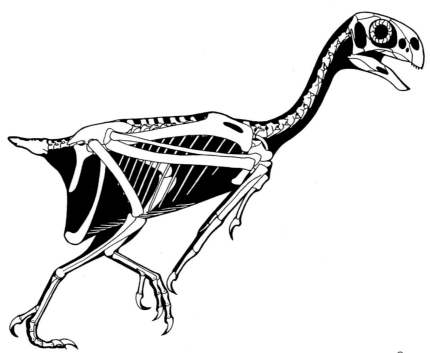

Sapeornis chaoyangensis

Sapeornis chaoyangensis

—OVIRAPTOROSAURS

SMALL TO LARGE HERBIVOROUS OR OMNIVOROUS AVEAIRFOILANS OF THE CRETACEOUS OF THE NORTHERN HEMISPHERE.

ANATOMICAL CHARACTERISTICS Fairly variable. Heads at least fairly short and deep, with parrot-like beak. Arms short to moderately long, fingers three to two. Pubes moderately retroverted to usually vertical or procumbent, legs moderately to very long.

HABITS Defense included biting with beaks, slashing with hand claws, and evasion.

NOTES Presence of large ossified sternal plates, ossified sternal ribs, ossified uncinate processes, tails short in most examples, reduction of outer finger in some examples, and pubic retroversion in basal examples strongly imply that the flightless oviraptorosaurs were the secondarily flightless relatives of known flying omnivoropterygids; this requires significant but not massive reversals of some sections of the skeleton to a nonavian condition; protarchaeopterygids, avimimids, caudipterids, and caenagnathoids may have evolved collectively and/or independently from earlier fliers. Alternatively, omnivoropterygids were the flying descendants of oviraptorosaurs. In the conventional scenario the two groups were not closely related, in which case the heads and hands evolved in a remarkably convergent manner despite the lack of a common flight heritage. Fragmentary remains may record presence in Australia.

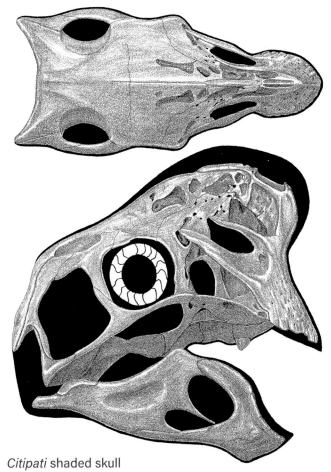

Citipati shaded skull

233

Conchoraptor muscle study

— OVIRAPTOROSAUR MISCELLANEA

Ningyuansaurus wangi
—— 1.7 m (6 ft) TL, 25 kg (50 lb)
FOSSIL REMAINS Majority of skull and skeleton.
ANATOMICAL CHARACTERISTICS Head somewhat
elongated, small teeth present. Tail fairly long. Arm short.
Leg long.
AGE Early Cretaceous, late Barremian.

DISTRIBUTION AND FORMATION/S Northeastern China;
Yixian.
HABITAT Well-watered volcanic highland forest-and-lake
district, winters chilly with snow.
HABITS Seeds present in gut.
NOTES Lack of detailed images hinders skeletal
restoration. Presence of crest and orientation of pubis
uncertain.

— PROTARCHAEOPTERYGIDS
SMALL OVIRAPTOROSAURS LIMITED TO THE EARLY CRETACEOUS OF ASIA.

ANATOMICAL CHARACTERISTICS Highly uniform. Head not as deep as in other oviraptorosaurs, subrectangular, roof
of mouths projects below rim of upper jaws, lower jaws shallow, frontmost teeth enlarged and well worn, rest of teeth
small, blunt, and unserrated, teeth absent from tip of lower jaw. Skeletons lightly built. Large sternal plates present,
arm long, three finger claws are large hooks. Pubes at least sometimes retroverted, legs long.
HABITS Divergence in tooth size and form is much greater than in other theropods. Incisor-like front teeth are
reminiscent of rodents and imply gnawing on some form of hard plant material. Both climbing and running
performance appear to be high, main defense climbing, high speed, and biting.
NOTES That these basal oviraptorosaurs are more birdlike than more-derived oviraptorosaurs is compatible with the
group being secondarily flightless.

Protarchaeopteryx (or Incisivosaurus) gauthieri
—— 0.8 m (2.7 ft) TL, 2 kg (5 lb)
FOSSIL REMAINS Almost complete skull and small
minority of skeleton.
ANATOMICAL CHARACTERISTICS Standard for group,
number of teeth differs from *P. robusta*.

AGE Early Cretaceous, latest Barremian.
DISTRIBUTION AND FORMATION/S Northeastern China;
lower Yixian.
HABITAT Well-watered volcanic highland forest-and-lake
district, winters chilly with snow.
NOTES May be direct ancestor of *P. robusta*.

composite protarchaeopterygid

Protarchaeopteryx robusta
— 0.7 m (2.3 ft) TL, 1.7 kg (3.7 lb)

FOSSIL REMAINS Majority of badly damaged skull and skeleton, some feathers.

ANATOMICAL CHARACTERISTICS Pubis appears to be at least somewhat retroverted. Tail feathers fairly long, vanes asymmetrical.

AGE Early Cretaceous, latest Barremian and/or earliest Aptian.

DISTRIBUTION AND FORMATION/S Northeastern China; middle Yixian.

HABITAT Well-watered volcanic highland forest-and-lake district, winters chilly with snow.

HABITS Arms not long enough and arm feathers too symmetrical for flight.

NOTES Misnamed as a closely related predecessor to the much earlier deinonychosaur *Archaeopteryx*. Pubic orientation ambiguous. May be direct ancestor of *P. ganqi*.

Protarchaeopteryx (or *Xingtianosaurus*) *ganqi*

—— 0.7 m (2.3 ft) TL, 1.7 kg (3.7 lb)

FOSSIL REMAINS Majority of skeleton.
ANATOMICAL CHARACTERISTICS Arms shorter than in *P. robusta*. Pubis moderately retroverted.
AGE Early Cretaceous, earliest Aptian.
DISTRIBUTION AND FORMATION/S Northeastern China; upper Yixian.

HABITAT Well-watered volcanic highland forest-and-lake district, winters chilly with snow.
NOTES Pubis clearly preserved in a retroverted posture; that pubic process of ilium is not retroverted is not incompatible with pubic retroversion because such is seen in some other dinosaurs. Elongated hand, pelvic characters and not highly elongated lower limb indicate is a protarchaeopterygid rather than a caudipterid.

—— CAUDIPTERIDS

SMALL OVIRAPTOROSAURS LIMITED TO THE EARLY CRETACEOUS OF ASIA.

ANATOMICAL CHARACTERISTICS Heads small, subtriangular, lower jaw shallow, a few procumbent, small, pointed teeth at front of upper jaw. Skeleton lightly built. Trunks short, uncinate processes on ribs. Ossified sternal plates and sternal ribs present, arm short, outer finger severely reduced so there are only two fully functional fingers, claws not large. Pelves very large, pubis vertical, leg very long and gracile, leg muscles exceptionally well developed, semireversed hallux small, so speed potential very high. Pennaceous feathers on hands and tail, propatagium present.
HABITS Climbing ability low or nonexistent, main defense high speed.
NOTES That these basal oviraptorosaurs are more birdlike in certain regards than more-derived oviraptorosaurs is compatible with the group being secondarily flightless.

Caudipteryx dongi

—— 0.5 m (1.75 ft) TL, 1.7 kg (3.5 lb)

FOSSIL REMAINS Majority of skeleton.
ANATOMICAL CHARACTERISTICS Similar to *C. zoui*.
AGE Early Cretaceous, latest Barremian.
DISTRIBUTION AND FORMATION/S Northeastern China; lower Yixian.
HABITAT Well-watered volcanic highland forest-and-lake district, winters chilly with snow.

DISTRIBUTION AND FORMATION/S Northeastern China; middle Yixian.
HABITAT Well-watered volcanic highland forest-and-lake district, winters chilly with snow.
HABITS Presence of some small, sharp teeth imply *Caudipteryx* may have caught small animals, but gizzard stones verify diet of plants that required grinding. Small hand and tail feather fans probably for display within the species.
NOTES May be ancestor of *C. yixianensis*.

Caudipteryx zoui

—— 0.65 m (2 ft) TL, 2.5 kg (5 lb)

FOSSIL REMAINS A number of complete skulls and skeletons, extensive feathers, bundles of gizzard stones.
ANATOMICAL CHARACTERISTICS Pygostyle not present. Well-developed feather fan on hand, possibly split fan at end of tail, latter showing pigment banding, large pennaceous feathers symmetrical, simpler feathers covering much of body.
AGE Early Cretaceous, latest Barremian and/or earliest Aptian.

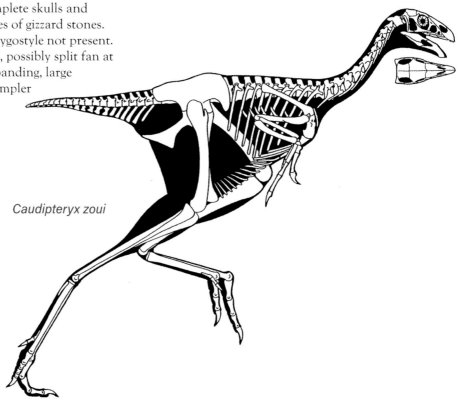

Caudipteryx zoui

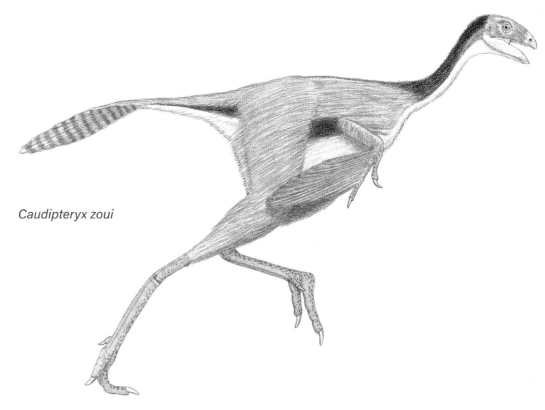

Caudipteryx zoui

Caudipteryx zoui

Caudipteryx (= Similicaudipteryx) yixianensis
—— 1 m (3 ft) TL, 9 kg (20 lb)
FOSSIL REMAINS Majority of poorly preserved skeleton.
ANATOMICAL CHARACTERISTICS Tail tipped with a small fused pygostyle.

AGE Early Cretaceous, early or middle Aptian.
DISTRIBUTION AND FORMATION/S Northeastern China; Jiufotang.
HABITAT Well-watered volcanic highland forest-and-lake district, winters chilly with snow.

—AVIMIMIDS

SMALL OVIRAPTOROSAURS LIMITED TO THE LATE LATE CRETACEOUS OF ASIA.

ANATOMICAL CHARACTERISTICS Heads short, deep, broad aft, top of aft head somewhat bulbous, lower jaw fairly deep, small teeth at front of upper jaw. Arms short. Pelves large, pubis procumbent, legs very long and its muscles exceptionally well developed, feet very long and strongly compressed from side to side, hallux absent and toes short, so speed potential very high.
HABITS Broad hips indicate large belly for processing plant material. Main defense high speed.

Avimimus portentosus
—— 1.2 m (3.5 ft) TL, 14 kg (30 lb)
FOSSIL REMAINS Partial skulls and skeletons.
ANATOMICAL CHARACTERISTICS Skeleton highly fused including upper hand. Pelvis very broad with ilial plate facing strongly upward, forward end rounded.

AGE Late Cretaceous, later Campanian or early Maastrichtian.
DISTRIBUTION AND FORMATION/S Mongolia; uncertain formation.
NOTES Tail dimensions not yet detailed. Possible formations of origin include Djadokhta, Barun Goyot, Nemegt.

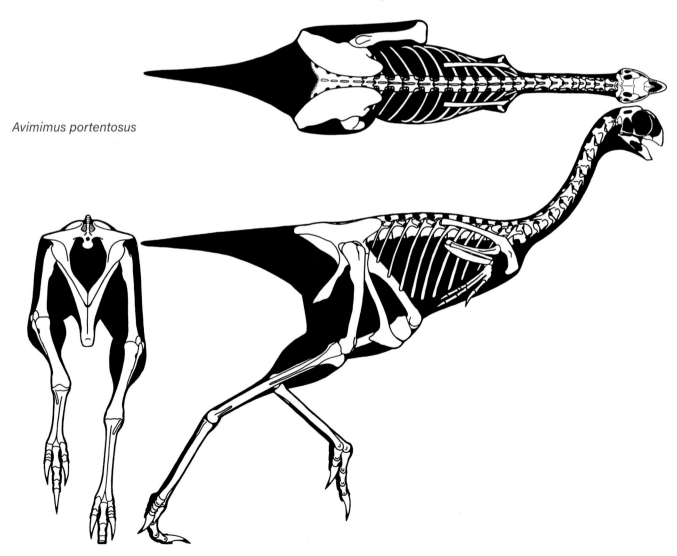

Avimimus portentosus

Unnamed genus *nemegtensis*

—— 1.2 m (3.5 ft) TL, 15 kg (30 lb)

FOSSIL REMAINS Partial skulls and skeletons in bone bed.

ANATOMICAL CHARACTERISTICS Pelvis breadth normal with ilial plate facing mainly sideways, forward prong projecting downward.

AGE Late Cretaceous, late Campanian and/or earliest Maastrichtian.

DISTRIBUTION AND FORMATION/S Mongolia; lower Nemegt.

HABITAT Temperate, well-watered, dense woodlands with seasonal rains and winter snow.

NOTES Placement in *Avimimus* incorrect because pelvis is so different.

—— CAENAGNATHOIDS

SMALL TO LARGE CAENAGATHOIDS OF THE CRETACEOUS OF THE NORTHERN HEMISPHERE.

ANATOMICAL CHARACTERISTICS Most or all adults with cassowary-like head crests, teeth absent, and blunt beaks present. Uncinate processes on ribs. Ossified sternal plates and sternal ribs present. Three finger claws well developed. Pubes vertical or procumbent. Olfactory bulbs reduced. Eggs highly elongated.

HABITS Pneumatic head crests too delicate for butting, for visual display within species. Defense included running, climbing, hand claws, and biting. Eggs formed and laid in pairs in flat, two-layered rings, partly exposed, probably by more than one female in each nest, brooded and incubated by adult sitting in empty center of nest with feathered arms and tail draped over eggs.

—— CAENAGNATHIDS

SMALL TO LARGE CAENAGATHOIDS OF THE CRETACEOUS OF THE NORTHERN HEMISPHERE.

ANATOMICAL CHARACTERISTICS Fairly uniform. Lower jaws not very deep. Arms and hands long. Legs fairly long.

Microvenator celer

—— Adult size uncertain

FOSSIL REMAINS Partial skeleton, juvenile.

ANATOMICAL CHARACTERISTICS Insufficient information.

AGE Early Cretaceous, middle Albian.

DISTRIBUTION AND FORMATION/S Montana; upper Cloverly.

HABITAT Short wet season, otherwise semiarid with floodplain prairies, open woodlands, and riverine forests.

Hagryphus giganteus

—— 2 m (6.5 ft) TL, 50 kg (100 lb)

FOSSIL REMAINS Small portion of skeleton.

ANATOMICAL CHARACTERISTICS Insufficient information.

AGE Late Cretaceous, late Campanian.

DISTRIBUTION AND FORMATION/S Utah; middle Kaiparowits.

Caenagnathasia martinsoni

—— 0.6 m (2 ft) TL, 1.5 kg (3 lb)

FOSSIL REMAINS Minority of two skulls.

ANATOMICAL CHARACTERISTICS Insufficient information.

AGE Late Cretaceous, Turonian.

DISTRIBUTION AND FORMATION/S Uzbekistan; Bissekty.

Chirostenotes pergracilis

—— 3 m (10 ft) TL, 200 kg (400 lb)

FOSSIL REMAINS Assorted fragmentary remains.

ANATOMICAL CHARACTERISTICS Lower jaw shallow.

AGE Late Cretaceous, late middle and/or early to middle late Campanian.

DISTRIBUTION AND FORMATION/S Alberta; Dinosaur Park, level uncertain.

HABITAT Well-watered, forested floodplain with coastal swamps and marshes, cool winters.

NOTES Fragmentary remains of varying ontogenetic ages and uncertain stratigraphy of original fossils leave taxonomy uncertain, genus and species may or may not include juvenile *Chirostenotes* (= *Citipati*) *elegans*, *Caenagnathus collinsi*, *Macrophalangia canadensis*. May include ancestor of *Chirostenotes pennatus*.

Chirostenotes (or *Apatoraptor*) *pennatus*

—— 1.5 m (5 ft) TL, 25 kg (50 lb)

FOSSIL REMAINS Minority of skull and skeleton.

ANATOMICAL CHARACTERISTICS Standard for group.

AGE Late Cretaceous, latest Campanian and/or earliest Maastrichtian.

DISTRIBUTION AND FORMATION/S Alberta; lower Horseshoe Canyon.

HABITAT Well-watered, forested floodplain with coastal swamps and marshes, cool winters.

NOTES Usual placement in much earlier *C. pergracilis* not correct. May be an ancestor of *Epichirostenotes curriei*.

Epichirostenotes? *curriei*

—— 2 m (6.5 ft) TL, 50 kg (100 lb)

FOSSIL REMAINS Fragmentary skull and skeleton.

Anzu wyliei

ANATOMICAL CHARACTERISTICS Insufficient information.
AGE Late Cretaceous, early Maastrichtian.
DISTRIBUTION AND FORMATION/S Alberta; upper Horseshoe Canyon.
HABITAT Well-watered, forested floodplain with coastal swamps and marshes, cool winters.
NOTES May be same genus as *Chirostenotes* (or *Apatoraptor*) *pennatus*. May be an ancestor of *Anzu wyliei*.

Anzu wyliei
—— 3.75 (12 ft) TL, 265 kg (580 lb)
FOSSIL REMAINS Majority of a few skulls and skeletons.
ANATOMICAL CHARACTERISTICS Tall, broad head crest. Leg long.
AGE Late Cretaceous, late Maastrichtian.
DISTRIBUTION AND FORMATION/S South Dakota; Hell Creek.

HABITAT Well-watered coastal woodlands.
NOTES Was usually included in much earlier *Chirostenotes pergracilis*.

Elmisaurus rarus
—— 1.7 m (5.5 ft) TL, 20 kg (40 lb)
FOSSIL REMAINS Partial skeleton and other remains.
ANATOMICAL CHARACTERISTICS Tail tipped with a short fused pygostyle.
AGE Late Cretaceous, late Campanian and early Maastrichtian?
DISTRIBUTION AND FORMATION/S Mongolia; at least lower Nemegt.
HABITAT Temperate, well-watered, dense woodlands with seasonal rains and winter snow.
NOTES That original fossil is from lower Nemegt suggests later remains are different species, probably includes *Nomingia gobiensis*.

—— OVIRAPTORIDS
SMALL, LIMITED TO THE CRETACEOUS OF ASIA.

ANATOMICAL CHARACTERISTICS Uniform. Highly pneumatic heads subrectangular, snouts short, somewhat parrot-like beak deep, nostrils above preorbital opening, blunt pair of pseudoteeth on strongly downward-projecting mouth roofs, eyes not especially large, lower jaw deep. Outer two fingers subequal in length and robustness, finger claws well developed. Pubes vertical or procumbent, legs not slender.
HABITS The downward-jutting pseudoteeth indicate a crushing action.
NOTES The number of genera named in this group appears excessive in part because fossils without crests may be juveniles or females of crested species. Head crests were probably enlarged by keratin coverings.

Luoyanggia liudianensis
—— 1.5 m (5 ft) TL, 20 kg (40 lb)
FOSSIL REMAINS Fragmentary skeleton.

ANATOMICAL CHARACTERISTICS Insufficient information.
AGE Early? Cretaceous, Aptian or Albian?

DISTRIBUTION AND FORMATION/S Eastern China; Haoling.
NOTES Age of formation not fully certain.

Anomalipes zhaoi
—— Adult size uncertain
FOSSIL REMAINS Fragmentary skeleton, possibly juvenile.
ANATOMICAL CHARACTERISTICS Insufficient information.
AGE Late Cretaceous.
DISTRIBUTION AND FORMATION/S Eastern China; Wangshi Group.

Beibeilong sinensis
—— Adult size uncertain
FOSSIL REMAINS Majority of hatchling skull and skeleton, eggs.
ANATOMICAL CHARACTERISTICS Insufficient information.
AGE Early Late Cretaceous.
DISTRIBUTION AND FORMATION/S Eastern China; Gaogou.
NOTES Very large size of eggs indicates adults were as large as *Gigantoraptor*.

Gigantoraptor erlianensis
—— 8 m (25 ft) TL, 2.4 tonnes
FOSSIL REMAINS Minority of skull and majority of skeleton.
ANATOMICAL CHARACTERISTICS Hand slender.
AGE Late Cretaceous, Santonian.
DISTRIBUTION AND FORMATION/S Northern China; Iren Dabasu.
HABITAT Seasonally wet-dry woodlands.
HABITS Another example of a large, high-browsing theropod similar to *Deinocheirus* and *Therizinosaurus*. Better able to defend itself against predators than were smaller oviraptors, also able to run away from predators.
NOTES Giant eggs up to 0.5 m (1.6 ft) long laid in enormous rings up to 3 m (10 ft) across, found in Asia, probably laid by big oviraptors such as *Gigantoraptor*.

Nankangia jiangxiensis
—— 2.5 m (8 ft) TL, 85 kg (190 kg)
FOSSIL REMAINS Minority of skull and majority of skeleton.
ANATOMICAL CHARACTERISTICS Insufficient information.
AGE Late Cretaceous, late Campanian and/or early Maastrichtian.
DISTRIBUTION AND FORMATION/S Southeastern China; Nanxiong.

Yulong mini
—— Adult size uncertain
FOSSIL REMAINS A few complete and partial juvenile skulls and skeletons.
ANATOMICAL CHARACTERISTICS Standard for group.
AGE Late Cretaceous.
DISTRIBUTION AND FORMATION/S Eastern China; Qiupa.

Oviraptor philoceratops
—— 1.6 m (5 ft) TL, 25 kg (50 lb)
FOSSIL REMAINS Majority of poorly preserved skull and minority of skeleton.
ANATOMICAL CHARACTERISTICS Head not as deep as in other oviraptorids, full extent of head crest uncertain. Hand large.
AGE Late Cretaceous, middle Campanian.
DISTRIBUTION AND FORMATION/S Mongolia; lower Djadokhta.
HABITAT Desert with dunes and oases.

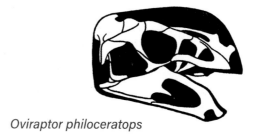

Oviraptor philoceratops

Gigantoraptor erlianensis

HABITS Presence of lizard skeleton in gut cavity of the skeleton indicates that the oviraptorid diet included at least some small animals.

NOTES Other oviraptorids were placed in *Oviraptor* until it was realized this is a very distinct genus.

Banji long
— 2.5 m (8 ft) TL, 85 kg (190 kg)

FOSSIL REMAINS Three nearly complete skulls and partial skeletons, adult to juvenile.

Banji long skulls growth series

ANATOMICAL CHARACTERISTICS Head crest low arced.

AGE Late Cretaceous, late Campanian and/or early Maastrichtian.

DISTRIBUTION AND FORMATION/S Southeastern China; Nanxiong.

NOTES *Huanansaurus ganzhouensis* and *Tongtianlong limosus* may be more mature male and females of this taxon. One fossil preserved in position suggests that it died stranded in mud.

Corythoraptor (or Citipati) jacobsi
— 2.5 m (8 ft) TL, 85 kg (190 kg)

FOSSIL REMAINS Nearly complete skull and majority of skeleton.

ANATOMICAL CHARACTERISTICS Well-developed vertical crest.

AGE Late Cretaceous, late Campanian and/or early Maastrichtian.

DISTRIBUTION AND FORMATION/S Southeastern China; Nanxiong.

Citipati osmolskae
— 2.5 m (8 ft) TL, 85 kg (190 kg)

FOSSIL REMAINS Several complete and partial skulls and skeletons from embryo to adult, completely known, nests, some with adults in brooding posture on complete nests.

ANATOMICAL CHARACTERISTICS Well-developed crest projects forward above upper beak. Tail tipped with a small fused pygostyle. Elongated eggs 18 cm (7 in) long.

AGE Late Cretaceous, late Campanian.

DISTRIBUTION AND FORMATION/S Mongolia; upper Djadokhta.

HABITAT Desert with dunes and oases.

HABITS Presence of remains of juvenile dinosaurs in some nests indicates that the oviraptorid diet included at least some small animals.

NOTES Showing that small, crestless *Khaan mckennai* is or is not the juvenile of this species requires examination of bone microstructure.

Rinchenia (or Citipati) mongoliensis
— 1.7 m (5.5 ft) TL, 25 kg (55 lb)

FOSSIL REMAINS Complete skull and minority of skeleton.

ANATOMICAL CHARACTERISTICS Subtriangular head crest very large.

AGE Late Cretaceous, latest Campanian and/or early Maastrichtian.

DISTRIBUTION AND FORMATION/S Mongolia; middle and/or upper Nemegt.

HABITAT Temperate, well-watered, dense woodlands with seasonal rains and winter snow.

Rinchenia (or *Citipati*) *mongoliensis*

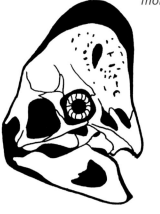
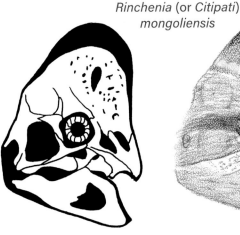

Wulatelong gobiensis
— 1.7 m (5.5 ft) TL, 25 kg (55 lb)

FOSSIL REMAINS Majority of poorly preserved skull and minority of skeleton.

ANATOMICAL CHARACTERISTICS Insufficient information.

AGE Late Cretaceous, Campanian.

DISTRIBUTION AND FORMATION/S Northwestern China; Ulansuhai.

Shixinggia oblita
— 2 m (7 ft) TL, 40 kg (85 lb)

FOSSIL REMAINS Minority of skeleton.

juvenile

Citipati osmolskae

ANATOMICAL CHARACTERISTICS Insufficient information.
AGE Late Cretaceous, Maastrichtian.
DISTRIBUTION AND FORMATION/S Southern China; Pingling.

Conchoraptor (= Ajancingenia) yanshini
—— 1.5 m (5 ft) TL, 19 kg (40 lb)
FOSSIL REMAINS Complete and partial skulls and skeletons, adult and juvenile.

ANATOMICAL CHARACTERISTICS Very large forward-pointed head crest. Tail deep in at least one morph, tipped with a short fused pygostyle. Thumb about as long as other fingers, hand robust at least in one morph.
AGE Late Cretaceous, late Campanian.
DISTRIBUTION AND FORMATION/S Mongolia; lower Barun Goyot.
HABITAT Semidesert with some dunes and oases.
HABITS Large thumb a weapon that may also have been used for feeding in some manner.

Citipati osmolskae including *Saurornithoides*

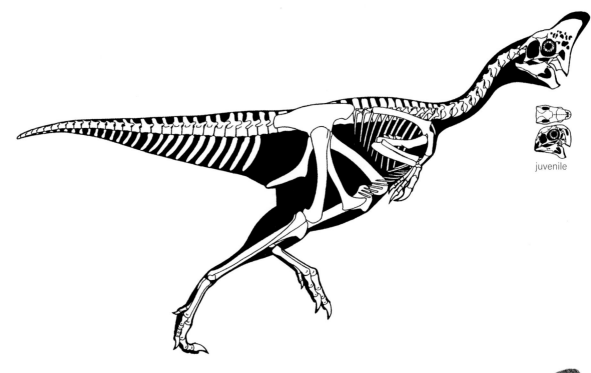

juvenile

Conchoraptor (= Ajancingenia) yanshini

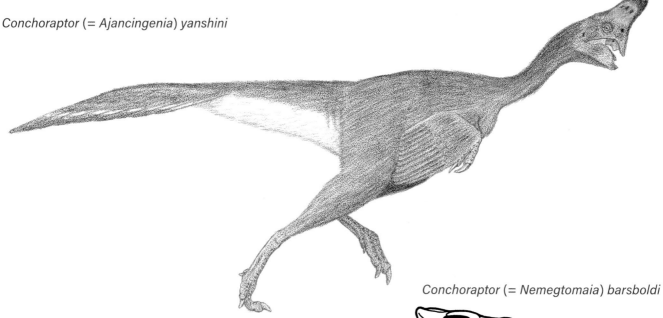

Conchoraptor (= Nemegtomaia) barsboldi

NOTES Showing that large, crested *C. gracilis* is or is not the adult of this species requires examination of bone microstructure. Original genus *Ingenia* preoccupied by an invertebrate. *Conchoraptor* has priority over more recent *Ajancingenia* if they are one genus. May be the ancestor of *C. barsboldi*.

Conchoraptor (= Nemegtomaia) barsboldi
—— 2 m (7 ft) TL, 45 kg (100 lb)
FOSSIL REMAINS Two poorly preserved complete skulls and minority of four skeletons.

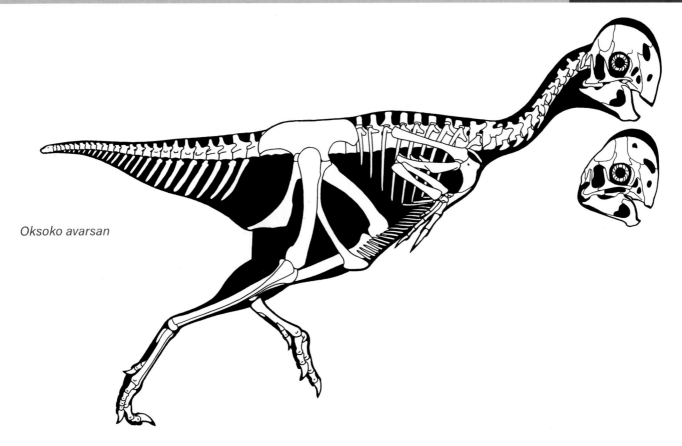

Oksoko avarsan

ANATOMICAL CHARACTERISTICS Large crest above upper beak.
AGE Late Cretaceous, latest Campanian and/or early Maastrichtian.
DISTRIBUTION AND FORMATION/S Mongolia; middle and/or upper Nemegt.
HABITAT Temperate, well-watered, dense woodlands with seasonal rains and winter snow.
NOTES Separation from the very similar *Conchoraptor* is not warranted. *Gobiraptor minutus* probably a juvenile of this species.

Conchoraptor (or *Heyuannia*) *huangi*
—— 1.5 m (5 ft) TL, 20 kg (45 lb)
FOSSIL REMAINS Partial skull and skeleton.
ANATOMICAL CHARACTERISTICS Insufficient information.
AGE Late Late Cretaceous.
DISTRIBUTION AND FORMATION/S Southern China; Dalangshan.

NOTES Available evidence insufficient to make separate genus from *Conchoraptor*.

Oksoko avarsan
—— Adult size uncertain
FOSSIL REMAINS Half a dozen large juvenile skulls and/or skeletons.
ANATOMICAL CHARACTERISTICS Head crest prominent. Tail tipped with a small fused pygostyle. Outer finger reduced to a splint, so two functional fingers.
AGE Late Cretaceous, early Maastrichtian.
DISTRIBUTION AND FORMATION/S Mongolia; Nemegt.
HABITAT Temperate, well-watered, dense woodlands with seasonal rains and winter snow.
NOTES Adults probably had more prominent crests. Reduction of fingers suggests that if not for the K/Pg extinction that many or all oviraptorids would have shifted to two-fingered hands.

ADDITIONAL READING

Brett-Surman, M., and J. Farlow. 2011. *The Complete Dinosaur*. 2nd ed. Bloomington: Indiana University Press.

Brusatte, S. 2018. *The Rise and Fall of the Dinosaurs*. Boston: Mariner Books.

Glut, D. 1997–2012. *Dinosaurs: The Encyclopedia* [including Supplements 1–7]. London: McFarland.

Molina-Perez, R., and A. Larramendi. 2019. *Dinosaur Facts & Figures: The Theropods and Other Dinosauriformes*. Princeton, NJ: Princeton University Press.

Paul, G. 2024. *The Princeton Field Guide to Dinosaurs*, 3rd ed. Princeton, NJ: Princeton University Press.

INDEX TO PREDATORY DINOSAUR TAXA

INDEX TO FORMATIONS

When a formation is cited more than once on a page, the number of times is indicated in parentheses.